CAMBRIDGE STUDIES IN
ADVANCED MATHEMATICS : 50

CLIFFORD ALGEBRAS AND THE CLASSICAL GROUPS

Postscript: On Saturday, 6th May, 1995, the 150th anniversary of the birth of William Kingdon Clifford (1845–1879), a Celebration of this life and that of his wife Lucy (1846–1929) was held at the University of Kent at Canterbury. There we learned that Clifford was not only theoretically but also athletically expert at rotations! Dave Chisholm's delightful portrayal of Clifford performing his 'corkscrew' at Cambridge in 1869 was a foil on that occasion to an entrancing exhibition of Victorian cartoons.

The Clifford algebras of real quadratic forms and their complexifications are studied here in detail, and those parts which are immediately relevant to theoretical physics are seen in the proper broad context.

Central to the work is the classification of the conjugation and reversion anti-involutions that arise naturally in the theory. It is of interest that all the classical groups play essential roles in this classification. Other features include detailed sections on conformal groups, the eight-dimensional non-associative Cayley algebra, its automorphism group, the exceptional Lie group G_2, and the triality automorphism of Spin 8.

The book is designed to be suitable for the last year of an undergraduate course or the first year of a postgraduate course.

Already published

1 W.M.L. Holcombe *Algebraic automata theory*
2 K. Petersen *Ergodic theory*
3 P.T. Johnstone *Stone spaces*
4 W.H. Schikhof *Ultrametric calculus*
5 J.-P. Kahane *Some random series of functions, 2nd edition*
6 H. Cohn *Introduction to the construction of class fields*
7 J. Lambek & P.J. Scott *Introduction to higher-order categorical logic*
8 H. Matsumura *Commutative ring theory*
9 C.B. Thomas *Characteristic classes and the cohomology of finite groups*
10 M. Aschbacher *Finite group theory*
11 J.L. Alperin *Local representation theory*
12 P. Koosis *The logarithmic integral I*
13 A. Pietsch *Eigenvalues and S-numbers*
14 S.J. Patterson *An introduction to the theory of the Riemann zeta-function*
15 H.J. Baues *Algebraic homotopy*
16 V.S. Varadarajan *Introduction to harmonic analysis on semisimple Lie groups*
17 W. Dicks & M. Dunwoody *Groups acting on graphs*
18 L.J. Corwin & F.P. Greenleaf *Representations of nilpotent Lie groups and their applications*
19 R. Fritsch & R. Piccinini *Cellular structures in topology*
20 H Klingen *Introductory lectures on Siegel modular forms*
21 P. Koosis *The logarithmic integral II*
22 M.J. Collins *Representations and characters of finite groups*
24 H. Kunita *Stochastic flows and stochastic differential equations*
25 P. Wojtaszczyk *Banach spaces for analysts*
26 J.E. Gilbert & M.A.M. Murray *Clifford algebras and Dirac operators in harmonic analysis*
27 A. Frohlich & M.J. Taylor *Algebraic number theory*
28 K. Goebel & W.A. Kirk *Topics in metric fixed point theory*
29 J.F. Humphreys *Reflection groups and Coxeter groups*
30 D.J. Benson *Representations and cohomology I*
31 D.J. Benson *Representations and cohomology II*
32 C. Allday & V. Puppe *Cohomological methods in transformation groups*
33 C. Soulé et al *Lectures on Arakelov geometry*
34 A. Ambrosetti & G. Prodi *A primer of nonlinear analysis*
35 J. Palis & F. Takens *Hyperbolicity and sensitive chaotic dynamics at homoclinic bifurcations*
36 M. Auslander, I. Reiten & S. Smalo *Representation theory of Artin algebras*
37 Y. Meyer *Wavelets and operators*
38 C. Weibel *An introduction to homological algebra*
39 W. Bruns & J. Herzog *Cohen-Macaulay rings*
40 V. Snaith *Explicit Brauer induction*
42 E.B. Davies *Spectral theory and differential operators*
43 J. Diestel, H. Jarchow & A. Tonge *Absolutely summing operators*
44 P. Mattila *Geometry of sets and measures in Euclidean spaces*
45 R. Pinsky *Positive harmonic functions and diffusion*
46 G. Tenenbaum *Introduction to analytic and probabilistic number theory*

Clifford Algebras and the Classical Groups

Ian R. Porteous
University of Liverpool

CAMBRIDGE
UNIVERSITY PRESS

[redacted]

CAMBRIDGE UNIVERSITY PRESS
Cambridge, New York, Melbourne, Madrid, Cape Town, Singapore, São Paulo, Delhi

Cambridge University Press
The Edinburgh Building, Cambridge CB2 8RU, UK

Published in the United States of America by Cambridge University Press, New York

www.cambridge.org
Information on this title: www.cambridge.org/9780521118026

First published 1995
Reprinted 2000
This digitally printed version 2009

A catalogue record for this publication is available from the British Library

ISBN 978-0-521-55177-9 hardback
ISBN 978-0-521-11802-6 paperback

[redacted]

Contents

Foreword

This book's parent *Topological Geometry* (Porteous (1969)), originally written in the 1960's to make propaganda for a basis-free approach to the differential calculus of functions of several variables, contained, almost by accident, a central section on Clifford algebras, a generalisation of quaternions that was at that time little known. This section was strengthened in the second edition (Porteous (1981)) by an additional chapter on the triality outer automorphism of the group $Spin(8)$, a feature which illuminates the structure of several of the other Spin groups and which is related to a property of six-dimensional projective quadrics first noticed almost a hundred years ago by Study in work on the rigid motions of three-dimensional space.

In recent years Clifford algebras have become a more popular tool in theoretical physics and it seems therefore appropriate to rework the original book, summarising the linear algebra and calculus required but expanding the Clifford algebra material. This seems the more worth while since it is clear that the central result of the old book, the classification of the conjugation anti-involution of the Clifford algebras $\mathbf{R}_{p,q}$ and their complexifications, was dealt with too briefly to be readily understood, and some of the more recent treatments of it elsewhere have been less than complete.

As in the previous version, the opportunity has been taken to give an exhaustive treatment of all the generalisations of the orthogonal and unitary groups known as the classical groups, since the full set plays a part in the Clifford algebra story. In particular, perhaps surprisingly, one learns to think of the general linear groups as unitary groups. Toward the end of the book the classical groups are presented as Lie groups and their Lie algebras are introduced. The exceptional Lie group G_2 also makes an appearance as the group of automorphisms of the Cayley algebra, a non-associative analogue of the quaternions that plays an essential role in the discussion of triality.

I owe a great debt not only to colleagues and students at the University of Liverpool over the years but also to new found friends at the by now regular international meetings on Clifford algebras and their applications to problems of mathematical physics, whose Proceedings have been published as Chisholm and Common (1986), Micali, Boudet and Helmstetter (1991) and Brackx, Delanghe and Serras (1993).

My interest in Clifford algebras and their use in physics was originally stimulated by discussions with my colleague at Liverpool Bob Boyer, tragically killed by a madman's bullet on the campus of the University of Austin, Texas, on August 1, 1966. Explicit classifications of both the conjugation and the reversion anti-involutions in the tables of Clifford algebras in Chapter 17 are in a Liverpool M.Sc. thesis by Tony Hampson (1969). On obtaining the answers Hampson and I wrote to my colleague Terry Wall, who was at that time on a visit to Mexico. He replied by drawing our attention to his paper (1968) which we had not read, and which presented the entire theory very succinctly! For the classical groups my main debt is to Prof. E. Artin's classic *Geometric Algebra* (1957). The observation that the Cayley algebra can be derived from one of the Clifford algebras I owe to Michael Atiyah, while the method adopted in Chapter 22 for constructing the Lie algebras of a Lie group was outlined to me by Frank Adams.

In preparing this fresh version of the material I am hugely indebted to Pertti Lounesto who has read much of the book in draft and over the years has kept me right on many points of detail. His knowledge of the history of the subject is unsurpassed. Much more recently, Chapter 23 on the conformal groups owes much to Jan Cnops, as is there acknowledged, and to the hospitality of Julius Ławrynowicz and the Banach Institute in Warsaw in 1994. An early version of some of the material of Chapter 17 has appeared as Porteous (1993).

Finally a disclaimer! I am no physicist, and therefore the reader will search in vain for particular applications to physics. On the other hand, works that are strongly biased toward applications frequently give only a fragmented and partial view of the subject. It is my belief that the subject only makes sense when the full picture is unfolded, and some of the otherwise confusing details are seen naturally to fall into place.

Ian Porteous, Liverpool, January, 1995

1
Linear spaces

In this chapter we recall briefly some salient facts about linear spaces and linear maps. Proofs for the most part are omitted.

Maps

Let X and Y be sets and $f : X \to Y$ a map. Then, for each $x \in X$ an element $f(x) \in Y$ is defined, the subset of Y consisting of all such elements being called the *image* of f, denoted by $\operatorname{im} f$. More generally $f : X \rightarrowtail Y$ will denote a map of an unspecified subset of X to Y, X being called the *source* of the map and the subset of X consisting of those points $x \in X$ for which $f(x)$ is defined being called the *domain* of f, denoted by $\operatorname{dom} f$. In either case the set Y is the *target* of f.

Given a map $f : X \rightarrowtail Y$ and a point $y \in Y$, the subset $f^{-1}\{y\}$ of X consisting of those points $x \in X$ such that $f(x) = y$ is called the *fibre* of f over y, this being non-null if and only if $y \in \operatorname{im} f$. The set of non-null fibres of f is called the *coimage* of f and the map

$$\operatorname{dom} f \to \operatorname{coim} f ; x \mapsto f^{-1}\{f(x)\}$$

the *partition* of $\operatorname{dom} f$ induced by f. The fibres of a map f are sometimes called the *level sets* or the *contours* of f, especially when the target of f is the field of real numbers **R**.

The *composite* gf of maps $f : X \rightarrowtail Y$ and $g : Y \rightarrowtail Z$ (read 'g following f') is the map $X \rightarrowtail Z ; x \mapsto g(f(x))$, with $\operatorname{dom} gf = f^{-1}(\operatorname{dom} g)$.

Proposition 1.1 *For any maps $f : W \to X$, $g : X \to Y$ and $h : Y \to Z$*

$$h(gf) = hgf = (hg)f.$$

Proof We traverse the *rebracketing pentagon* as follows:– for any $w \in W$

$$((h(gf))(w) = h((gf)(w)) = h(g(f(w))) = (hg)(f(w)) = ((hg)f)(w).$$

□

For any set X the *identity map* $X \to X : x \mapsto x$ will be denoted by 1_X, maps $f : X \to Y$ and $g : Y \to X$ such that $gf = 1_X$ and $fg = 1_Y$ being said to be *inverses* of each other, with $g = f^{-1}$ and $f = g^{-1}$. Given invertible maps $f : X \to Y$ and $g : Y \to Z$ then $(gf)^{-1} = f^{-1}g^{-1}$.

To any map $f : X \to Y$ there is associated an equation $f(x) = y$. The map f is said to be *surjective* or a *surjection* if, for each $y \in Y$, there is some $x \in X$ such that $f(x) = y$. It is said to be *injective* or an *injection* if, for each $y \in Y$, there is at most one element $x \in X$, though possibly none, such that $f(x) = y$. The map fails to be surjective if there exists an element $y \in Y$ such that the equation $f(x) = y$ has no solution $x \in X$, and fails to be injective if there exist $x, x' \in X$ such that $f(x') = f(x)$. A map that is both injective and surjective is said to be *bijective* or a *bijection*. A map is bijective if and only if it is invertible.

If maps $f : X \to Y$ and $g : Y \to X$ are such that $fg = 1_Y$ then, by Exercise 1.1, f is surjective and g is injective. The injection g is said to be a *section* of the surjection f. It selects for each $y \in Y$ a *single* $x \in X$ such that $f(x) = y$. It is assumed that any surjection $f : X \to Y$ has a section $g : Y \to X$, this assumption being known as the *axiom of choice*.

Linear spaces and maps

A *linear space* (of vectors), X, over a field (of scalars), $\mathbf{K}$, is an additive abelian group X, with zero element O (or 0) and furnished with a scalar multiplication $\mathbf{K} \times X \to X; (x, \lambda) \mapsto x\lambda = \lambda x$ satisfying both distributive laws, with $\lambda(\mu x) = (\lambda\mu)x$, for any $\lambda, \mu \in \mathbf{K}$ and any $x \in X$. Moreover, $1x = x$, for any $x \in X$, implying that $0x = O$ and $(-1)x = -x$. Also, for any $\lambda \in \mathbf{K}$, $\lambda O = O$. For us the field $\mathbf{K}$ will normally be either the real field $\mathbf{R}$ or the complex field $\mathbf{C}$ and the linear spaces will normally be *finite-dimensional*, a *basis* for such a space X being a finite set of vectors that are linearly independent and that span X. The number of vectors in any basis is independent of the basis and is called the *dimension* of X.

A *linear map* is a map between linear spaces that respects the linear structures; that is $f : X \to Y$ between linear spaces X and Y is *linear* if, for any $a, b \in X$, $\lambda, \mu \in \mathbf{K}$, $f(\lambda a + \mu b) = \lambda f(a) + \mu f(b)$. Such a map $f : X \to Y$ is uniquely determined by the action of f on any basis for X, and any assignment of f on the elements of a basis for X extends to a linear map of the whole of X to Y.

It is easily verified that the composite of any two composable linear maps is linear, while it follows as a corollary of Exercise 1.2 that the inverse of a linear bijection is linear.

A *linear subspace* of a linear space X is a subset W that acquires a linear structure by the restriction to W of the linear structure for X.

The *kernel* of a linear map $f : X \to Y$ is the set $\{x \in X : f(x) = 0\}$, written $\ker f$.

Proposition 1.2 *For any linear map* $f : X \to Y$, $\ker f$ *is a linear subspace of* X *and* $\operatorname{im} f$ *is a linear subspace of* Y.

Proposition 1.3 *A linear map* f *is injective if and only if* $\ker f = \{0\}$.

The *rank* of a linear map between finite-dimensional linear spaces is the dimension of the image of the map, this image being a linear subspace of the target space. The *kernel rank* or *nullity* of the map is the dimension of its kernel. The rank of the linear map $f : X \to Y$ will be denoted by $\operatorname{rk} f$ and the kernel rank by $\operatorname{kr} f$.

Proposition 1.4 *Let* $f : X \to Y$ *be a linear map,* X *and* Y *being finite-dimensional linear spaces. Then* $\operatorname{rk} f + \operatorname{kr} f = \dim X$.

In the case that X is a finite-dimensional linear space a linear map $f : X \to X$ is injective if and only if it is surjective, it then being an *automorphism* or self-isomorphism of X. More generally if X and Y are linear spaces of the same finite dimension then any linear injection $X \to Y$ is an isomorphism.

A linear space X is said to be the *direct sum* $X_0 \oplus X_1$ of linear subspaces X_0 and X_1 if $X_0 \cap X_1 = \{0\}$ and $X_0 + X_1 = X$, each of the subspaces then being a *linear complement* of the other. Then $\dim X_0 + \dim X_1 = \dim X$. Associated to any direct sum decomposition $X = X_0 \oplus X_1$ there are *projection maps* $X \to X_0$ with kernel X_1 and $X \to X_1$ with kernel X_0.

The *direct product* of linear spaces X and Y is the Cartesian product $X \times Y$, with the sum $(X \times Y)^2 \to X \times Y; ((x, y), (x', y')) \mapsto (x + x', y + y')$ and scalar product $\mathbf{K} \times (X \times Y) \to X \times Y; (\lambda, (x, y)) \mapsto (\lambda x, \lambda y)$, the linear space $\mathbf{K}^n$ being the n-fold direct power of the field $\mathbf{K}$.

The choice of a basis for an n-dimensional $\mathbf{K}$-linear space induces an isomorphism with the linear space $\mathbf{K}^n$, any linear map $\mathbf{K}^n \to \mathbf{K}^m$ being represented by its *matrix*, an array of real numbers with m rows and n

columns, whose columns are the images of the vectors of the standard ordered basis for $\mathbf{K}^n$, vectors in this context being represented by column matrices.

A non-zero kernel vector of a linear map $\mathbf{K}^n \to \mathbf{K}^m$ 'is' a linear dependence relation between the columns of the matrix of the map.

The *transpose* of a matrix with m rows and n columns is the matrix with n rows and m columns whose ith row is the ith column of the original matrix, for each i. Transposition will be denoted by τ, the transpose of a matrix a being denoted by a^τ.

A map $\beta : X \times Y \to Z$ is said to be *bilinear* if for any $a \in X$ and $b \in Y$ the maps $X \to Z; (x, b) \mapsto \beta(x, b)$ and $Y \to Z; y \mapsto \beta(a, y)$ are both linear. Scalar multiplication is an example of a bilinear map.

The set $L(X, Y)$ of linear maps between linear spaces X and Y, of dimensions n and m say, has a natural linear structure of dimension mn. In particular the linear space $L(X, \mathbf{K})$ of linear maps from X to $\mathbf{K}$, also of dimension n, is called the *dual* of X and will be denoted by X^L. The map

$$X \to (X^L)^L; \; x \mapsto \epsilon_x,$$

where $\epsilon_x(f) = f(x)$, is easily proved to be injective and so is an isomorphism.

The *dual* of a linear map $f : X \to Y$ between finite-dimensional linear spaces X and Y is the linear map

$$f^L : Y^L \to X^L; \; \omega \mapsto \omega f,$$

where ωf denotes the composite of the map $\omega : Y \to \mathbf{R}$ following the map $f : X \to Y$. Clearly, for composable linear maps $f : X \to Y$ and $g : W \to X$ we have $(fg)^L = g^L f^L$.

Proposition 1.5 *Let α and β be elements of the dual space X^L of a finite-dimensional real or complex linear space X such that $\ker \alpha = \ker \beta$. Then there exists a non-zero scalar λ such that $\beta = \lambda\,\alpha$.*

Proof Either $\ker \alpha = \ker \beta = X$, in which case $\alpha = \beta = 0$ and λ can be any non-zero element of $\mathbf{R}$, or both α and β are surjective. Then any element of X is of the form $a + \mu b$, where $\alpha(a) = 0$ and $\alpha(b) = 1$, and $\beta(a + \mu b) = \mu\,\beta(b) = \alpha(a + \mu b)\lambda$, where $\lambda = \beta(b)$; that is, $\beta = \lambda\,\alpha$. □

The *dual annihilator* $W^@$ of a linear subspace W of a finite-dimensional linear space X is the kernel of the map $\imath^L : X^L \to W^L$ dual to the inclusion map $\imath : W \to X$. Its dimension is equal to $\dim X - \dim W$.

A linear map $f : X \to X$ is said to be an *endomorphism* of the linear space X and the linear space $L(X, X)$ of all such endomorphisms of X is also denoted by $\operatorname{End} X$. As previously mentioned, a linear isomorphism $f : X \to X$ is said to be an *automorphism* of the linear space X and the set of all such automorphisms of X is denoted either by $GL(X)$ ('GL' standing for 'general linear') or by $\operatorname{Aut} X$. The linear space of all $n \times n$ matrices representing the elements of $L(\mathbf{K}^n, \mathbf{K}^n)$ will be denoted by $\mathbf{K}(n)$.

A *linear involution* of a linear space X is a linear map $t : X \to X$ such that $t^2 = 1_X$.

$^2\mathbf{K}$-*modules and maps*

Let Λ be a commutative and associative ring with unit element. Then a Λ-*module* X is a linear space over the ring Λ, the terminology *linear space* being reserved mainly for the case that Λ is a field.

Consider the case that Λ is the *double field* $^2\mathbf{K}$ consisting of the $\mathbf{K}$-linear space $\mathbf{K}^2$ assigned the product $(a, b)(c, d) = (ac, bd)$. A direct sum decomposition $X_0 \oplus X_1$ of a $\mathbf{K}$-linear space X may be regarded as a $^2\mathbf{K}$-module structure for X by setting

$$(\lambda, \mu)x = \lambda x_0 + \mu x_1, \quad \text{for all } x \in X \text{ and } (\lambda, \mu) \in {}^2\mathbf{K}.$$

Conversely, any $^2\mathbf{K}$-module structure for X determines a direct sum decomposition $X_0 \oplus X_1$ of X as a $\mathbf{K}$-linear space in which $X_0 = (1, 0)X$ $(= \{(1, 0)x : x \in X\})$ and $X_1 = (0, 1)X$.

Proposition 1.6 *Let $t : X \to X$ be a linear involution of the* $\mathbf{K}$*-linear space X. Then a* $^2\mathbf{K}$*-module structure, and therefore a direct sum decomposition, is defined for X by setting, for any $x \in X$,*

$$(1, 0)x = \frac{1}{2}(x + t(x)) \text{ and } (0, 1)x = \frac{1}{2}(x - t(x)).$$

$^2\mathbf{K}$-*module maps* and $^2\mathbf{K}$-*submodules* are defined in the obvious ways.

In working with a $^2\mathbf{K}$-module map $t : X \to Y$ it is often convenient to represent X and Y each as the *product* of its components and then to use notations associated with maps between products, as, for example, in the next proposition.

Proposition 1.7 *Let $t : X \to Y$ be a* $^2\mathbf{K}$*-module map. Then t is of the form* $\begin{pmatrix} a_0 & 0 \\ 0 & a_1 \end{pmatrix}$, *where $a_0 \in L(X_0, Y_0)$ and $a_1 \in L(X_1, Y_1)$. Conversely any map of this form is a* $^2\mathbf{K}$*-module map.*

A ${}^2\mathbf{K}$-module X such that $X_0 = (1, 0)X$ and $X_1 = (0, 1)X$ are isomorphic will be called a ${}^2\mathbf{K}$-*linear space* and a ${}^2\mathbf{K}$-module map $X \to Y$ between ${}^2\mathbf{K}$-linear spaces X and Y will be called a ${}^2\mathbf{K}$-*linear map*.

Affine spaces and maps

An *affine space* is a linear space with its origin deleted – it acquires a unique linear structure so soon as a point is chosen as origin, and the transfer from any one linear structure to any other is by a translation. An *affine map* is a map $f : X \to Y$ between affine spaces X and Y that becomes linear so soon as a point a of X is chosen as origin for X and $f(a)$ is chosen as origin for Y. An *affine map* between *linear spaces* is the sum of a linear map and a constant map. An *affine subspace* of a linear space is a translate or parallel of a linear subspace.

Determinants

We assume that the reader is familiar with the basic properties of determinants. Briefly, to any element a of $\mathbf{K}(n)$ there is a unique real number, the *determinant* of a, $\det a$, such that

(i) if any column of a matrix a is multiplied by a scalar λ then the determinant is multiplied by λ,
(ii) if any column of a matrix a is added to another then the determinant remains unaltered,
(iii) the determinant of the identity is 1.

The map is defined, for all $a \in \mathbf{K}(n)$, by the formula

$$\det a = \sum_{\pi \in \mathbf{n}!} \operatorname{sgn} \pi \prod_{j \in \mathbf{n}} a_{\pi(j),j} ,$$

where $\mathbf{n}!$ denotes the set of permutations of the set $\mathbf{n}$ of all natural numbers m such that $0 \leq m < n$. Moreover,

(iv) for any $a, b \in \mathbf{K}(n), \det b\, a = \det b \det a$,
(v) for any invertible $a \in \mathbf{K}(n), \det a^{-1} = (\det a)^{-1}$,
(vi) for any $a \in \mathbf{K}(n)$, a is invertible if and only if $\det a$ is invertible, that is, if and only if $\det a = 0$.

Any linear isomorphism $a : \mathbf{K}^n \to X$ induces a map

$$\operatorname{End} X = L(X,X) \to \mathbf{K};\ f \mapsto \det(a^{-1} f a)$$

called the *determinant* on $\operatorname{End} X$ and also denoted by det. This map is easily seen to be independent of the isomorphism a.

Proposition 1.8 *Any invertible matrix $a \in \mathbf{K}(n)$ is reducible by a series of column operations to a matrix with all entries on the main diagonal equal to* 1 *except for one which is equal to* $\det a$*, and all entries off the main diagonal equal to zero.*

Note that each of the column operations may be performed by multiplying the matrix on the right by a matrix all of whose entries on the main diagonal are equal to 1 and all entries off the main diagonal except one are equal to zero.

Linear groups

For any linear space X the set $\operatorname{Aut} X$ has a natural group structure. The group $\operatorname{Aut} \mathbf{K}^n$ is usually denoted by $GL(n;\mathbf{K})$ and called the *general linear group of degree n.* The subgroup of $GL(n;\mathbf{K})$ consisting of all automorphisms of $\mathbf{K}^n$ of determinant 1 is called the *special linear group of degree n*, denoted by $SL(n;\mathbf{K})$.

For any *real* linear space X there is a map $\zeta : \operatorname{End} X \to \{-1,0,1\}$, taking the value 1 if the determinant is positive, the value -1 if the determinant is negative and the value 0 if the determinant is zero. Automorphisms for which the value of ζ is equal to 1 are said to be *orientation-preserving*, while those for which the value is equal to -1 are said to be *orientation-reversing*. For any finite-dimensional linear space X the restriction of ζ to $\operatorname{Aut} X$ is a group isomorphism with the multiplicative group $S^0 = \{\pm 1\}$. The orientation-preserving automorphisms of X form a subgroup of $\operatorname{Aut} X$ which we shall denote by $\operatorname{Aut}^+ X$.

The question of orientation does not arise for complex linear spaces since there is no notion of a *positive* complex number.

Exercises

1.1 Let $f : X \to Y$ and $g : Y \to X$ be such that $fg = 1_Y$. Prove that f is surjective and that g is injective.

1.2 Let W, X and Y be linear spaces and let $t : X \to Y$ and $u : W \to X$ be maps whose composite $tu : W \to Y$ is linear.

Then

(a) if t is a linear injection, u is linear,
(b) if u is a linear surjection, t is linear.

1.3 Let X and Y be $\mathbf{K}$-linear spaces. Prove that

$$X \times Y = (X \times \{0\}) \oplus (\{0\} \times Y).$$

1.4 Let $\mathbf{K}$ be either the field $\mathbf{R}$ or the field $\mathbf{C}$ and consider the product $\mathbf{K}^{2n} \times \mathbf{K}^{2n} \to \mathbf{K};\ (x, y) \mapsto x \wedge y$ defined by

$$x \wedge y = \sum_{i \in \mathbf{n}} (x_i y_{n+i} - x_{n+i} y_i).$$

Verify that the product is bilinear and that, for every $x, y \in \mathbf{K}^{2n}$, $y \wedge x = -x \wedge y$.

Now define $\theta : \mathbf{K}(2n) \to \mathbf{K}$ by the formula

$$\theta(a) = \frac{1}{n!2^n} \sum_{\pi \in \mathbf{2n}!} \operatorname{sgn} \pi \prod_{i \in \mathbf{n}} a_{\pi(i)} \wedge a_{\pi(n+i)},$$

where $\mathbf{2n}!$ denotes the set of permutations of all the natural numbers m such that $0 \le m < 2n$.

Verify that this is an alternating $2n$-linear map on the columns of the matrix a, with $\theta({}^{2n}1) = 1$, where ${}^{2n}1$ denotes the unit $2n \times 2n$ matrix, and therefore that $\theta(a) = \det a$.

This exercise will be of use in Proposition 6.11.

2
Real and complex algebras

A *linear algebra* over the field of real numbers $\mathbf{R}$ is, by definition, a linear space A over $\mathbf{R}$ together with a bilinear map $A^2 \to A$, the *algebra product*.

Examples include $\mathbf{R}$ itself, the field of *complex numbers*, $\mathbf{C}$, consisting of the linear space $\mathbf{R}^2$ with the product $(a,b)(c,d) = (ac - bd, ad + bc)$, the *double field* ${}^2\mathbf{R}$ consisting of the linear space $\mathbf{R}^2$ with the product $(a,b)(c,d) = (ac, bd)$, and the *full matrix algebra* $\mathbf{R}(n)$ of all $n \times n$ matrices with real entries, with matrix multiplication as the product.

An algebra A may, or may not, have a unit element, and the product need be neither commutative nor associative, though it is usual to mention explicitly any failure of associativity. The unit element, if it exists, will normally be denoted by $1_{(A)}$ or simply, where no confusion need arise, by 1, the map $\mathbf{R} \to A; \lambda \mapsto \lambda 1_{(A)}$ being injective. (The notation 1_A is reserved for the identity map on A.)

All the above examples are associative and have a unit element, and all are commutative, with the exception of the matrix algebra $\mathbf{R}(n)$, with $n > 1$. The double field ${}^2\mathbf{R}$ is often identified with the subalgebra of $\mathbf{R}(2)$ consisting of the diagonal 2×2 matrices, the unit element being denoted by ^{21}. Likewise, for any n the n-fold power ${}^n\mathbf{R}$ of $\mathbf{R}$ may be identified with the subalgebra of the algebra $\mathbf{R}(n)$ consisting of the diagonal $n \times n$ matrices, the unit matrix in $\mathbf{R}(n)$ similarly being denoted by ${}^n1$.

Examples of non-associative algebras include the Cayley algebra and Lie algebras, discussed in later chapters.

Concepts defined in the obvious ways include not only *subalgebras* but *algebra maps*, *algebra-reversing maps*, and in particular *algebra isomorphisms* and *algebra anti-isomorphisms*, the latter for example being a linear isomorphism f of one algebra A to another B that reverses the order of multiplication, that is, for all $x, y \in A$, and all $\lambda, \mu \in \mathbf{R}$,

$$f(\lambda x + \mu y) = \lambda f(x) + \mu f(y), \qquad \text{and} \qquad f(xy) = f(y)f(x).$$

The *centre* of a real algebra A is the set of all those elements of A that commute with each element of A. For example the centre of $\mathbf{R}(n)$ consists of all real multiples of the identity ${}^n1$. The centre of an algebra A is a subalgebra of A.

Analogous definitions hold for linear algebras not only over any field, in particular the field $\mathbf{C}$ of complex numbers, but also over the double fields ${}^2\mathbf{R}$ and ${}^2\mathbf{C}$. The part of a ${}^2\mathbf{R}$-linear space is played by an $\mathbf{R}$-linear space A with a prescribed direct sum decomposition $A_0 \oplus A_1$, where A_0 and A_1 are isomorphic linear subspaces of A (so, in the case that A is finite-dimensional, $\dim A_0 = \dim A_1$) with scalar multiplication defined by

$$((\lambda, \mu), (x_0 + x_1)) \mapsto (\lambda x_0 + \mu x_1).$$

For example, the elements of ${}^2\mathbf{R}(2)$ may be represented as matrices either of the form $\begin{pmatrix} a_0 & 0 & c_0 & 0 \\ 0 & a_1 & 0 & c_1 \\ b_0 & 0 & d_0 & 0 \\ 0 & b_1 & 0 & d_1 \end{pmatrix}$ or of the form $\begin{pmatrix} a_0 & c_0 & 0 & 0 \\ b_0 & d_0 & 0 & 0 \\ 0 & 0 & a_1 & c_1 \\ 0 & 0 & b_1 & d_1 \end{pmatrix}$.

Our preference is for the second form, where $\mathbf{R}^4 = \mathbf{R}^2 \times \{0\} \oplus \{0\} \times \mathbf{R}^2$ is thought of as $\mathbf{R}^2 \times \mathbf{R}^2$.

For any k with $0 \le k < n$ the kth and $(n+k)$th columns of ${}^2\mathbf{R}(n)$ will be said to be *partners*. Thus in the example the columns $\begin{pmatrix} c_0 \\ d_0 \\ 0 \\ 0 \end{pmatrix}$ and $\begin{pmatrix} 0 \\ 0 \\ c_1 \\ d_1 \end{pmatrix}$ are partners.

Similar remarks apply to the algebra ${}^2\mathbf{C}(n)$.

Any algebra over $\mathbf{C}$, ${}^2\mathbf{R}$ or ${}^2\mathbf{C}$ may also be regarded as an algebra over $\mathbf{R}$, while any algebra over ${}^2\mathbf{C}$ may be regarded as an algebra over $\mathbf{C}$. In particular, as the corollary to the next proposition shows, the real algebra $\mathbf{C}$ is isomorphic to a subalgebra of the algebra $\mathbf{R}(2)$ of 2×2 real matrices.

Proposition 2.1 *Let* $\mathbf{C}$ *be identified with* $\mathbf{R}^2$ *in the standard way. Then, for any* $c = a + \mathrm{i}\,b \in \mathbf{C}$, *the real linear map* $\mathbf{C} \to \mathbf{C};\ z \mapsto c\,z$ *has matrix* $\begin{pmatrix} a & -b \\ b & a \end{pmatrix}$.

Corollary 2.2 *The set of matrices* $\left\{ \begin{pmatrix} a & -b \\ b & a \end{pmatrix} : (a, b) \in \mathbf{R} \right\}$ *is a subalgebra of the algebra* $\mathbf{R}(2)$, *isomorphic to* $\mathbf{C}$.

Note that the determinant of the real matrix $\begin{pmatrix} a & -b \\ b & a \end{pmatrix}$ is equal to $a^2 + b^2 = (\overline{a + \mathrm{i}\,b})(a + \mathrm{i}\,b) = |a + \mathrm{i}\,b|^2$.

Proposition 2.3 *Let* X *be a finite-dimensional complex linear space and let* $X_{\mathbf{R}}$ *be the underlying real linear space. Then, if* $t : X \to X$ *is a complex linear map,*

$$\det_{\mathbf{R}} t = |\det_{\mathbf{C}} t|^2,$$

where $\det_{\mathbf{C}} t$ *is the determinant of* t *regarded as a complex linear map and* $\det_{\mathbf{R}} t$ *is the determinant of* t *regarded as a real linear map.*

Proof By Proposition 1.8 the matrix of t with respect to any basis for X is reducible by elementary column operations, all of determinant 1 in either the real or the complex sense, to a diagonal complex $n \times n$ matrix all of whose entries except one are equal to 1, and for such a matrix the statement is clearly true, by the above remark. □

Corollary 2.4 *Let* X *be as in Proposition 2.3, and let* $t : X \to X$ *be a complex linear map. Then* $\det_{\mathbf{R}} t \geq 0$.

Minimal ideals

Let $A = \mathbf{K}(n)$, where $\mathbf{K} = \mathbf{R}$ or $\mathbf{C}$. A *left ideal* $\mathscr{I}$ of A is a linear subspace $\mathscr{I}$ of A such that, for all $x \in \mathscr{I}$ and all $a \in A$, $ax \in \mathscr{I}$. *Right ideals* are similarly defined.

Proposition 2.5 *Let* X *be a* $\mathbf{K}$*-linear space, and let* $t \in \operatorname{End} X$. *Then the subset*

$$\mathscr{I}(t) = \{at \in \operatorname{End} X : a \in \operatorname{End} X\}$$

is a left ideal of $\operatorname{End} X$.

A left ideal $\mathscr{I}$ of A is said to be *minimal* if the only proper subset of $\mathscr{I}$ which is a left ideal of A is $\{0\}$.

Theorem 2.6 *Let X be a finite-dimensional $\mathbf{K}$-linear space. The minimal left ideals of* $\operatorname{End} X$ *are those of the form*

$$\mathscr{I}(t) = \{at : a \in \operatorname{End} X\},$$

where $t \in \operatorname{End} X$ and $\operatorname{rk} t = 1$.

Proof Suppose first that $\mathscr{I}$ is a minimal left ideal of $\operatorname{End} X$. Then, for any $t \in \mathscr{I}$, $\mathscr{I}(t)$ is a left ideal of $\operatorname{End} X$ and a subset of $\mathscr{I}$. Since $\mathscr{I}$ is minimal, it follows that $\mathscr{I} = \mathscr{I}(t)$, for any non-zero $t \in \mathscr{I}$.

Suppose that $\operatorname{rk} t > 1$. Then, for any $s \in \operatorname{End} X$ with $\operatorname{rk} st = 1$, $\mathscr{I}(st)$ is a proper subset of $\mathscr{I}(t)$. Since there is such an s, it follows that $\mathscr{I}(t)$ is not minimal.

So $\mathscr{I}(t)$ is minimal if and only if $\operatorname{rk} t = 1$. □

The minimal left ideals remain the same even if $\operatorname{End} X$ is regarded as an algebra over any subfield of the field $\mathbf{K}$.

We shall generally index the columns of an $m \times n$ matrix by the integers $0, 1, 2, 3, \ldots, n-1$, a set we denote by $\mathbf{n}$.

Proposition 2.7 *The matrices in $\mathbf{K}(n)$ all of whose columns except the zeroth are zero form a minimal left ideal of the algebra $\mathbf{K}(n)$.*

For this reason it is possible to think of the linear space $\mathbf{K}^n$ on which the matrix algebra $\mathbf{K}(n)$ acts as a minimal left ideal of that algebra. Similar remarks may be made about minimal right ideals.

Algebra maps

In practice one often wishes to construct an algebra map or an algebra-reversing map of one algebra, A, to another, B, and such a map, in so far as it must be linear, will be determined by its restriction to any linear basis for A. However, the converse is no longer true – we are not free to assign arbitrarily the values in B of a map of some basis for A to B, and then to extend this to an algebra map of the whole of A to B. In general such an extension will not be possible.

In fact what one normally starts with is a subset S of A that generates A as a $\mathbf{K}$-algebra, the subset S being said to *generate* A if each element of A is expressible, possibly in more than one way, as a linear combination of a finite sequence of elements of A each of which is the product of a finite sequence of elements of S. For example, the set of matrices $\left\{\begin{pmatrix} 1 & 0 \\ 0 & -1 \end{pmatrix}, \begin{pmatrix} 0 & 1 \\ 1 & 0 \end{pmatrix}\right\}$ generates $\mathbf{R}(2)$ as an $\mathbf{R}$-algebra. The following is then true.

Proposition 2.8 *Let A and B be algebras over a field $\mathbf{K}$ and let S be a subset of A that generates A. Then any algebra or algebra-reversing map $t : A \to B$ is uniquely determined by its restriction to S.*

There will be much interest later in certain automorphisms and anti-automorphisms of linear algebras. An *automorphism* of an algebra A is an algebra isomorphism $A \to A$, while an *anti-automorphism* of A is an algebra anti-isomorphism $A \to A$. An automorphism or anti-automorphism f of A such that $f^2 = 1_A$ is said to be, respectively, an *involution* or an *anti-involution* of A.

In the sequel it will often be convenient to denote an automorphism ϕ of an algebra A by $x \mapsto x^{\phi}$, rather than by $x \mapsto \phi(x)$. The composite $\phi\psi$ of two automorphisms of A will then be denoted by $x \mapsto x^{\phi\psi}$, and not by $x \mapsto x^{\psi\phi}$. In fact, however, in most of the uses of the notation that we have in mind, when we have two automorphisms of the same algebra they will commute.

Proposition 2.9 *The only automorphism of* $\mathbf{R}$ *is the identity* $1_{\mathbf{R}}$.

Proof Let f be an automorphism of $\mathbf{R}$. Necessarily f sends 0 to 0 and 1 to 1, from which it follows by an easy argument that f sends each rational number to itself. Also the order of the elements of $\mathbf{R}$ is determined by the field structure. So f also respects order and, in particular, limits of sequences. Since each real number is the limit of a convergent sequence of rational numbers and since the limit of a convergent sequence on $\mathbf{R}$ is unique, it follows that f sends each element of $\mathbf{R}$ to itself. □

Proposition 2.10 *The only algebra automorphisms of* $\mathbf{C}$, *regarded as a real algebra, are the identity,* $1_{\mathbf{C}}$ *and conjugation.*

However, unlike $\mathbf{R}$, the *field* $\mathbf{C}$ has many automorphisms. (See Segre (1947) and also Fuchs (1963), page 122.) Conjugation is the only field automorphism other than the identity that sends each real number to itself, but a field isomorphism of $\mathbf{C}$ need not send each real number to itself. It is true that each *rational* number must be sent to itself, but the remainder of the proof in the real case is no longer applicable. Indeed one of the non-standard automorphisms of **C** sends $\sqrt{2}$ to $-\sqrt{2}$. What is implied by these remarks is that the real subfield of **C** is not uniquely determined by the field structure of **C** alone. The field injection or inclusion $\mathbf{R} \to \mathbf{C}$ is an additional piece of structure. In practice the additional structure is usually taken for granted. It is unusual to say so explicitly.

Proposition 2.11 *The only algebra automorphisms of the real algebra* $^2\mathbf{R}$ *are the identity and* swap $\sigma : {}^2\mathbf{R} \to {}^2\mathbf{R};\ (\lambda, \mu) \mapsto (\mu, \lambda)$, *the analogy with conjugation on* **C** *being made evident if we remark that*

$$(\lambda(1,1) + \mu(1,-1))^{\sigma} = \lambda(1,1) - \mu(1,-1).$$

There seems to be no recognised notation for swap. In Porteous (1969) I used the clumsy notation hb, this being an abbreviation for *hyperbolic*, and suggested by the observation that the set $\{(\lambda, \mu) \in {}^2\mathbf{R} : (\lambda, \mu)^{\sigma}(\lambda, \mu) = 1\}$ is just the rectangular hyperbola $\{(\lambda, \mu) : \lambda\,\mu = 1\}$.

Two automorphisms or anti-automorphisms β, γ of an algebra A are said to be *similar* if there is an automorphism α of A such that $\gamma\alpha = \alpha\beta$. If no such exists then β and γ are said to be *dissimilar*.

Proposition 2.12 *The identity and conjugation on* **C** *are dissimilar.*

Proposition 2.13 *The identity and swap on* $^2\mathbf{R}$ *are dissimilar.*

Proposition 2.14 *The involutions* $\sigma : {}^2\mathbf{C} \to {}^2\mathbf{C} : (\lambda, \mu) \to (\mu, \lambda)$ *and* $\overline{\sigma} : {}^2\mathbf{C} \to {}^2\mathbf{C} : (\lambda, \mu) \to (\overline{\mu}, \overline{\lambda})$ *are* similar.

Proof Define $\phi : {}^2\mathbf{C} \to {}^2\mathbf{C}; (\lambda,\ \mu) \mapsto (\lambda,\ \overline{\mu})$. Then $\phi\,\sigma = \overline{\sigma}\,\phi$; since, for any $(\lambda,\ \mu) \in {}^2\mathbf{C}$,

$$\phi\,\sigma(\lambda,\ \mu) = \phi(\mu,\ \lambda) = (\mu,\ \overline{\lambda}) = \overline{\sigma}(\lambda,\ \overline{\mu}) = \overline{\sigma}\,\phi(\lambda,\ \mu).$$

□

Irreducible automorphisms and anti-automorphisms

An *idempotent* of an algebra A is an element a of A such that $a^2 = a$. (The word means that any *power* of the element is the *same* as the element itself.) A *primitive idempotent* of A is an idempotent of A that is not the sum of two non-zero idempotents of A.

Proposition 2.15 *Any automorphism or anti-automorphism of a real algebra A permutes the primitive idempotents of A.*

Proposition 2.16 *Let* $\mathbf{K}$ *be any field. Then, for any positive n, the elements of the standard basis for* $\mathbf{K}^n$ *are the primitive idempotents of the n-fold power of* $\mathbf{K}$, ${}^n\mathbf{K}$.

A permutation π of a finite set S is said to be *reducible* if there is a proper subset T of S such that $\pi(T) = T$, and an automorphism or anti-automorphism of an algebra A is said to be *reducible* if the induced permutation of the primitive idempotents of A is reducible. A permutation or automorphism or anti-automorphism that is not reducible is said to be *irreducible.*

Proposition 2.17 *Let n be a positive number such that* ${}^n\mathbf{K}$ *admits an irreducible involution. Then* $n = 1$ *or* 2.

Proposition 2.18 *An automorphism of* ${}^2\mathbf{K}$ *is reducible if and only if it is of the form*

$$ {}^2\mathbf{K} \to {}^2\mathbf{K};\ (\lambda, \mu) \mapsto (\lambda^\chi, \mu^\phi), $$

where $\chi, \phi : \mathbf{K} \to \mathbf{K}$ *are both automorphisms of* $\mathbf{K}$. *It is an involution of* ${}^2\mathbf{K}$ *if and only if both* χ *and* ϕ *are involutions of* $\mathbf{K}$.

Such an automorphism is denoted by $\chi \times \phi$.

More interesting are the irreducible automorphisms of ${}^2\mathbf{K}$.

Proposition 2.19 *An automorphism of* ${}^2\mathbf{K}$ *is irreducible if and only if it is of the form*

$$ {}^2\mathbf{K} \to {}^2\mathbf{K};\ (\lambda, \mu) \mapsto (\mu^\phi, \lambda^\chi), $$

where $\chi, \phi : \mathbf{K} \to \mathbf{K}$ *are both automorphisms of* $\mathbf{K}$. *An involution of* ${}^2\mathbf{K}$ *is irreducible if and only if it is of that form with, moreover,* $\chi = \phi^{-1}$.

Thus, for example, for any field $\mathbf{K}$, swap is an irreducible involution of ${}^2\mathbf{K}$.

There is the following generalisation of Proposition 2.14.

Proposition 2.20 *For any automorphisms* ϕ *and* χ *of a field* $\mathbf{K}$ *the involutions* $(\phi \times \phi^{-1})\sigma$ *and* $(\chi \times \chi^{-1})\sigma$ *of* ${}^2\mathbf{K}$ *are similar.*

Proof Let $\alpha = \chi^{-1}\phi$. Then

$$(\chi \times \chi^{-1})\sigma(1 \times \alpha) = (1 \times \alpha)(\phi \times \phi^{-1})\sigma,$$

since, for any $(\lambda, \mu) \in {}^2\mathbf{K}$,

$$\begin{aligned}(\chi \times \chi^{-1})\sigma(1 \times \alpha)\begin{pmatrix}\lambda & 0\\ 0 & \mu\end{pmatrix} &= (\chi \times \chi^{-1})\sigma\begin{pmatrix}\lambda & 0\\ 0 & \mu^{\alpha}\end{pmatrix}\\ &= \begin{pmatrix}\mu^{\chi\alpha} & 0\\ 0 & \lambda^{\chi^{-1}}\end{pmatrix}\end{aligned}$$

and

$$\begin{aligned}(1 \times \alpha)(\phi \times \phi^{-1})\sigma\begin{pmatrix}\lambda & 0\\ 0 & \mu\end{pmatrix} &= (1 \times \alpha)\begin{pmatrix}\mu^{\phi} & 0\\ 0 & \lambda^{\phi^{-1}}\end{pmatrix}\\ &= \begin{pmatrix}\mu^{\phi} & 0\\ 0 & \lambda^{\alpha\phi^{-1}}\end{pmatrix}.\end{aligned}$$

□

Exercises

2.1 An element s of the matrix algebra $\mathbf{R}(n)$ is said to be *skew* if $s^{\tau} = -s$. Show that if s is a skew element of $\mathbf{R}(n)$ then, for any $x \in \mathbf{R}^n$, $x^{\tau}s\,x = 0$. Deduce that for such an element s the linear map $1 - s : \mathbf{R}^n \to \mathbf{R}^n;\ x \mapsto x - s(x)$ is injective and therefore invertible.

2.2 Let s be a skew element of $\mathbf{R}(n)$. Prove that, for all $u, v \in \mathbf{R}^n$, $(1 - s)v = (1 + s)u \Rightarrow v^{\tau}v = u^{\tau}u$.

2.3 Prove that ${}^2\mathbf{R}$ is isomorphic to the subalgebra of $\mathbf{R}(2)$ consisting of all matrices of the form $\begin{pmatrix}a & b\\ b & a\end{pmatrix}$, the map

$$a(1,1) + b(1,-1) \mapsto \begin{pmatrix}a & b\\ b & a\end{pmatrix}$$

being an isomorphism.

2.4 Prove that

$$\mathbf{R}(2) \to \mathbf{R}(2);\ \begin{pmatrix}a & c\\ b & d\end{pmatrix} \mapsto \begin{pmatrix}a & -b\\ -c & d\end{pmatrix}$$

and

$$\mathbf{R}(2) \to \mathbf{R}(2);\ \begin{pmatrix}a & c\\ b & d\end{pmatrix} \mapsto \begin{pmatrix}d & c\\ b & a\end{pmatrix}$$

are similar anti-involutions of $\mathbf{R}(2)$.

2.5 Prove that

$$\mathbf{R}(2) \to \mathbf{R}(2);\ \begin{pmatrix} a & c \\ b & d \end{pmatrix} \mapsto \begin{pmatrix} a & b \\ c & d \end{pmatrix},$$

$$\mathbf{R}(2) \to \mathbf{R}(2);\ \begin{pmatrix} a & c \\ b & d \end{pmatrix} \mapsto \begin{pmatrix} a & -b \\ -c & d \end{pmatrix}$$

and

$$\mathbf{R}(2) \to \mathbf{R}(2);\ \begin{pmatrix} a & c \\ b & d \end{pmatrix} \mapsto \begin{pmatrix} d & -c \\ -b & a \end{pmatrix}$$

are dissimilar anti-involutions of $\mathbf{R}(2)$.

Hint: Track what happens to the matrices

$$\begin{pmatrix} 1 & 0 \\ 0 & 1 \end{pmatrix}, \begin{pmatrix} 1 & 0 \\ 0 & -1 \end{pmatrix}, \begin{pmatrix} 0 & 1 \\ 1 & 0 \end{pmatrix}, \text{ and } \begin{pmatrix} 0 & 1 \\ -1 & 0 \end{pmatrix},$$

all of which have squares in $\mathbf{R} = \mathbf{R}\,1_{\mathbf{R}(2)}$.

2.6 Prove that the primitive idempotents of $\mathbf{R}(n)$ are those matrices similar to u, where u has a single non-zero entry on the main diagonal, all off-diagonal entries being zero, that is, is of the form aua^{-1}, where a is any invertible element of $\mathbf{R}(n)$.

3
Exact sequences

The properties of exact sequences of linear maps and group maps are summarised and left-coset space representations are introduced.

Exact sequences of linear maps

Let s and t be linear maps such that the target of s is also the source of t. Such a pair of maps is said to be *exact* if $\operatorname{im} s = \ker t$. Note that this is stronger than the assertion that $ts = 0$, which is equivalent to the condition $\operatorname{im} s \subset \ker t$ only. A possibly doubly infinite sequence of linear maps such that the target of each map coincides with the source of its successor is said to be *exact* if each pair of adjacent maps is exact.

Proposition 3.1 *Let W be a linear subspace of a linear space X. Then the sequence of linear maps*

$$\{0\} \longrightarrow W \xrightarrow{\iota} X \xrightarrow{\pi} X/W \longrightarrow \{0\},$$

where ι is the inclusion and π is the partition, is exact.

An exact sequence of linear maps of the form

$$\{0\} \to W \xrightarrow{s} X \xrightarrow{t} Y \to \{0\}$$

is said to be a *short exact sequence*. The exactness of the sequence at W is equivalent to the injectivity of s and the exactness at Y to the surjectivity of t. Given such a sequence one thinks of W as a subspace of X and of Y as the quotient space X/W. The fibres of t are the parallels in X of the linear subspace which is the image of s.

The following proposition is a first example of the technique known as *diagram-chasing*.

Proposition 3.2 *Let*

$$\begin{array}{ccccccccc} \{0\} & \longrightarrow & W & \xrightarrow{s} & X & \xrightarrow{t} & Y & \longrightarrow & \{0\} \\ & & \downarrow\alpha & & \downarrow\beta & & & & \\ \{0\} & \longrightarrow & W' & \xrightarrow{s'} & X' & \xrightarrow{t'} & Y' & \longrightarrow & \{0\} \end{array}$$

be a diagram of linear maps such that the rows are exact and $\beta s = s'\alpha$, *that is the square formed by these maps is* commutative. *Then there exists a unique linear map* $\gamma : Y \to Y'$ *such that* $\gamma t = t'\beta$, *and if* α *and* β *are isomorphisms then* γ *also is an isomorphism.*

Proof *Uniqueness of* γ: Suppose that γ exists. By hypothesis $\gamma t(x) = t'\beta(x)$, for all $x \in X$. Since for each $y \in Y$ there exists x in $t^{-1}\{y\}$, $\gamma(y) = t'\beta(x)$ for any such x; that is, γ is uniquely determined.

Existence of γ: Let $y \in Y$ and let $x, x_1 \in t^{-1}\{y\}$. Then $x_1 - x \in \ker t = \operatorname{im} s$ and so $x_1 = x + s(w)$, for some $w \in W$. Then

$$\begin{aligned} t'\beta(x_1) &= t'\beta(x + s(w)) = t'\beta(x) + t'\beta\, s(w) \\ &= t'\beta(x) + t's'\alpha(w) \\ &= t'\beta(x), \text{ since } t's' = 0. \end{aligned}$$

The prescription $\gamma(y) = t'\beta(x)$, for any x in $t^{-1}\{y\}$, does therefore determine a map $\gamma : Y \to Y'$ such that $\gamma t = t'\beta$. Since $t'\beta$ is linear and t is a linear surjection, γ is linear, by Exercise 1.2.

Now suppose that α and β are isomorphisms, and let $\eta : Y \to Y'$ be the unique linear map such that $\eta t' = t\beta^{-1}$. Then applying the uniqueness part of the proposition to the diagram

$$\begin{array}{ccccccccc} \{0\} & \longrightarrow & W & \xrightarrow{s} & X & \xrightarrow{t} & Y & \longrightarrow & \{0\} \\ & & \downarrow 1_W & & \downarrow 1_X & & \downarrow \eta\gamma & & \\ \{0\} & \longrightarrow & W & \xrightarrow{s} & X & \xrightarrow{t} & Y & \longrightarrow & \{0\} \end{array}$$

yields $\eta\gamma = 1_Y$. Similarly $\gamma\eta = 1_{Y'}$. That is, $\eta = \gamma^{-1}$ and so γ is an isomorphism. □

Proposition 3.3 is an important special case.

Proposition 3.3 *Let* W, X *and* Y *be finite-dimensional linear spaces, and let*

$$\{0\} \to W \xrightarrow{s} X \xrightarrow{t} Y \to \{0\}$$

be a short exact sequence. Then the dual sequence

$$\{0\} \to Y^L \xrightarrow{t^L} X^L \xrightarrow{s^L} W^L \to \{0\}$$

is exact.

Analogues for group maps

The *definition* of an exact sequence goes over without change to sequences of group maps, as does the definition of a short exact sequence, a *group map* being synonymous with a *group homomorphism*, a map between groups that respects the group products. In work with multiplicative groups the symbol $\{1\}$ is usually used in place of $\{0\}$ to denote the one-element group. Of course a subgroup of a group need not be normal, and in many of the situations that we shall encounter it will not be. This possibility prompts the following definition, extending the notion of short exact sequence.

Let F and G be groups, let H be a set, and let

$$F \xrightarrow{s} G \xrightarrow{t} H$$

be a pair of maps such that s is a group injection and t is a surjection whose fibres are the left cosets on the image of s in G, the set of such left cosets being denoted by G/F, the specific injection s being understood. The pair will then be said to be *left-coset exact*, and the induced bijection $G/F \to H$ will be called a *(left-) coset space representation* of the set H. Numerous examples will be given later.

Proposition 3.4 *Let F, G, F' and G' be groups, let H, H', M and N be sets, and let*

$$\begin{array}{ccccc} F & \xrightarrow{s} & G & \xrightarrow{t} & H \\ \downarrow{\scriptstyle\alpha} & & \downarrow{\scriptstyle\beta} & & \\ F' & \xrightarrow{s'} & G' & \xrightarrow{t} & H' \\ \downarrow{\scriptstyle\mu} & & \downarrow{\scriptstyle\nu} & & \\ M & & N & & \end{array}$$

be a commutative diagram of maps whose rows and columns are left-coset exact. Then if there is a (necessarily unique) bijection $u : M \to N$ such that $u\mu = \nu s'$ there is a unique bijection $\gamma : H \to H'$ such that $\gamma t = t' \beta$. If, moreover, H and H' are groups and if t and t' are group maps, then γ is a group isomorphism.

Group actions

Let G be a group and X a set. Then a map $G \times X \to X$; $g \mapsto g\,x$ is said to be a *(left) action* of G on X if, for all $x \in X$, and $g, g' \in G$,

$$(g'\,g)x = g'(g\,x) \quad \text{with} \quad 1_{(G)}\,x = x.$$

For any $a \in X$ the subset $G_a = \{g \in G : g\,a = a\}$ is then a subgroup of G called the *isotropy subgroup* of the action of G at a or the *stabiliser* of a in G. It is easy to verify that the relation $\sim$ on X defined by

$$x \sim x' \Leftrightarrow \text{for some } g \in G,\ x' = g\,x$$

is an equivalence. Each equivalence set is said to be an *orbit* of the action. If there is only a single orbit, namely the whole of X, then the action of G on X is said to be *transitive*. In this case, for any $a \in X$, the sequence

$$\begin{array}{ccccc} G_a & \xrightarrow{\iota} & G & \xrightarrow{a_R} & X \\ & & g & \longmapsto & g\,a \end{array}$$

is left-coset exact.

Similar remarks apply to group actions on the *right*. Consider, in particular, the case of a subgroup G of a group G' acting on G' on the right by the map $G' \times G \to G'$; $(g', g) \mapsto g'\,g$. In this case the set of orbits of the action coincides with the set of *left* cosets on G in G', G'/G. For this reason the set of orbits of a group G acting on a set X on the *right* may without confusion be denoted by X/G.

Exercise

3.1 Let $G \times X \to X$; $(g, x) \mapsto gx$ be a left action of a group G on a set X. Prove that the subset $G_a = \{g \in G : ga = a\}$ is a subgroup of G and that the relation $\sim$ on X defined by $x \sim x' \Leftrightarrow$ for some $g \in G$, $x' = gx$ is an equivalence on X. Prove also that if the action is transitive then, for any $a \in X$, the sequence $G_a \xrightarrow{\iota} G \xrightarrow{a_R} X$ is left-coset exact, while, for any $a, b \in X$ with $b = ha$, where $h \in G$, and for any $g \in G_a$, then $hgh^{-1} \in G_b$, the map $G_a \to G_b$; $g \mapsto hgh^{-1}$ being a group isomorphism.

4
Real quadratic spaces

A *real quadratic space* is a finite-dimensional real linear space with an assigned quadratic form.

Quadratic forms

A quadratic form on a real linear space X is most conveniently introduced in terms of a *symmetric scalar product* on X. This, by definition, is a bilinear map

$$X^2 \to \mathbf{R};\ (a, b) \mapsto a \cdot b$$

such that, for all $a, b \in X$, $b \cdot a = a \cdot b$. The map

$$X \to \mathbf{R};\ a \mapsto a \cdot a$$

is called the *quadratic form* of the scalar product, $a^{(2)} = a \cdot a$ being called the *square* or *quadratic norm* of a. (The notation a^2 is reserved for later use.) Since, for each $a, b \in X$,

$$2a \cdot b = a^{(2)} + b^{(2)} - (a - b)^{(2)},$$

the scalar product is uniquely determined by its quadratic form. In particular, the scalar product is the zero map if and only if its quadratic form is the zero map.

The following are examples of scalar products on $\mathbf{R}^2$:

$$((x, y), (x'y')) \mapsto 0,\ xx',\ xx' + yy',\ -xx' + yy' \text{ and } xy' + yx',$$

their respective quadratic forms being

$$(x, y) \mapsto 0,\ x^2,\ x^2 + y^2,\ -x^2 + y^2 \text{ and } 2xy.$$

A finite-dimensional real linear space with symmetric scalar product will be called a *(real) quadratic space*, any linear subspace W of a quadratic space X being tacitly assigned the restriction to W^2 of the scalar product for X.

A quadratic space X is said to be *positive-definite* if, for all non-zero $a \in X$, $a^{(2)} > 0$, and to be *negative-definite* if, for all non-zero $a \in X$, $a^{(2)} < 0$. An example of a positive-definite space is the linear space $\mathbf{R}^2$ with the scalar product

$$((x, y), (x', y')) \mapsto xx' + yy'.$$

A quadratic space whose scalar product is the zero map is said to be *null* or *isotropic* and a quadratic space of dimension $2n$ that is the direct sum of two null subspaces of dimension n is said to be *neutral*.

It is convenient to have short notations for the quadratic spaces that most commonly occur in practice. The linear space $\mathbf{R}^{p+q}$ with the scalar product

$$(a, b) \mapsto -\sum_{0 \le i < p} a_i b_i + \sum_{0 \le j < q} a_{p+j} b_{p+j}$$

will therefore be denoted by $\mathbf{R}^{p,q}$, while the linear space $\mathbf{R}^{2n}$ with the scalar product

$$(a, b) \mapsto \sum_{0 \le i < n} (a_i b_{n+i} + a_{n+i} b_i)$$

will be denoted by $\mathbf{R}^{2n}_{\text{hb}}$, the quadratic space $\mathbf{R}^2_{\text{hb}}$ being called the *standard hyperbolic plane*.

The linear space underlying $\mathbf{R}^{p,q}$ will frequently be identified with $\mathbf{R}^p \times \mathbf{R}^q$ and the linear space underlying $\mathbf{R}^{2n}_{\text{hb}}$ with $\mathbf{R}^n \times \mathbf{R}^n$. The linear subspaces $\mathbf{R}^n \times \{0\}$ and $\{0\} \times \mathbf{R}^n$ of $\mathbf{R}^n \times \mathbf{R}^n$ are null spaces of $\mathbf{R}^{2n}_{\text{hb}}$. This orthogonal space is therefore neutral. The quadratic spaces $\mathbf{R}^{0,n}$ and $\mathbf{R}^{n,0}$ are, respectively, positive-definite and negative-definite.

The dot notation for the scalar product derives from its traditional use on $\mathbf{R}^{0,n}$, that is $\mathbf{R}^n$ with its standard positive-definite scalar product, in which context vectors a and b are said to be *orthogonal* if $a \cdot b = 0$, subsets A and B of the space being said to be *mutually orthogonal* if, for each $a \in A$, $b \in B$, $a \cdot b = 0$. It will be convenient to use the same terminology without further comment for any quadratic space, whether or not the assigned scalar product is positive-definite. Alternative notations will be introduced later in cases where more than one scalar product is assigned at the same time to the same linear space.

An element a of a quadratic space X is said to be *invertible* if $a^{(2)} \neq 0$, the element $a^{(-1)} = (a^{(2)})^{-1}a$ being called the *inverse* of a. Every non-null real quadratic space X possesses invertible elements, since the quadratic form on X is zero only if the scalar product is zero.

Proposition 4.1 *Let a be an invertible element of a quadratic space X. Then, for some $\lambda \in \mathbf{R}, (\lambda a)^{(2)} = \pm 1$.*

Proof Since $(\lambda a) \cdot (\lambda a)^{(2)} = \lambda^2 a^{(2)}$ and since $a^{(2)} \neq 0$ we may choose $\lambda = (\sqrt{(|a^{(2)}|)})^{-1}$. □

Proposition 4.2 *If a and b are invertible elements of a quadratic space X with $a^{(2)} = b^{(2)}$, than $a + b$ and $a - b$ are mutually orthogonal and either $a + b$ or $a - b$ is invertible.*

Proof Since $a^{(2)} = b^{(2)}$ it follows that $(a + b) \cdot (a - b) = 0$, while $(a + b)^{(2)} + (a - b)^{(2)} = 4a^{(2)} \neq 0$. □

The elements $a + b$ and $a - b$ need not both be invertible even when $a \neq \pm b$. Consider, for example, $\mathbf{R}^{1,2}$. In this case we have $(1, 1, 1)^{(2)} = (1, 1, -1)^{(2)} = 1$, and $(0, 0, 2) = (1, 1, 1) - (1, 1, -1)$ is invertible, since $(0, 0, 2)^{(2)} = 4 \neq 0$. However, $(2, 2, 0) = (1, 1, 1) + (1, 1, -1)$ is non-invertible, since $(2, 2, 0)^{(2)} = 0$.

For a sequel to Proposition 4.2 see Proposition 5.14.

Linear correlations

Let X be any finite-dimensional real linear space with dual space X^L. Any linear map $\xi : X \to X^L;\ x \mapsto x^\xi = \xi(x)$ is said to be a *linear correlation* on X. An example of a linear correlation on $\mathbf{R}^n$ is *transposition*:

$$\tau : \mathbf{R}^n \to (\mathbf{R}^n)^L;\ x \mapsto x^\tau.$$

A correlation ξ is *symmetric* if, for all $a, b \in X$, $a^\xi(b) = b^\xi(a)$. A symmetric correlation ξ induces a scalar product $(a, b) \mapsto a^\xi b = a^\xi(b)$. Conversely, any scalar product $(a, b) \mapsto a \cdot b$ is so induced by a unique symmetric correlation, namely the map $a \mapsto (a\cdot)$, where, for all $a, b \in X, (a\cdot)(b) = a \cdot b$. A real linear space X with a correlation ξ will be called a real *correlated (linear) space.* By the above remarks any real quadratic space may be thought of as a symmetric real correlated space, and conversely.

Non-degenerate spaces

Let X be a real quadratic space, with correlation ξ. For such a space ξ is injective if and only if it is bijective, in which case X, its scalar product, its quadratic form and its correlation are all said to be *non-degenerate.*

The kernel of ξ, $\ker\xi$, is also called the *kernel* of X and denoted by $\ker X$. An element $a \in X$ belongs to $\ker X$ if and only if, for all $x \in X, a \cdot x = a^{\xi}x = 0$, that is if and only if a is orthogonal to each element of X. From this it at once follows that a positive-definite space is non-degenerate. The rank of ξ is also called the *rank* of X and denoted by $\operatorname{rk} X$. The space X is non-degenerate if and only if $\operatorname{rk} X = \dim X$.

Proposition 4.3 *Let A be a finite set of mutually orthogonal invertible elements of a real quadratic space X, with correlation ξ. Then the linear subspace of X spanned by A is a non-degenerate subspace of X.*

Proof Let λ be any set of coefficients for A such that $\sum_{a \in A} \lambda_a a \neq 0$. Then, by the orthogonality condition,

$$(\sum_{a \in A} \lambda_a a)^{\xi}(\sum_{a \in A} \lambda_a a^{(-1)}) = \sum_{a \in A} \lambda_a^2 > 0.$$

Therefore the kernel of the linear subspace spanned by A is $\{0\}$. □

Corollary 4.4 *For any finite p, q, the quadratic space $\mathbf{R}^{p,q}$ is non-degenerate.*

Proposition 4.5 *Let X be a real quadratic space and X' a linear complement in X of* $\ker X$. *Then X' is a non-degenerate subspace of X.*

Orthogonal maps

As always, there is interest in the maps preserving a given structure.

Let X and Y be real quadratic spaces, with correlations ξ and η, respectively. A map $f : X \to Y$ is said to be an *orthogonal map* if it is linear and, for all $a, b \in X$,

$$f(a)^{\eta} f(b) = a^{\xi} b$$

or, informally, in terms of the dot notation,

$$f(a) \cdot f(b) = a \cdot b.$$

This condition may be re-expressed in terms of a commutative diagram involving the linear dual f^L of f, as follows.

Proposition 4.6 *Let X, Y, ξ and η be as above. Then a linear map $f : X \to Y$ is orthogonal if and only if $f^L \eta f = \xi$, that is if and only if the diagram*

$$\begin{array}{ccc} X & \xrightarrow{f} & Y \\ \xi\downarrow & & \downarrow\eta \\ X^L & \xleftarrow{f^L} & Y^L \end{array}$$

commutes.

Corollary 4.7 *If the quadratic space X is non-degenerate then any orthogonal map $f : X \to Y$ is injective.*

Proof Let f be such a map. Then $(f^L\eta)f = \xi$ is injective and so f is injective. □

Proposition 4.8 *Let X and Y be real quadratic spaces with correlations ξ and η, respectively. A linear map $f : X \to Y$ such that, for all $a \in X$,*

$$f(a)^\eta f(a) = a^\xi a$$

is orthogonal.

Proposition 4.9 *Let W, X and Y be real quadratic spaces and let $f : X \to Y$ and $g : W \to X$ be orthogonal maps. Then 1_X is orthogonal, fg is orthogonal and, if f is invertible, f^{-1} is orthogonal.*

An invertible orthogonal map $f : X \to Y$ will be called an *orthogonal isomorphism*, and two quadratic spaces X and Y so related will be said to be *isomorphic*.

Proposition 4.10 *For any finite n the quadratic spaces $\mathbf{R}^{n,n}$ and $\mathbf{R}^{2n}_{\mathrm{hb}}$ are isomorphic.*

Any two-dimensional quadratic space isomorphic to the standard hyperbolic plane $\mathbf{R}^2_{\mathrm{hb}}$ will be called a *hyperbolic plane*.

Proposition 4.11 *Let X be a real quadratic space. Then any two linear complements in X of* $\ker X$ *are isomorphic as quadratic spaces.*

Orthogonal groups

An invertible orthogonal map $f : X \to X$ will be called an *orthogonal automorphism* of the quadratic space X. By Corollary 4.7 any orthogonal transformation of a non-degenerate quadratic space is an orthogonal automorphism.

For quadratic spaces X and Y the set of orthogonal maps $f : X \to Y$ will be denoted by $O(X, Y)$ and the *group* of orthogonal automorphisms $f : X \to X$ will be denoted by $O(X)$. That this is a group is immediate by Proposition 4.9. For any finite p, q, n the groups $O(\mathbf{R}^{p,q})$ and $O(\mathbf{R}^{0,n})$ will also be denoted, respectively, by $O(p, q; \mathbf{R})$ and $O(n; \mathbf{R})$ or, more briefly, by $O(p, q)$ and $O(n)$.

An orthogonal transformation of a quadratic space X may or may not preserve the orientations of X. An orientation-preserving orthogonal automorphism of X is said to be a *special* orthogonal automorphism, or a *rotation*, of X. The subgroup of $O(X)$ consisting of the special orthogonal automorphisms of X is denoted by $SO(X)$, the groups $SO(\mathbf{R}^{p,q})$ and $SO(\mathbf{R}^{0,n})$ also being denoted, respectively, by $SO(p, q)$ and $SO(n)$. These are the *special orthogonal groups.*

An orthogonal automorphism of X that reverses the orientations of X will be called an *anti-rotation* of X.

Proposition 4.12 *For any finite p, q the groups $O(p, q)$ and $O(q, p)$ are isomorphic, as are the groups $SO(p, q)$ and $SO(q, p)$.*

Adjoints

Suppose now that $f : X \to Y$ is a linear map of a non-degenerate quadratic space X, with correlation $\xi : X \to X^L$, to a quadratic space Y, with correlation $\eta : Y \to Y^L$. Since ξ is bijective there is a unique linear map $f^* : Y \to X$ such that $\xi f^* = f^L \eta : Y \to X^L$, that is, such that, for any $x \in X$, $y \in Y$,

$$f^*(y) \cdot x = y \cdot f(x).$$

The map $f^* = \xi^{-1} f^L \eta : Y \to X$ is called the *adjoint* of f with respect to the correlations ξ and η.

Proposition 4.13 *Let X be a non-degenerate real quadratic space. Then the map*

$$\operatorname{End} X \to \operatorname{End} X;\ f \mapsto f^{*}$$

is an anti-involution of the real algebra $\operatorname{End} X$.

Proposition 4.14 *Let $f : X \to Y$ be a linear map of a non-degenerate real quadratic space X to a quadratic space Y. Then f is orthogonal if and only if $f^{*}f = 1_X$.*

Proof Let ξ and η be the correlations on X and Y respectively. Since ξ is bijective,

$$f^{-1}\eta f = \xi \Leftrightarrow f^{*}f = \xi^{-1}f^{L}\eta f = 1_X.$$

□

Corollary 4.15 *A linear automorphism f of a non-degenerate real quadratic space X is orthogonal if and only if $f^{*} = f^{-1}$.*

Proposition 4.16 *Let f be a linear endomorphism of a non-degenerate real quadratic space X. Then $x \cdot f(x) = 0$ for all $x \in X$ if and only if $f^{*} = -f$.*

Proof

$$\begin{aligned} & x \cdot f(x) = 0, \text{ for all } x \in X \\ \Leftrightarrow\ & a \cdot f(a) + b \cdot f(b) - (a-b) \cdot f(a-b) \\ & \quad = a \cdot f(b) + b \cdot f(a) = 0, \text{ for all } a, b \in X, \\ \Leftrightarrow\ & f(b) \cdot a + f^{*}(b) \cdot a = 0, \text{ for all } a, b \in X, \\ \Leftrightarrow\ & (f + f^{*})(b) = 0, \text{ for all } x' \in X, \text{ since } \ker X = 0, \\ \Leftrightarrow\ & f + f^{*} = 0. \end{aligned}$$

□

Corollary 4.17 *Let f be an orthogonal endomorphism of the space X. Then $x \cdot f(x) = 0$ for all $x \in X$ if and only if $f^2 = -1_X$.*

Examples of adjoints

The next few propositions show what the adjoint of a linear map looks like in several important cases. It is convenient throughout these examples to use the same letter to denote not only a linear map $t : \mathbf{R}^p \to \mathbf{R}^q$ but also its $q \times p$ matrix over $\mathbf{R}$. Elements of $\mathbf{R}^p$ are identified with column matrices and elements of $(\mathbf{R}^p)^L$ with row matrices. For any linear map $t : \mathbf{R}^p \to \mathbf{R}^q$, t^τ denotes both the transpose of the matrix of t and also the linear map $\mathbf{R}^q \to \mathbf{R}^p$ represented by this matrix.

Proposition 4.18 *Let $t : \mathbf{R}^{0,p} \to \mathbf{R}^{0,q}$ be a linear map. Then $t^* = t^\tau$.*

Proof For any $x \in \mathbf{R}^p$, $y \in \mathbf{R}^q$,

$$y \cdot t(x) = y^\tau t x = (t^\tau y)^\tau x = t^\tau(y) \cdot x.$$

Now $\mathbf{R}^{0,p}$ is non-degenerate, implying that the adjoint of t is unique. So $t^* = t^\tau$. □

The case $(p,q) = (0,2)$ is worth considering in detail.

Proposition 4.19 *Let $t : \mathbf{R}^{0,2} \to \mathbf{R}^{0,2}$ be a linear map, with matrix*

$$\begin{pmatrix} a & c \\ b & d \end{pmatrix}.$$

Then t^* has matrix $\begin{pmatrix} a & b \\ c & d \end{pmatrix}$, and t is therefore orthogonal if and only if

$$\begin{pmatrix} a & b \\ c & d \end{pmatrix}\begin{pmatrix} a & c \\ b & d \end{pmatrix} = \begin{pmatrix} 1 & 0 \\ 0 & 1 \end{pmatrix},$$

that is, if and only if $a^2 + b^2 = c^2 + d^2 = 1$ and $ac + bd = 0$, from which it follows that the matrix is either of the form $\begin{pmatrix} a & -b \\ b & a \end{pmatrix}$ or of the form $\begin{pmatrix} a & b \\ b & -a \end{pmatrix}$, with $a^2 + b^2 = 1$. The map in the first case is a rotation and in the second case an anti-rotation, as can be verified by examination of the sign of the determinant.

To simplify notations in the next two propositions $\mathbf{R}^{p,q}$ and $\mathbf{R}^{2n}_{\text{hb}}$ are identified, as linear spaces, with $\mathbf{R}^p \times \mathbf{R}^q$ and $\mathbf{R}^n \times \mathbf{R}^n$, respectively. The entries in the matrices are linear maps.

Proposition 4.20 *Let $t : \mathbf{R}^{p,q} \to \mathbf{R}^{p,q}$ be linear, and let $t = \begin{pmatrix} a & c \\ b & d \end{pmatrix}$. Then*

$$t^* = \begin{pmatrix} a^\tau & -b^\tau \\ -c^\tau & d^\tau \end{pmatrix}.$$

Proof For all $(x, y), (x', y') \in \mathbf{R}^{p,q}$,

$$\begin{aligned} &(x, y) \cdot (ax' + cy', bx' + dy') \\ &= -x^\tau(ax' + cy') + y^\tau(bx' + dy') \\ &= -(a^\tau x)^\tau x' - (c^\tau x)^\tau y' + (b^\tau y)^\tau x' + (d^\tau y)^\tau y' \\ &= -(a^\tau x - b^\tau y)^\tau x' + (-c^\tau x + d^\tau u)^\tau y' \\ &= (a^\tau x - b^\tau y, -c^\tau x + d^\tau y) \cdot (x', y'). \end{aligned}$$

□

Corollary 4.21 *For such a linear map t, $\det t^* = \det t$, and, if t is orthogonal, $(\det t)^2 = 1$.*

Proposition 4.22 *Let $t : \mathbf{R}^{2n}_{\text{hb}} \to \mathbf{R}^{2n}_{\text{hb}}$ be linear, where $t = \begin{pmatrix} a & c \\ b & d \end{pmatrix}$. Then $t^* = \begin{pmatrix} d^\tau & c^\tau \\ b^\tau & a^\tau \end{pmatrix}$.*

Exercises

4.1 Let $t : X \to X$ be a linear transformation of a finite-dimensional quadratic space X, and suppose that t^2 is orthogonal. Discuss whether or not t is orthogonal. Discuss, in particular, the case where X is positive-definite.

4.2 Let X be a finite-dimensional real linear space. Prove that the map

$$(X \times X^L)^2 \to \mathbf{K};\ ((x, t), (y, u)) \mapsto t(y) + u(x))$$

is a neutral non-degenerate scalar product on the linear space $X \times X^L$.

4.3 Prove that $\mathbf{R}(2)$ with the quadratic form

$$\mathbf{R}(2) \to \mathbf{R};\ t \mapsto \det t$$

is isomorphic as a real quadratic space with the space $\mathbf{R}^{2,2}$, the subset $\left\{\begin{pmatrix} 1 & 0 \\ 0 & 1 \end{pmatrix}, \begin{pmatrix} 1 & 0 \\ 0 & -1 \end{pmatrix}, \begin{pmatrix} 0 & -1 \\ 1 & 0 \end{pmatrix}, \begin{pmatrix} 0 & 1 \\ 1 & 0 \end{pmatrix}\right\}$ being an orthonormal basis. Verify that $t \in \mathbf{R}(2)$ is invertible with respect to the quadratic form if and only if it is invertible as an element of the algebra $\mathbf{R}(2)$.

4.4 For any $\begin{pmatrix} a & c \\ b & d \end{pmatrix} \in \mathbf{R}(2)$, define $\begin{pmatrix} a & c \\ b & d \end{pmatrix}^{-} = \begin{pmatrix} d & -c \\ -b & a \end{pmatrix}$, the space $\mathbf{R}(2)$ being assigned the determinant quadratic form, and let any $\lambda \in \mathbf{R}$ be identified with $\begin{pmatrix} \lambda & 0 \\ 0 & \lambda \end{pmatrix} \in \mathbf{R}(2)$. Verify that, for any $t \in \mathbf{R}(2)$, $t^{-}t = t^{(2)}$ and that the subset $T = \{t \in \mathbf{R}(2) : t + t^{-} = 0\}$ is a quadratic subspace of $\mathbf{R}(2)$ isomorphic to $\mathbf{R}^{2,1}$.

4.5 Let $u \in \mathbf{R}(2)$ and let $t \in T$, where T is as in Exercise 4.4. Suppose also that t is orthogonal to $u - u^{-}$. Show that $tu \in T$. Hence prove that any element of $\mathbf{R}(2)$ is expressible as the product of two elements of T.

4.6 With T as in Exercise 4.4, prove that, for any $u \in SL(2; \mathbf{R})$, the map $T \to T;\ t \mapsto -utu^{-1}$ is reflection in the plane $(\mathbf{R}\{u\})^{\perp}$.

4.7 Prove that, for any $u \in SL(2; \mathbf{R})$, the maps $\mathbf{R}(2) \to \mathbf{R}(2);\ t \mapsto ut$ and $t \mapsto tu$ are rotations of $\mathbf{R}(2)$. (It has to be shown not only that the quadratic form is preserved but also that orientations are preserved.)

4.8 Find linear injections $\alpha : \mathbf{R} \to \mathbf{R}(2)$ and $\beta : \mathbf{R}^{1,1} \to \mathbf{R}(2)$ such that, for all $x \in \mathbf{R}^{1,1}$, $(\beta(x))^2 = -\alpha(x^{(2)})$.

5

The classification of real quadratic spaces

Central to this chapter are the basis theorem (Theorem 5.9) and the signature theorem (Theorem 5.22). Also important in the sequel are the factorisations of orthogonal automorphisms as composites of hyperplane reflections (Theorem 5.15) and a characterisation of neutral spaces (Theorem 5.28).

Orthogonal annihilators

Let X be a real quadratic space with correlation ξ and let W be a linear subspace of X. The dual annihilator $W^@$ of W has been defined as the subspace of X^L annihilating W, namely

$$\{\beta \in X^L : \text{for all } w \in W,\ \beta(w) = 0\}.$$

Moreover $\dim W^@ = \dim X - \dim W$. We now define $W^\perp = \xi^{-1}(W^@)$. That is,

$$W^\perp = \{a \in X : a \cdot w = 0, \text{ for all } w \in W\}.$$

This linear subspace is called the *orthogonal annihilator* of W in X. Its dimension is not less that $\dim X - \dim W$, being equal to this when ξ is bijective, that is, when X is non-degenerate.

A linear complement Y of W in X that is also a linear subspace of $W^\perp$ is said to be *an orthogonal complement* of W in X. The direct sum decomposition $W \oplus Y$ of X is then said to be *an orthogonal decomposition* of X.

Proposition 5.1 *Let W be a linear subspace of a real quadratic space X. Then* $\ker W = W \cap W^\perp$.

Proposition 5.2 *Let W be a linear subspace of a non-degenerate real quadratic space X. Then $X = W \oplus W^{\perp}$ if and only if W is non-degenerate, $W^{\perp}$, in this case, being the* unique *orthogonal complement of W in X.*

Corollary 5.3 *Let a be a non-zero element of a non-degenerate real quadratic space X and let $\mathbf{R}\{a\}$ denote the line in X spanned by a. Then $X = \mathbf{R}\{a\} \oplus (\mathbf{R}\{a\})^{\perp}$ if and only if a is invertible.*

Proposition 5.4 *Let W be a linear subspace of a non-degenerate real quadratic space X. Then $(W^{\perp})^{\perp} = W$.*

Proposition 5.5 *Let V and W be linear subspaces of a real quadratic space X. Then $V \subset W^{\perp} \Leftrightarrow W \subset V^{\perp}$.*

A first application of the orthogonal annihilator is to null subspaces.

Proposition 5.6 *Let W be a linear subspace of a real quadratic space X. Then W is null if and only if $W \subset W^{\perp}$.*

Corollary 5.7 *Let W be a null subspace of a non-degenerate real quadratic space X. Then* $\dim W \leq \frac{1}{2} \dim X$.

By this corollary it is only just possible for a necessarily even-dimensional non-degenerate real quadratic space to be neutral. As we noted earlier, $\mathbf{R}^{2n}_{\text{hb}}$, and therefore also $\mathbf{R}^{n,n}$, is such a space.

The basis theorem

Let W be a linear subspace of a real quadratic space X. Then, by Proposition 5.2, $X = W \oplus W^{\perp}$ if and only if W is non-degenerate. Moreover, if W is non-degenerate, then, by Proposition 5.4, $W^{\perp}$ also is non-degenerate.

These remarks lead to the basis theorem, which we take in two stages.

Theorem 5.8 *An n-dimensional non-degenerate real quadratic space, with $n > 0$, is expressible as the direct sum of n non-degenerate mutually orthogonal lines.*

A linearly independent set S of a real quadratic space X such that any two distinct elements of S are mutually orthogonal, with the square of any element of S equal to $0, -1$ or 1, is said to be an *orthonormal subset* of X. If S also spans X then S is said to be an *orthonormal basis* for X.

Theorem 5.9 (The basis theorem.) *Any real quadratic space X has an orthonormal basis.*

Proof Let X' be a linear complement in X of $\ker X$. Then X' is a non-degenerate subspace of X, by Proposition 4.5, and so has an orthonormal basis, B, say, by Theorem 5.8 and Proposition 4.1. Let A be any basis for $\ker X$. Then $A \cup B$ is an orthonormal basis for X. □

As a corollary we have

Theorem 5.10 (The first part of the classification theorem, the second part of which is the signature theorem, Theorem 5.22.) *Any non-degenerate real quadratic space X is isomorphic to $\mathbf{R}^{p,q}$ for some finite p, q.*

Proposition 5.11 *Let f be an automorphism of a non-degenerate real quadratic space X. Then* $(\det f)^2 = 1$, *and, for any rotation f of X,* $\det f = 1$.

Proof Apply Theorem 5.10 and Corollary 4.21. □

Reflections

Proposition 5.12 *Let $W \oplus Y$ be an orthogonal decomposition of a real quadratic space X. Then the map*

$$X \to X;\ w + y \mapsto w - y,$$

where $w \in W$ and $y \in Y$, is orthogonal.

Such a map is said to be a *reflection* of X in W. When $Y = W^\perp$ the map is said to be *the* reflection of X in W. A reflection of X in a linear hyperplane W is said to be a *hyperplane reflection* of X. Such a reflection exists if $\dim X > 0$, for the hyperplane can be chosen to be an orthogonal complement of $\mathbf{R}(a)$, where a is either an element of $\ker X$ or an invertible element of X.

Proposition 5.13 *A hyperplane reflection of a real quadratic space X is an anti-rotation of X.*

Let a be an invertible element of a real quadratic space X. Then the reflection of X in the hyperplane $(\mathbf{R}(a))^\perp$ will be denoted by ρ_a.

Proposition 5.14 *Suppose that a and b are invertible elements of a real quadratic space X, such that $a^{(2)} = b^{(2)}$. Then a may be mapped to b either by a single hyperplane reflection of X or by the composite of two hyperplane reflections of X.*

Proof By Proposition 4.2 either $a - b$ or $a + b$ is invertible, $a - b$ and $a + b$ being in any case mutually orthogonal. In the first case, ρ_{a-b} exists and

$$\rho_{a-b}(a) = \rho_{a-b}(\tfrac{1}{2}(a-b) + \tfrac{1}{2}(a+b)) = -\tfrac{1}{2}(a-b) + \tfrac{1}{2}(a+b) = b.$$

In the second case, ρ_{a+b} exists and

$$\rho_b \rho_{a+b}(a) = \rho_b(-b) = b.$$

□

Theorem 5.15 *Any orthogonal automorphism t of a non-degenerate real quadratic space X is expressible as the composite of a finite number of hyperplane reflections of X, the number being not greater than $2 \dim X$, or, if X is positive-definite, $\dim X$.*

Proof This is a straightforward induction based on Proposition 5.14. The details are left as an exercise. □

By Proposition 5.13 the number of reflections composing t is even when t is a rotation and odd when t is an anti-rotation. The following corollaries are important. To simplify notations we write $\mathbf{R}^2$ for $\mathbf{R}^{0,2}$ and $\mathbf{R}^3$ for $\mathbf{R}^{0,3}$.

Corollary 5.16 *Any anti-rotation of $\mathbf{R}^2$ is a reflection in some line of $\mathbf{R}^2$.*

Corollary 5.17 *Any rotation of $\mathbf{R}^3$ is the composite of two plane reflections.*

Proposition 5.18 *The only rotation of $\mathbf{R}^2$ leaving a non-zero point of $\mathbf{R}^2$ fixed is the identity.*

Proposition 5.19 *Any rotation of $\mathbf{R}^3$, other than the identity, leaves fixed each point of a unique line in $\mathbf{R}^3$.*

The line left fixed is called the *axis* of the rotation.

A common mistake is to assume that every anti-rotation of $\mathbf{R}^3$ is a reflection, for example in referring loosely to the orthogonal automorphisms of $\mathbf{R}^3$ as the 'rotations and reflections' of $\mathbf{R}^3$. This is clearly wrong, any reflection of $\mathbf{R}^3$ being a linear *involution* of $\mathbf{R}^3$.

Theorem 5.15 has an important part to play in Chapter 16.

Signature

Proposition 5.20 *Let $U \oplus V$ and $U' \oplus V'$ be orthogonal decompositions of a real quadratic space X such that, for all non-zero $u' \in U'$, $u'^{(2)} < 0$, and, for all $v \in V$, $v^{(2)} \geq 0$. Then the projection of X on to U with kernel V maps U' injectively to U.*

Proof Let $u' = u + v$, where $u \in U$, $v \in V$, be any element of U'. Then $u'^{(2)} = u^{(2)} + v^{(2)}$, so that, if $u = 0$, $u'^{(2)} = v^{(2)}$, implying that $u'^{(2)} = 0$ and therefore that $u' = 0$. □

Corollary 5.21 *If also $u^{(2)} < 0$, for all non-zero $u \in U$, and $v'(2) \geq 0$, for all $v' \in V'$, then* $\dim U = \dim U'$.

Proof By Proposition 5.20, $\dim U' \leq \dim U$. By a similar argument $\dim U \leq \dim U'$. □

As a further corollary of Proposition 5.20 we then have the following.

Theorem 5.22 (The signature theorem.) *The real quadratic spaces $\mathbf{R}^{p,q}$ and $\mathbf{R}^{p',q'}$ are isomorphic if and only if $p = p'$ and $q = q'$.*

Yet a further corollary of Proposition 5.20 is the following proposition.

Proposition 5.23 *Let $\begin{pmatrix} a & c \\ b & d \end{pmatrix}$: $\mathbf{R}^{p,q} \to \mathbf{R}^{p,q}$ be an orthogonal automorphism of $\mathbf{R}^{p,q}$, $\mathbf{R}^{p,q}$ being identified as usual with $\mathbf{R}^p \times \mathbf{R}^q$. Then $a : \mathbf{R}^p \to \mathbf{R}^p$ and $d : \mathbf{R}^q \to \mathbf{R}^q$ are linear isomorphisms.*

The orthogonal automorphism $\begin{pmatrix} a & c \\ b & d \end{pmatrix}$ of $\mathbf{R}^{p,q}$ will be said to be *semi-orientation-preserving* if a and d are orientation-preserving. We shall prove later (Proposition 22.48) that the set $SO^+(p,q)$ of all semi-orientation-preserving orthogonal automorphisms of $\mathbf{R}^{p,q}$ is a normal subgroup of $SO(p,q)$, with quotient group isomorphic to ± 1.

It follows from Theorem 5.10 that the quadratic form $x \mapsto x^{(2)}$ of a real quadratic space X may be represented as a *sum of squares*

$$x = \sum_{0 \le i < n} x_i e_i \mapsto \sum_{0 \le i < n} \zeta_i x_i^2$$

with respect to some suitable orthonormal basis $\{e_i : 0 \le i < n\}$ for X, with $\zeta_i = e_i^{(2)} = 0, -1$ or 1, for each $i \in n$. By Theorem 5.22 the number p of negative squares and the number q of positive squares are each independent of the basis chosen and $\dim(\ker X) + p + q = \dim X$; that is, $\operatorname{rk} X = p + q$. The pair of numbers (p, q) will be called the *signature* of the quadratic form and of the quadratic space. The number $\inf\{p, q\}$ will be called the *index* of the form and the space.

The definitions of 'signature' and 'index' are not standard in the literature, and almost all possibilities occur. For example, the signature is frequently defined to be $-p + q$ and the index to be p. The number we have called the index is sometimes called the *Witt index* of the quadratic space.)

The geometrical significance of the index is brought out in the next proposition.

Proposition 5.24 *Let W be a null subspace of the real quadratic space $\mathbf{R}^{p,q}$. Then $\dim W \le \inf\{p, q\}$.*

Proof Clearly $\mathbf{R}^{p,q} = X \oplus Y$, where $X \cong \mathbf{R}^{p,0}$, and $Y \cong \mathbf{R}^{0,q}$. As in the proof of Proposition 5.20, the restrictions to W of the projections of $\mathbf{R}^{p,q}$ on to X and Y, with kernels Y and X respectively, are injective. Hence the result. □

The bound is attained, since there is a subspace of $\mathbf{R}^{p,q}$ isomorphic to $\mathbf{R}^{r,r}$, where $r = \inf\{p, q\}$, and $\mathbf{R}^{r,r}$ is neutral.

Witt decompositions

Let X be a non-degenerate finite-dimensional real quadratic space. A *Witt decomposition* of X is a direct sum decomposition of X of the form $W \oplus W' \oplus (W \oplus W')^{\perp}$, where W and W' are null subspaces of X. (Some authors restrict the use of the term to the case where $\dim W = \operatorname{index} X$.)

Proposition 5.25 *Let X be a non-degenerate finite-dimensional real quadratic space with a one-dimensional null subspace W. Then there exists* *another one-dimensional null subspace W' such that the plane spanned by W and W' is a hyperbolic plane.*

Proof Let w be a non-zero element of W. Since X is non-degenerate, there exists $x \in X$ such that $w \cdot x \neq 0$ and x may be so chosen so that $w \cdot x = 1$. Then, for any $\lambda \in \mathbf{R}$, $(x + \lambda w)^{(2)} = x^{(2)} + 2\lambda$, this being zero if $\lambda = -\frac{1}{2}x^{(2)}$. Let W' be the line spanned by $w' = x - \frac{1}{2}x^{(2)}w$. The line is null since $(w')^{(2)} = 0$, and the plane spanned by w and w' is isomorphic to $\mathbf{R}^{1,1}$ since $w \cdot w' = w' \cdot w = 1$, and therefore, for any $a, b \in \mathbf{R}$, $(a w + b w') \cdot w = b$ and $(a w + b w') \cdot w' = a$, both being zero only if $a = b = 0$. □

Corollary 5.26 *Let W be a null subspace of a non-degenerate finite-dimensional real quadratic space X. Then there exists a null subspace W' such that $X = W \oplus W' \oplus (W \oplus W')^{\perp}$ (a Witt decomposition of X).*

Corollary 5.27 *Any non-degenerate finite-dimensional real quadratic space may be expressed as the direct sum of a finite number of hyperbolic planes and a positive- or negative-definite subspace, any two components of the decomposition being mutually orthogonal.*

The number of hyperbolic planes in the decomposition is of course equal to the (Witt) index of the space.

Neutral spaces

By Proposition 5.24 a non-degenerate real quadratic space is neutral if and only if its signature is (n, n), for some finite number n, that is, if and only if it is isomorphic to $\mathbf{R}^{2n}_{\mathrm{hb}}$, or, equivalently, $\mathbf{R}^{n,n}$, for some n.

The following theorem sometimes provides a quick method of detecting whether or not a real quadratic space is neutral.

Theorem 5.28 *A non-degenerate real quadratic space X is neutral if and only if there exists a linear map $t : X \to X$ such that $t^{*}t = -1$.*

Proof It is enough to remark that

$$t^{*}t = -1 \Leftrightarrow \text{for all } x \in X, (t(x))^{(2)} = -x^{(2)}.$$

□

The next proposition has an important role to play in Chapter 15.

Proposition 5.29 *Let W be a possibly degenerate n-dimensional real quadratic space. Then W is isomorphic to a subspace of the neutral quadratic space $\mathbf{R}^{n,n}$.*

Euclidean spaces

By Theorem 5.10 any finite-dimensional positive-definite quadratic space is isomorphic to $\mathbf{R}^{0,n}$ for some n. Throughout the remainder of this chapter this quadratic space will be denoted simply by $\mathbf{R}^n$, and any such space is said to be a *euclidean space.*

Let X be a euclidean space. For all $b \in X$ the *norm* of a is, by definition $|a| = \sqrt{(a^{(2)})}$, defined for all $a \in X$ since $a^{(2)} = a \cdot a \geq 0$, and the *distance* of a from b or the *length* of the line-segment $[a, b]$ is, by definition, $|a - b|$. In particular, for all $\lambda \in \mathbf{R}, \lambda^{(2)} = \lambda^2$, and $|\lambda| = \sqrt{(\lambda^2)}$ is the usual absolute value.

Proposition 5.30 *Let $a, b, c \in \mathbf{R}$. Then the quadratic form $(x, y) \mapsto ax^2 + 2bxy = cy^2$ is positive-definite if and only if $a > 0$ and $ac - b^2 > 0$.*

Proposition 5.31 *Let X be a euclidean space. Then, for all $a, b \in X$, $\lambda \in \mathbf{R}$,*

(i) $|a| \geq 0$,
(ii) $|a| = 0$ *if and only if* $a = 0$,
(iii) $|a - b| = 0$ *if and only if* $a = b$,
(iv) $|\lambda a| = |\lambda|\,|a|$,
(v) *a and b are collinear with* 0 *if and only if* $|b|a = \pm|a|b$,
(vi) *a and b are collinear with* 0 *and not on opposite sides of* 0 *if and only if* $|b|\,a = |a|\,b$,
(vii) $|a \cdot b| \leq |a|\,|b|$,
(viii) $a \cdot b \leq |a|\,|b|$,
(ix) $|a + b| \leq |a| + |b|$ (*the* triangle inequality),
(x) $||a| - |b|| \leq |a - b|$,

with equality in (vii) if and only if a and b are collinear with 0 *and in (viii), (ix) and (x) if and only if a and b are collinear with* 0 *and not on opposite sides of* 0.

Inequality (vii) of Proposition 5.31 is known as the *Cauchy-Schwarz* inequality. It follows from this inequality that, for all non-zero $a, b \in X$,

$$-1 \leq \frac{a \cdot b}{|a|\,|b|} \leq 1,$$

the quotient being equal to 1 if and only if b is a positive multiple of a and equal to -1 if and only if b is a negative multiple of a. The *absolute angle* between the line segments $[0, a]$ and $[0, b]$ is defined by

$$\cos\theta = \frac{a \cdot b}{|a|\,|b|}, \quad 0 \le \theta \le \pi,$$

with $a \cdot b = 0$ if and only if $\cos\theta = 0$ if and only if $\theta = \pi/2$, this being consistent with the ordinary usage of the word 'orthogonal'.

A map $t : X \to Y$ between euclidean spaces X and Y is said to *preserve scalar product* if, for all $a, b \in X$, $t(a) \cdot t(b) = a \cdot b$, to *preserve norm* if, for all $a \in X$, $|t(a)| = |a|$, to *preserve distance* if, for all $a, b \in X$, $|t(a) - t(b)| = |a - b|$, and to *preserve zero* if $t(0) = 0$.

According to our earlier definition, t is *orthogonal* if it is linear and preserves scalar product.

Of the various definitions of an orthogonal map given in Proposition 5.32 (iii), is probably the most natural from a practical point of view, being closest to our intuition of a rotation or anti-rotation.

Proposition 5.32 *Let $t : X \to Y$ be a map between euclidean spaces X and Y. Then the following are equivalent:*

(i) *t is orthogonal,*
(ii) *t is linear and preserves norm,*
(iii) *t preserves distance and zero,*
(iv) *t preserves scalar product.*

Let X be a euclidean space and let $a \in X$ and r be a positive real number. Then the set $\{x \in X : |x - a| = r\}$ is called the *sphere* with *centre* a and *radius* r in X, the sphere $\{x \in X : |x| = 1\}$ being called the *unit sphere* in X. The unit sphere in $\mathbf{R}^{n+1}$ is usually denoted by S^n and called the *unit n-sphere*. In particular

$$S^0 = \{x \in \mathbf{R} : x^2 = 1\} = \{-1, 1\},$$
$$S^1 = \{(x, y) \in \mathbf{R}^2;\ x^2 + y^2 = 1\}, \text{ the } \textit{unit circle},$$
and
$$S^2 = \{(x, y, z) \in \mathbf{R}^3 : x^2 + y^2 + z^2 = 1\}, \text{ the } \textit{unit sphere}.$$

In studying S^n it is often useful to identify $\mathbf{R}^{n+1}$ with $\mathbf{R}^n \times \mathbf{R}$. The points $(0, 1)$ and $(0, -1)$ are then referred to, respectively, as the *North* and *South poles* of S^n.

Proposition 5.33 *Let S be the unit sphere in a euclidean space X and let $t : X \to X$ be a linear transformation of X such that $t(S) \subset S$. Then t is orthogonal, and $t(S) = S$.*

Proposition 5.34 *For any non-negative integer n the map*

$$S^n \rightarrowtail \mathbf{R}^n;\ (u,v) \mapsto \frac{u}{1-v},$$

undefined only at the North pole, (0, 1), *is invertible.*

Proof Since $(u, v) \in S^n$, $|u|^2 + v^2 = 1$. So, if $x = \dfrac{u}{1-v}$,

$$|x|^2 = \frac{1+v}{1-v} = \frac{2}{1-v} - 1,$$

and v, and therefore u, is uniquely determined by x. □

The map defined in Proposition 5.34 is said to be the *stereographic projection* of S^n from the North pole on to its equatorial plane $\mathbf{R}^n \times \{0\}$, identified with $\mathbf{R}^n$. For, since $(u, v) = (1-v)(\dfrac{u}{1-v}, 0) + v(0, 1)$, the three points $(0, 1)$, (u, v) and $(\dfrac{u}{1-v}, 0)$ are collinear.

Similarly the map $S^n \rightarrowtail \mathbf{R}^n;\ (u, v) \mapsto \dfrac{u}{1+v}$, undefined only at the South pole, is invertible. This is stereographic projection from the South pole to the equatorial plane.

Exercises

5.1 Prove Proposition 5.23 in the case that $p = q = 1$.

(Show first that any element of $SO^+(1,1)$ may be written in the form

$$\begin{pmatrix} \cosh u & \sinh u \\ \sinh u & \cosh u \end{pmatrix}, \text{where } u \in \mathbf{R}.$$

Note that it has to be proved that $SO^+(1,1)$ is a subset of $SO(1,1)$.)

5.2 Let $t, u : \mathbf{R}^{n+1} \to \mathbf{R}^{n+1}$ be linear transformations of $\mathbf{R}^{n+1}$. Prove that the following statements are equivalent:

(a) For each $x \in S^n$, $(x, t(x), u(x))$ is an orthonormal 3-frame;

(b) t and u are orthogonal, $t^2 = u^2 = -1_{n+1}$ and $u\,t = -t\,u$.

(Use Propositions 4.16 and 5.33 and Corollary 4.15.)

5.3 Let i and j : $\mathbf{R}^n \to S^n$ be the 'inverses' of the stereographic projection of S^n, the unit sphere in $\mathbf{R}^{n+1} = \mathbf{R}^n \times \mathbf{R}$, from its North and South poles, respectively, on to its equatorial plane $\mathbf{R}^n = \mathbf{R}^n \times \{0\}$. Prove that, for all $x \in \mathbf{R}^n \backslash \{0\}$,

$$i^{-1}j(x) = j^{-1}i(x) = x^{(-1)}.$$

5.4 Let $\mathscr{S}(\mathbf{R}^{p,q+1}) = \{x \in \mathbf{R}^{p,q+1} : x^{(2)} = 1\}$. Prove that, for any $(x, y) \in \mathbf{R}^p \times S^q$, $(x, \sqrt{(1 + x^{(2)})}y) \in \mathscr{S}(\mathbf{R}^{p,q+1})$, and that the map

$$\mathbf{R}^p \times S^q \to \mathscr{S}(\mathbf{R}^{p,q+1});\ (x, y) \mapsto (x, \sqrt{(1 + x^{(2)})}y)$$

is bijective.

5.5 Determine whether or not the point-pairs (or 0-spheres)

$$\{x \in \mathbf{R} : x^2 - 8x - 12 = 0\} \quad \text{and} \quad \{x \in \mathbf{R} : x^2 - 10x + 7 = 0\}$$

are linked.

5.6 Determine whether or not the circles

$$\{(x, y, z) \in \mathbf{R}^3 : x^2 + y^2 + z^2 = 5 \quad \text{and} \quad x + y - 1 = 0\}$$

and

$$\{(x, y, z) \in \mathbf{R}^3 : x^2 + y^2 + z^2 + 2y - 4z = 0 \quad \text{and} \quad x - z + 1 = 0\}$$

are linked.

5.7 Let A and B be mutually disjoint circles which link in $\mathbf{R}^3$. Prove that the map

$$A \times B \to S^2 : (a, b) \mapsto (b - a)/|b - a|$$

is surjective, and describe the fibres of the map.

5.8 What was your definition of *linked* in the preceding three exercises? Can two circles in $\mathbf{R}^4$ be linked, according to your definition? Extend your definition to cover point pairs on S^1 or circles on S^3.

5.9 Show that any two mutually disjoint great circles on S^3 are linked.

5.10 Where can one find a pair of linked 3-spheres?

6
Anti-involutions of $\mathbf{R}(n)$

We saw in Chapter 4 how a non-degenerate scalar product on $\mathbf{R}^n$ induces an anti-automorphism of $\mathbf{R}(n)$, the adjoint $\mathbf{R}(n) \to \mathbf{R}(n)$; $t \mapsto t^*$, this being an anti-involution in the particular case that the scalar product is symmetric. For example, for the standard scalar product $(x, y) \mapsto x^\tau y$, the adjoint of a matrix t is its transpose t^τ. What can be said in the reverse direction?

The answer is that any anti-involution of $\mathbf{R}(n)$ is the adjoint of some symmetric or skew scalar product on $\mathbf{R}^n$. An intermediate role is played by *reflexive* scalar products.

Reflexive scalar products

We are already familiar with symmetric scalar products. The scalar product induced by a correlation ξ on a finite-dimensional real linear space X is said to be *skew* if, for all $a, b \in X$, $b^\xi a = -a^\xi b$. A scalar product $(a, b) \mapsto a^\xi b$ on X is said to be *reflexive* if, for all $a, b \in X$,

$$a^\xi b = 0 \Leftrightarrow b^\xi a = 0.$$

Not all correlations are reflexive. For example, the $\mathbf{R}$-bilinear product of $\mathbf{R}^2$,

$$\mathbf{R}^2 \times \mathbf{R}^2 \to \mathbf{R};\ ((a,b),\ a',b')) \mapsto aa' + ab' + bb',$$

is induced by a correlation that is not reflexive.

We prove first that any reflexive product is either symmetric or skew and then show how any anti-involution of $\mathbf{R}(n)$ may be induced by a non-degenerate reflexive scalar product.

Proposition 6.1 *Any non-zero reflexive correlation ξ on $\mathbf{R}^n$, where $n > 1$, is either symmetric or skew.*

Proof Since $b^\xi a = 0 \Leftrightarrow a^\xi b = 0$ for all $a, b \in \mathbf{R}^n$, it follows that for any non-zero $a \in \mathbf{R}^n$ the kernels of the linear maps $b \mapsto b^\xi a$ and a^ξ coincide. Accordingly, by Proposition 1.5, there is a non-zero real number λ_a such that, for all $b \in \mathbf{R}^n$, $b^\xi a = \lambda_a a^\xi b$.

Now λ_a does not depend on the vector a. For let a' be any other non-zero vector. Since $n > 1$ there exists a vector c, independent both of a and of a' (separately!). Then, since

$$b^\xi a + b^\xi c = b^\xi(a + c),$$

it follows that

$$\lambda_a a^\xi b + \lambda_c c^\xi b = \lambda_{a+c}(a + c)^\xi b,$$

for all $b \in \mathbf{R}^n$. So

$$\lambda_a a + \lambda_c c = \lambda_{a+c}(a + c).$$

But a and c are linearly independent. So

$$\lambda_a = \lambda_{a+c} = \lambda_c.$$

Similarly, $\lambda_{a'} = \lambda_c$. So $\lambda_{a'} = \lambda_a$. That is, there exists $\lambda \in \mathbf{R}$, non-zero, such that, for all $a, b \in \mathbf{R}^n, b^\xi a = \lambda a^\xi b$.

There are two cases.

Suppose first that $a^\xi a = 0$, for all $a, b \in \mathbf{R}^n$. Then, since

$$2(a^\xi b + b^\xi a) = a^\xi a + b^\xi b - (a - b)^\xi(a - b),$$

$$b^\xi a = \lambda a^\xi b = -\lambda b^\xi a,$$

for all $a, b \in \mathbf{R}^n$. Now, since ξ is non-degenerate, there exist a, b such that $b^\xi a \neq 0$. So $\lambda = -1$. That is, the correlation is skew.

The alternative is that, for some $a \in \mathbf{R}^n$, $a^\xi a \neq 0$, implying that the correlation is symmetric. □

We then have the following extension of Proposition 4.13.

Proposition 6.2 *For any non-degenerate symmetric or skew correlation on $\mathbf{R}^n$ the adjoint endomorphism of $\mathbf{R}(n)$ is an anti-involution.*

The following proposition is required early in the proof of the converse to Proposition 6.2

Proposition 6.3 *Let α be an anti-automorphism of $\mathbf{R}(n)$ and let $t \in \mathbf{R}(n)$, with $\operatorname{rk} t = 1$. Then $\operatorname{rk} t^\alpha = 1$.*

Proof By Theorem 2.6 the map t generates a minimal left ideal of $\mathbf{R}(n)$. Since α is an anti-automorphism of $\mathbf{R}(n)$ the image of this ideal by α is a minimal right ideal of $\mathbf{R}(n)$. This ideal is generated by t^α; so, by Theorem 2.6 again, or, rather, its analogue for right ideals, $\operatorname{rk} t^\alpha = 1$. □

Now for the converse to Proposition 6.2.

Theorem 6.4 *Any anti-involution α of $\mathbf{R}(n+1)$ is representable as the adjoint anti-involution induced by a non-degenerate reflexive correlation on $\mathbf{R}^{n+1}$.*

Proof Throughout the proof we think of $\mathbf{R}^{n+1}$ as $\mathbf{R}^n \times \mathbf{R}$.

Let $w = \begin{pmatrix} 0 & 0 \\ 0 & 1 \end{pmatrix}^\alpha$. Then $w^2 = w$, while, by Proposition 6.3, w has rank 1. Accordingly $w = vu$, where $u : \mathbf{R}^n \to \mathbf{R}$ and $v : \mathbf{R} \to \mathbf{R}^n$ are linear maps, u surjective and v injective. Now $w^2 = w$, so $vuvu = vu$, implying that $uv = 1 = uwv$, while, for all $\begin{pmatrix} c \\ d \end{pmatrix} \in \mathbf{R}^n \times \mathbf{R}$,

$$vu\begin{pmatrix} 0 & c \\ 0 & d \end{pmatrix}^\alpha = \begin{pmatrix} 0 & 0 \\ 0 & 1 \end{pmatrix}^\alpha \begin{pmatrix} 0 & c \\ 0 & d \end{pmatrix}^\alpha = \begin{pmatrix} 0 & c \\ 0 & d \end{pmatrix}^\alpha.$$

The map

$$\psi : \mathbf{R} \to \mathbf{R};\ \lambda \mapsto u\begin{pmatrix} 0 & c \\ 0 & d \end{pmatrix}^\alpha v$$

is the identity, since $uwv = 1$.

Now define

$$\xi : \mathbf{R}^n \times \mathbf{R} \to (\mathbf{R}^n \times \mathbf{R})^L;\ \begin{pmatrix} c \\ d \end{pmatrix} \mapsto u\begin{pmatrix} 0 & c \\ 0 & d \end{pmatrix}^\alpha.$$

Clearly ξ is linear. Moreover it is injective, for if $u\begin{pmatrix} 0 & c \\ 0 & d \end{pmatrix}^\alpha = 0$, for any $\begin{pmatrix} c \\ d \end{pmatrix} \in \mathbf{R}^n \times \mathbf{R}$, then $\begin{pmatrix} 0 & c \\ 0 & d \end{pmatrix}^{\alpha} = vu \begin{pmatrix} 0 & c \\ 0 & d \end{pmatrix}^{\alpha} = 0$, implying that $\begin{pmatrix} c \\ d \end{pmatrix}^{\alpha} = 0$, since $\alpha^2 = 1$. So ξ is a non-degenerate correlation on $\mathbf{R}^n \times \mathbf{R}$.

This correlation is reflexive, since, for all $\begin{pmatrix} c \\ d \end{pmatrix}, \begin{pmatrix} c' \\ d' \end{pmatrix} \in \mathbf{R}^n \times \mathbf{R}$,

$$
\begin{aligned}
\left(\begin{pmatrix} c' \\ d' \end{pmatrix}^{\xi} \begin{pmatrix} c \\ d \end{pmatrix}\right)^{\psi} &= u \left(u \begin{pmatrix} 0 & c' \\ 0 & d' \end{pmatrix}^{\alpha} \begin{pmatrix} c \\ d \end{pmatrix}\right)^{\alpha} v \\
&= u \left(\begin{pmatrix} 0 \\ u \end{pmatrix} \begin{pmatrix} 0 & c' \\ 0 & d' \end{pmatrix}^{\alpha} \begin{pmatrix} 0 & c \\ 0 & d \end{pmatrix}\right)^{\alpha} v \\
&= u \begin{pmatrix} 0 & c \\ 0 & d \end{pmatrix}^{\alpha} \begin{pmatrix} 0 & c' \\ 0 & d' \end{pmatrix} \begin{pmatrix} 0 \\ u \end{pmatrix}^{\alpha} v \\
&= u \begin{pmatrix} 0 & c \\ 0 & d \end{pmatrix}^{\alpha} \begin{pmatrix} c' \\ d' \end{pmatrix} \begin{pmatrix} 1 & 0 \end{pmatrix} \begin{pmatrix} 0 \\ u \end{pmatrix}^{\alpha} v \\
&= \begin{pmatrix} c \\ d \end{pmatrix}^{\xi} \begin{pmatrix} c' \\ d' \end{pmatrix} \mu,
\end{aligned}
$$

where $\mu = \begin{pmatrix} 1 & 0 \end{pmatrix} \begin{pmatrix} 0 \\ u \end{pmatrix}^{\alpha} v \in \mathbf{R}$, from which it follows that ξ is reflexive.

Finally, for any $\begin{pmatrix} c \\ d \end{pmatrix}, \begin{pmatrix} c' \\ d' \end{pmatrix} \in \mathbf{R}^n \times \mathbf{R}$, and any $t \in \mathbf{R}(n+1)$,

$$
\left(t^{\alpha} \begin{pmatrix} c' \\ d' \end{pmatrix}\right)^{\xi} \begin{pmatrix} c \\ d \end{pmatrix} = \begin{pmatrix} c' \\ d' \end{pmatrix}^{\xi} t \begin{pmatrix} c \\ d \end{pmatrix},
$$

each side being equal to $u \begin{pmatrix} 0 & c' \\ 0 & d' \end{pmatrix}^{\alpha} t \begin{pmatrix} c \\ d \end{pmatrix}$, since $t^{\alpha\alpha} = t$. That is, t^{α} is the adjoint of t with respect to the correlation ξ. □

Real symplectic spaces

A *real symplectic space* is a finite-dimensional real linear space with a skew correlation ξ.

Let X and Y be real symplectic spaces, with correlations ξ and η respectively. A linear map $f : X \to Y$ is said to be *symplectic* if, for all $a, b \in X$,

$$f(a)^{\eta} f(b) = a^{\xi} b.$$

Proposition 6.5 *The real linear space* $\mathbf{R}^2$ *with the bilinear product*

$$\mathbf{R}^2 \times \mathbf{R}^2 \to \mathbf{R};\ \left(\begin{pmatrix} a \\ b \end{pmatrix}, \begin{pmatrix} a' \\ b' \end{pmatrix}\right) \mapsto a\,b' - a\,b'$$

is a non-degenerate neutral symplectic space.

That the space is neutral follows at once from the remark that any one-dimensional subspace of a symplectic space is null.

The symplectic space of Proposition 6.5 is called the *standard real symplectic plane* and denoted by $\mathbf{R}^2_{\mathrm{sp}}$. Any symplectic space isomorphic to this one is called a *real symplectic plane.*

Proposition 6.6 *Let X be a non-degenerate real symplectic space and let W be any one-dimensional subspace of X, automatically null. Then there exists a one-dimensional subspace W', necessarily distinct from W, such that the plane spanned by W and W' is a real symplectic plane.*

Theorem 6.7 (The classification theorem for real symplectic spaces.) *Let X be a non-degenerate real symplectic space. Then X is isomorphic to* $(\mathbf{R}^{2k}_{\mathrm{sp}}) = (\mathbf{R}^2_{\mathrm{sp}})^k$, *where* $2k = \dim X$, $\dim X$ *necessarily being even.*

Proposition 6.8 *Let* $t = \begin{pmatrix} a & c \\ b & d \end{pmatrix}$ *be an endomorphism of the symplectic space* $\mathbf{R}^{2k}_{\mathrm{sp}}$. *Then the adjoint of t is* $t^{\bullet} = \begin{pmatrix} d^{\tau} & -c^{\tau} \\ -b^{\tau} & a^{\tau} \end{pmatrix}$.

The group of symplectic automorphisms $f : X \to X$ will be denoted by $Sp(X;\mathbf{R})$. For any finite $n = 2k$ the groups $Sp(\mathbf{R}^{2k}_{\mathrm{sp}})$ will also be denoted either by $Sp(2k;\mathbf{R})$ or by $Sp(n;\mathbf{R})$. These are the *real symplectic groups.*

Proposition 6.9 *An endomorphism t of a real symplectic space X with correlation ξ is symplectic if and only if* $t^{\bullet}t = 1$.

Corollary 6.10 *Any column of the matrix of an automorphism of the symplectic space* $\mathbf{R}^{2k}_{\mathrm{sp}}$ *is orthogonal in the ordinary sense to every column of the matrix except one, its partner.*

Proposition 6.11 *The determinant of a symplectic automorphism of a non-degenerate real symplectic space X is equal to* 1.

Proof Use Proposition 6.8, Corollary 6.10 and Exercise 1.4, considering first the case that $\dim X = 4$. □

7

Anti-involutions of $\mathbf{C}(n)$

The classification of the anti-involutions of $\mathbf{C}(n)$, regarded as a real algebra, parallels the work of the last chapter, but also extends it with the introduction of *hermitian products* and *unitary groups*.

Complex quadratic spaces

Much of the last three chapters extends at once to complex quadratic spaces, or indeed to quadratic spaces over any commutative field $\mathbf{K}$ (of characteristic not equal to 2), $\mathbf{R}$ being replaced simply by $\mathbf{C}$, or by $\mathbf{K}$, in the definitions, propositions and theorems. An exception is the signature theorem, Theorem 5.22, which is false.

The main classification theorem for complex quadratic spaces is the following.

Theorem 7.1 *Let X be a non-degenerate n-dimensional complex quadratic space. Then X is isomorphic to $\mathbf{C}^n$ with its standard complex scalar product*

$$\mathbf{C}^n \times \mathbf{C}^n \to \mathbf{C}^n : (a, b) \mapsto \sum_{0 \le i < n} a_i b_i.$$

As in the real case, a neutral non-degenerate quadratic space is even-dimensional, but in the complex case we can say more.

Proposition 7.2 *Let X be any non-degenerate complex quadratic space of even dimension 2n. Then X is neutral, being isomorphic not only to $\mathbf{C}^{2n}$ but also to $\mathbf{C}^{n,n}$ and to ${}^2\mathbf{C}^n$.*

For any complex quadratic space X the group of orthogonal automorphisms of X will be denoted by $O(X; \mathbf{C})$, the group $O(\mathbf{C}^n; \mathbf{C})$ being normally denoted by $O(n; \mathbf{C})$. These are the *complex orthogonal groups.*

As in the real case, the determinant of any element of $O(n;\mathbf{C})$ is equal to $+1$ or -1, the subgroup of elements of determinant 1 being denoted by $SO(n;\mathbf{C})$. Such groups are the *special complex orthogonal groups.*

Complex symplectic spaces

A *complex symplectic space* is a finite-dimensional complex linear space with a skew correlation ξ.

Let X and Y be complex symplectic spaces, with correlations ξ and η respectively. A linear map $f : X \to Y$ is said to be *symplectic* if, for all $a, b \in X$,

$$f(a)^\eta f(b) = a^\xi b.$$

Proposition 7.3 *The complex linear space* $\mathbf{C}^2$ *with the bilinear product*

$$\mathbf{C}^2 \times \mathbf{C}^2 \to \mathbf{C};\ \left(\begin{pmatrix} a \\ b \end{pmatrix}, \begin{pmatrix} a' \\ b' \end{pmatrix}\right) \mapsto a\,b' - a\,b'$$

is a non-degenerate neutral symplectic space.

That the space is neutral follows at once from the remark that any one-dimensional subspace of a symplectic space is null.

The symplectic space of Proposition 7.3 is called the *standard complex symplectic plane* and denoted by $\mathbf{C}^2_{\mathrm{sp}}$. Any symplectic space isomorphic to this one is called a *complex symplectic plane.*

Theorem 7.4 (Classification theorem for complex symplectic spaces.) *Let* X *be a non-degenerate complex symplectic space. Then* X *is isomorphic to* $(\mathbf{C}^{2k}_{\mathrm{sp}}) = (\mathbf{C}^2_{\mathrm{sp}})^k$, *where* $2k = \dim X$, $\dim X$ *necessarily being even.*

Proposition 7.5 *Let* $t = \begin{pmatrix} a & c \\ b & d \end{pmatrix}$ *be an endomorphism of the symplectic space* $\mathbf{C}^{2k}_{\mathrm{sp}}$. *Then the adjoint of* t *is* $t^\bullet = \begin{pmatrix} d^\tau & -c^\tau \\ -b^\tau & a^\tau \end{pmatrix}$.

The group of symplectic automorphisms $f : X \to X$ will be denoted by $Sp(X;\mathbf{C})$. For any finite $n = 2k$ the groups $Sp(\mathbf{C}^{2k}_{\mathrm{sp}})$ will also be denoted either by $Sp(2k;\mathbf{C})$ or by $Sp(n;\mathbf{C})$.

Proposition 7.6 *An endomorphism* t *of a complex symplectic space* X *with correlation* ξ *is symplectic if and only if* $t^\bullet t = 1$.

Proposition 7.7 *The determinant of a symplectic automorphism of a non-degenerate complex symplectic space X is equal to* 1.

The proof follows the same route as the proof of Proposition 6.11

Hermitian spaces

Complex quadratic spaces are not to be confused with *hermitian spaces,* finite-dimensional complex linear spaces that carry a hermitian form. These arise when the field $\mathbf{C}$ is regarded as a real algebra, assigned complex conjugation as an anti-involution, $\overline{\mathbf{C}}$ then denoting $\mathbf{C}$ with this assignment.

Let X and Y be complex linear spaces. Then an $\mathbf{R}$-linear map $f : X \to Y$ is said to be *semi-linear* over $\mathbf{C}$ if for all $x \in X, \lambda \in \mathbf{C}$ either $f(\lambda x) = \lambda f(x)$ or $f(\lambda x) = \overline{\lambda} f(x)$. In the former case the map is $\mathbf{C}$-linear. In the latter case it will be said to be $\overline{\mathbf{C}}$-linear. A semi-linear map t uniquely determines the automorphism ψ of $\mathbf{C}$ with respect to which it is semi-linear, unless $t = 0$. On occasion it is convenient to refer directly to ψ, the map t then being said to be $\mathbf{C}^\psi$*-linear* (*not* '$\mathbf{C}^\psi$*-semi-linear*' since, when $\psi = 1_\mathbf{C}$, $\mathbf{C}^\psi$ is usually abbreviated to $\mathbf{C}$ and the term '$\mathbf{C}$-semi-linear' could therefore be ambiguous).

The composite of any pair of composable semi-linear maps is semi-linear and the inverse of an invertible semi-linear map is semi-linear. An invertible semi-linear map is said to be a *semi-linear isomorphism*.

Proposition 7.8 *Let $f : X \to Y$ be a semi-linear map over* $\mathbf{C}$. *Then* $\operatorname{im} f$ *is a* $\mathbf{C}$*-linear subspace of Y and* $\ker f$ *is a* $\mathbf{C}$*-linear subspace of X.*

Rank and *kernel rank* are defined for semi-linear maps as for linear maps.

Proposition 7.9 *Let $f : X \to Y$ be a* $\mathbf{C}^\psi$*-linear map, where X and Y are finite-dimensional* $\mathbf{C}$*-linear spaces, and ψ is complex conjugation. Then, for any $\gamma \in Y^L, \psi\gamma f \in X^L$.*

Proof The map $\psi\gamma f$ is certainly $\mathbf{R}$-linear. it remains to consider its interaction with conjugation. However, for each $x \in X$, $\lambda \in \mathbf{C}$,

$$\psi\gamma f(\lambda x) = \overline{\gamma f(\lambda x)} = \overline{\gamma \overline{\lambda} f(x)} = \overline{\overline{\lambda} \gamma f(x)} = \lambda \overline{\gamma f(x)} = \lambda \psi\gamma f(x).$$

□

The map $f^L : Y^L \to X^L$, defined, for all $\gamma \in Y^L$, by the formula $f^L(\gamma) = \psi\gamma f$, is called the *dual* of f.

Proposition 7.10 *The dual f^L of a $\overline{\mathbf{C}}$-linear map $f : X \to Y$ is $\overline{\mathbf{C}}$-linear.*

Many properties of the duals of **R**-linear maps carry over to semi-linear maps over **C**.

A *correlation* on a finite-dimensional **C**-linear space X is a semi-linear map $\xi : X \to X^L;\ x \mapsto x^\xi$. The map $X \times X \to (a,b) \mapsto a^\xi b = a^\xi(b)$ is the *product* induced by the correlation, and the map $X \to \mathbf{C};\ a \mapsto a^\xi a$ the *form* induced by the correlation. Such a product is **R**-bilinear, but not, in general, **C**-bilinear, for although the map

$$X \to \mathbf{C};\ x \mapsto a^\xi x$$

is **C**-linear, for any $a \in X$, the map

$$X \to \mathbf{C};\ x \mapsto x^\xi b,$$

for any $b \in X$, is, in general, not linear but only semi-linear over **C**. Such a product is said to be *sesqui-linear*, the prefix being derived from a Latin word meaning 'one and a half times'.

Let ψ denote either of the involutions of **C**. Then a $\mathbf{C}^\psi$-correlation $\xi : X \to X^L$ and the induced product on the **C**-linear space X are said to be, respectively, *symmetric* or *skew* with respect to ψ or *over* $\mathbf{C}^\psi$ according as, for each $a, b \in X$,

$$b^\xi a = (a^\xi b)^\psi \quad \text{or} \quad -(a^\xi b)^\psi.$$

Symmetric products over $\overline{\mathbf{C}}$ are called *hermitian* products, and their forms are called *hermitian* forms. For example, the product

$$\mathbf{C}^2 \times \mathbf{C}^2 \to \mathbf{C};\ ((a,b),(a',b')) \mapsto \bar{a}a' + \bar{b}b'$$

is hermitian.

A *hermitian space* is a finite-dimensional **C**-linear space assigned a $\overline{\mathbf{C}}$-correlation.

An invertible correlation is said to be *non-degenerate*.

Semi-linear correlations $\xi, \eta : X \to X^L$ on a **C**-linear space X are said to be *equivalent* if, for some invertible $\lambda \in \mathbf{C}, \eta = \lambda\xi$. This is clearly an equivalence on any set of semi-linear correlations on X.

Proposition 7.11 *Any skew $\overline{\mathbf{C}}$-correlation on a* **C***-linear space X is equivalent to a symmetric $\overline{\mathbf{C}}$-correlation on X, and conversely.*

Proof Let ξ be a skew $\overline{\mathbf{C}}$-correlation on X. Then $\mathrm{i}\xi$ is a $\overline{\mathbf{C}}$-correlation on X since, for all $x \in X$, $\lambda \in \mathbf{C}$,

$$(\mathrm{i}\xi)(x\lambda) = \mathrm{i}(\xi(x\lambda)) = \mathrm{i}\overline{\lambda}\xi(x) = \overline{\lambda}(\mathrm{i}\xi)(x).$$

Moreover, for all $a, b \in X$,

$$b^{\mathrm{i}\xi}a = \mathrm{i}b^{\xi}a = (-\overline{\mathrm{i}})(-\overline{a^{\xi}b}) = \overline{a^{\mathrm{i}\xi}b}.$$

That is, $\mathrm{i}\xi$ is symmetric over $\overline{\mathbf{C}}$.

Similarly, if ξ is symmetric over $\overline{\mathbf{C}}$, then $\mathrm{i}\xi$ is skew over $\overline{\mathbf{C}}$. □

Equivalent semi-linear correlations ξ and η on $\mathbf{C}^n$ induce the same adjoint anti-automorphism of $\mathbf{C}(n)$.

Suppose now that $f : X \to Y$ is a linear map of a non-degenerate hermitian space X, with correlation ξ, to a hermitian space Y, with correlation η. Since ξ is bijective there is a unique $\mathbf{C}$-linear map $t^* : Y \to X$ such that $\xi t^* = \xi^{-1}t^L\eta$, that is, such that, for any $x \in X, y \in Y$, $t^*(y) \cdot x = y \cdot t(x)$. The map $t^* = \xi^{-1}t^L\eta$ is called the *adjoint* of t with respect to ξ and η.

The adjoint of a linear map $u : X \to X$ with respect to a non-degenerate correlation ξ on a complex linear space X will be denoted by u^{ξ}. The map u is said to be *self-adjoint* if $u^{\xi} = u$ and *skew-adjoint* if $u^{\xi} = -u$. The real linear subspaces $\{u \in \operatorname{End} X : u^{\xi} = u\}$ and $\{u \in \operatorname{End} X : u^{\xi} = -u\}$ of the real linear space $\operatorname{End} X = L(X, X)$ will be denoted by $\operatorname{End}_+(X, \xi)$ and $\operatorname{End}_-(X, \xi)$, respectively.

Proposition 7.12 *Let X be a non-degenerate hermitian space. Then the map*

$$\operatorname{End} X \to \operatorname{End} X;\ f \mapsto f^*$$

is an anti-involution of End X, *regarded as a real algebra.*

Proposition 7.13 *Any non-zero reflexive correlation ξ on a hermitian space X of dimension greater than one is either symmetric or skew.*

Proof This is just a re-run of the proof of Proposition 6.1, but taking account of conjugation.

For all $a, b \in X$, $\overline{b^{\xi}a} = 0 \Leftrightarrow b^{\xi}a = 0 \Leftrightarrow a^{\xi}b = 0$. That is, for any non-zero $a \in X$, the kernels of the surjective $\mathbf{C}$-*linear* maps $b \mapsto \overline{b^{\xi}a}$ and a^{ξ} coincide. Therefore, by Proposition 1.5, there is a non-zero complex number λ_a such that, for all $b \in X$,

$$\overline{b^{\xi}a} = \lambda_a a^{\xi} b.$$

To prove that λ_a is independent of a one then considers $c \in X$ linearly independent of a. Then, since

$$b^\xi a + b^\xi c = b^\xi(a+c),$$

it follows that

$$\lambda_a a^\xi b + \lambda_c c^\xi b = \lambda_{a+c}(a+c)^\xi b,$$

for all $b \in X$. So

$$\overline{\lambda_a} a + \overline{\lambda_c} c = \overline{\lambda_{a+c}}(a+c).$$

But a and c are linearly independent. So

$$\overline{\lambda_a} = \overline{\lambda_{a+c}} = \overline{\lambda_c},$$

implying that $\lambda_a = \lambda_c$. So, as before, there exists $\lambda \in \mathbf{C}$ such that, for all $a, b \in X, \overline{b^\xi a} = \lambda a^\xi b$.

Again there are apparently two cases.

However, if $a^\xi a = 0$, for all $a \in X$, it follows that $\overline{b^\xi a} = -b^\xi a$ for all $a, b \in X$, clearly not the case.

The alternative is that, for some $x \in X$, $x^\xi x \neq 0$, implying that, for some invertible $\mu \in \mathbf{C}$, $\overline{\mu^{-1}} = \lambda \mu^{-1}$, or, equivalently, $\overline{\mu} = \overline{\lambda}\mu$. Then, for all $a, b \in X$,

$$b^{\mu\xi} a = \mu b^\xi a = \mu \overline{\lambda a^\xi b} = \mu \overline{\lambda}\, \overline{a^\xi b} = \overline{\mu}\, \overline{a^\xi b} = \overline{a^{\mu\xi} b}.$$

That is, the correlation $\mu\,\xi$, equivalent to ξ, is symmetric. □

There is a converse of Proposition 7.12, analogous to the converse of Proposition 6.2.

Theorem 7.14 *Any anti-involution α of the* real *algebra* $\mathbf{C}(n+1)$ *is representable as the adjoint anti-involution induced by a non-degenerate reflexive correlation on* $\mathbf{C}^{n+1}$.

Proof This follows the proof of Theorem 6.4. At the end of the first stage of the argument one proves that the map ψ is an automorphism of $\mathbf{C}$, so is either the identity or conjugation. The correlation ξ is defined as before and proved to be $\mathbf{C}^\psi$-linear, and injective, and so is a non-degenerate $\mathbf{C}^\psi$-linear correlation on $\mathbf{C}^n \times \mathbf{C}$.

The remainder of the proof then goes through without change. □

Theorem 7.15 (The basis theorem for hermitian spaces.) *Each hermitian space has an orthonormal basis.*

Theorem 7.16 (The classification theorem for hermitian spaces.) *Let X be a hermitian space of finite dimension n. Then there exists a unique pair of numbers (p,q), with $p+q=n$, such that X is isomorphic to $\overline{\mathbf{C}}^{p,q}$, this being the $\mathbf{C}$-linear space $\mathbf{C}^{p+q}$, with the hermitian product*

$$(\mathbf{C}^{p+q})^2 \to \mathbf{C};\ (a,b) \mapsto -\sum_{0\le i<p} \overline{a_i}\, b_i + \sum_{0\le j<q} \overline{a_{p+j}}\, b_{p+j}.$$

The pair of numbers (p,q) is called the *signature* of the hermitian space X and of its form. The space and its form are said to be *positive-definite* if $p=0$.

Unitary groups

Proposition 7.17 *Let $t = \begin{pmatrix} a & c \\ b & d \end{pmatrix}$ be an endomorphism of the hermitian space $\overline{\mathbf{C}}^{p,q}$. Then the adjoint of t is $t^\xi = \begin{pmatrix} \overline{a}^\tau & -\overline{b}^\tau \\ -\overline{c}^\tau & \overline{d}^\tau \end{pmatrix}$.*

Let X and Y be hermitian spaces, with correlations ξ and η respectively. A linear map $f : X \to Y$ is said to be *unitary* if, for all $a, b \in X$,

$$f(a)^\eta f(b) = a^\xi b.$$

The group of unitary automorphisms $f : X \to X$ will be denoted by $U(X)$. For any finite p, q, n the groups $U(\overline{\mathbf{C}}^{p,q})$ and $U(\overline{\mathbf{C}}^{0,n})$ will also be denoted, respectively, by $U(p,q)$ and $U(n)$. These are the *unitary groups*.

Proposition 7.18 *An endomorphism t of a hermitian space X with correlation ξ is unitary if and only if $t^\xi t = 1$.*

Proposition 7.19 *The determinant of a unitary automorphism of a hermitian space X has modulus* 1.

Proof Use Theorem 7.15 and Proposition 7.17. □

A unitary automorphism of a hermitian space X of determinant 1 is called a *special unitary* automorphism. The special unitary automorphisms of X form a subgroup $SU(X)$ of the group $U(X)$, the notations $SU(p,q)$ and $SU(n)$ being in common use for the special unitary groups of the hermitian spaces $\overline{\mathbf{C}}^{p,q}$ and $\overline{\mathbf{C}}^{0,n}$.

Proposition 7.20 *For any finite n, and $p + q = n$, with a rather obvious definition of $O(p, q; \mathbf{C})$,*

$$\begin{aligned} O(n) &= O(n;\mathbf{C}) \cap GL(n;\mathbf{R}) &= O(n;\mathbf{C}) \cap U(n), \\ Sp(2n;\mathbf{R}) &= Sp(2n;\mathbf{C}) \cap GL(2n;\mathbf{R}), \\ O(p,q) &= O(p, q;\mathbf{C}) \cap U(p, q). \end{aligned}$$

Clearly $U(1) = S^1$, the *circle group* of complex numbers of modulus 1. Moreover $U(1) \cong SO(2)$.

8
Quaternions

Since for any $z = x + iy \in \mathbf{C}$, $|z|^2 = \bar{z}\,z = x^2 + y^2$, the field of complex numbers $\mathbf{C}$, when identified with the linear space $\mathbf{R}^2$, can also in a natural way be identified with the real quadratic space $\mathbf{R}^{0,2}$. Moreover, since, for any $z \in \mathbf{C}$,

$$|z| = 1 \Leftrightarrow x^2 + y^2 = 1,$$

the subgroup of the group of invertible complex numbers $\mathbf{C}^{\bullet}$, consisting of all complex numbers of absolute value 1, may be identified with the unit circle S^1 in $\mathbf{R}^2$.

In what follows the identification of $\mathbf{C}$ with $\mathbf{R}^2 = \mathbf{R}^{0,2}$ is taken for granted.

We have already remarked in Proposition 2.1 that for any complex number $c = a + ib$, the map $\mathbf{C} \to \mathbf{C};\ z \mapsto c\,z$ may be regarded as the real linear map $\mathbf{R}^2 \to \mathbf{R}^2$ with matrix $\begin{pmatrix} a & -b \\ b & a \end{pmatrix}$. Moreover, $|c|^2 = \bar{c}\,c = \det\begin{pmatrix} a & -b \\ b & a \end{pmatrix} = 1 \Leftrightarrow |c| = 1$, while, for all $c \in \mathbf{C}$, with $|c| = 1$, and all $z \in \mathbf{C}$, $|c\,z| = |z|$. Thus, (cf. Proposition 4.19) multiplication by a complex number of modulus 1 induces a rotation of $\mathbf{R}^2$, and any rotation of $\mathbf{R}^2$ may be so represented by a point of the unit circle, S^1.

The following statement sums this all up.

Proposition 8.1 $SO(2) \cong S^1$.

The group S^1 is called the *circle group.* Anti-rotations of $\mathbf{R}^2$ also can be handled by $\mathbf{C}$, for conjugation is an anti-rotation, and any other anti-rotation can be regarded as the composite of conjugation with a rotation.

The algebra of quaternions, **H**, the subject of this chapter, is analogous in many ways to the algebra of complex numbers **C**. For example, it has application to the description of the groups $O(3)$ and $O(4)$. The letter H is the initial letter of the surname of Sir William Hamilton, who first studied quaternions and gave them their name (1844).

The algebra H

Let 1, i, j and k denote the elements of the standard basis for $\mathbf{R}^4$. The *quaternion product* on $\mathbf{R}^4$ is then the **R**-bilinear product

$$\mathbf{R}^4 \times \mathbf{R}^4 \to \mathbf{R}^4;\ (a, b) \mapsto ab$$

with unit element 1, defined by the formulae

$$\mathrm{i}^2 = \mathrm{j}^2 = \mathrm{k}^2 = -1$$

$$\text{and } \mathrm{i\,j} = \mathrm{k} = -\mathrm{j\,i},\ \mathrm{j\,k} = \mathrm{i} = -\mathrm{k\,j},\ \mathrm{k\,i} = \mathrm{j} = -\mathrm{i\,k}.$$

Proposition 8.2 *The quaternion product is associative.*

On the other hand the quaternion product is not commutative. For example, $\mathrm{j\,i} \neq \mathrm{i\,j}$. Moreover it does not necessarily follow that, if $a^2 = b^2$, then $a = \pm b$. For example, $\mathrm{i}^2 = \mathrm{j}^2$, but $\mathrm{i} \neq \pm\mathrm{j}$.

The linear space $\mathbf{R}^4$, with the quaternion product, is a real algebra **H** known as the *algebra of quaternions*. In working with **H** it is usual to identify **R** with $\mathbf{R}\{1\}$ and $\mathbf{R}^3$ with $\mathbf{R}\{\mathrm{i},\mathrm{j},\mathrm{k}\}$, the first identification having been anticipated by our use of the symbol 1 to denote the unit element of the algebra. The subspace $\mathbf{R}\{\mathrm{i},\mathrm{j},\mathrm{k}\}$ is known as the subspace of *pure quaternions*. Each quaternion q is uniquely expressible in the form $\mathrm{re}\,q + \mathrm{pu}\,q$, where $\mathrm{re}\,q \in \mathbf{R}$ and $\mathrm{pu}\,q \in \mathbf{R}^3$, $\mathrm{re}\,q$ being called the *real part* of q and $\mathrm{pu}\,q$ the *pure part* of q.

Proposition 8.3 *A quaternion is real if and only if it commutes with every quaternion. That is,* **R** *is the centre of* **H**.

Proof $\Rightarrow$: Clear.

$\Leftarrow$: Let $q = a + b\mathrm{i} + c\mathrm{j} + d\mathrm{k}$, where a, b, c, d are real, be a quaternion commuting with i and j. Since q commutes with i,

$$a\mathrm{i} - b + c\mathrm{k} - d\mathrm{j} = \mathrm{i}\,q = q\,\mathrm{i} = a\mathrm{i} - b - c\mathrm{k} + d\mathrm{j},$$

implying that $2(c\mathrm{k} - d\mathrm{j}) = 0$. So $c = d = 0$. Similarly, since q commutes with j, $b = 0$. So $q = a$, and is real. □

Proposition 8.4 *The ring structure of* **H** *induces the real linear structure, and any ring automorphism or anti-automorphism of* **H** *is a real linear automorphism of* **H**, *and therefore also a real algebra automorphism or anti-automorphism of* **H**.

Proof By Proposition 8.3 the injection of **R** in **H**, and hence the real scalar multiplication $\mathbf{R} \times \mathbf{H} \to \mathbf{H};\ (\lambda, q) \mapsto \lambda q$, is determined by the ring structure.

Also, again by Proposition 8.3, any automorphism or anti-automorphism t of **H** maps **R** to **R**, this restriction being an automorphism of **R** and therefore the identity, by Proposition 2.9. Therefore t not only respects addition and respects or reverses multiplication but also, for any $\lambda \in \mathbf{R}$ and $q \in \mathbf{H}$, $t(\lambda q) = t(\lambda)\, t(q) = \lambda\, t(q)$. □

This result is to be contrasted with the more involved situation for the field of complex numbers described following Proposition 2.10. The automorphisms and anti-automorphisms of **H** are discussed in more detail below.

Proposition 8.5 *A quaternion is pure if and only if its square is a non-positive real number.*

Proof ⇒ : Consider the pure quaternion $q = b\mathrm{i} + c\mathrm{j} + d\mathrm{k}$, where $b, c, d \in \mathbf{R}$. Then $q^2 = -(b^2 + c^2 + d^2)$, which is real and non-positive.

⇐ : Consider $q = a + b\mathrm{i} + c\mathrm{j} + d\mathrm{k}$, where $a, b, c, d \in \mathbf{R}$. Then

$$q^2 = a^2 - b^2 - c^2 - d^2 + 2a(b\mathrm{i} + c\mathrm{j} + d\mathrm{k}).$$

If q^2 is real, either $a = 0$ and q is pure, or $b = c = d = 0$ and $a \leq 0$, in which case q^2 is positive. So, if q^2 is real and non-positive, q is pure. □

Proposition 8.6 *The direct sum decomposition of* **H**, *with components the real and pure subspaces, is induced by the ring structure for* **H**.

The *conjugate* $\overline{q}$ of a quaternion q is defined to be the quaternion $\operatorname{re} q - \operatorname{pu} q$.

Proposition 8.7 *Conjugation* : $\mathbf{H} \to \mathbf{H};\ q \mapsto \overline{q}$ *is a real algebra anti-involution. That is, for all* $a, b \in \mathbf{H}$ *and all* $\lambda \in \mathbf{R}$,

$$\overline{a + b} = \overline{a} + \overline{b},\ \overline{\lambda a} = \lambda\,\overline{a},\ \overline{\overline{a}} = a \text{ and } \overline{a\, b} = \overline{b}\,\overline{a}.$$

Moreover, $a \in \mathbf{R} \Leftrightarrow \bar{a} = a$ *and* $a \in \mathbf{R}^3 \Leftrightarrow \bar{a} = -a$, *while* $\operatorname{re} a = \frac{1}{2}(a + \bar{a})$ *and* $\operatorname{pu} a = \frac{1}{2}(a - \bar{a})$.

Now let $\mathbf{H}$ be assigned the standard positive-definite scalar product on $\mathbf{R}^4$, denoted as usual by $\cdot$.

Proposition 8.8 *For all* $a, b \in \mathbf{H}$, $a \cdot b = \operatorname{re}(\bar{a}\, b) = \frac{1}{2}(\bar{a}\, b + \bar{b}\, a)$. *In particular, for any* $a \in \mathbf{H}$, $\bar{a}\, a = a \cdot a$, *so* $\bar{a}\, a$ *is non-negative.*

In particular also, for all $a, b \in \mathbf{R}^3$, $a \cdot b = -\frac{1}{2}(a\, b + b\, a) = -\operatorname{re}(a\, b)$, *with* $a^{(2)} = a \cdot a = -a^2$ *and with* $a \cdot b = 0$ *if and only if* a *and* b *anti-commute.*

The non-negative number $|a| = \sqrt{\bar{a}\, a}$ is called the *norm* or *modulus* or *absolute value* of the quaternion a.

Proposition 8.9 *Let* x *be a non-real element of* $\mathbf{H}$. *Then* $\mathbf{R}\{1, x\}$ *is a subalgebra of* $\mathbf{H}$ *isomorphic to* $\mathbf{C}$. *In particular,* $\mathbf{R}\{1, x\}$ *is commutative.*

Proof It is enough to remark that

$$x^2 + b\, x + c = 0,$$

where $b = -(x + \bar{x})$, and $c = \bar{x}\, x$, b and c both being real. □

Proposition 8.10 *Each non-zero* $a \in \mathbf{H}$ *is invertible, with* $a^{-1} = |a|^{-2}\, \bar{a}$ *and with* $|a^{-1}| = |a|^{-1}$.

Note that the quaternion inverse of a is the conjugate of the scalar product inverse of a, $a^{(-1)} = |a|^{-2} a$.

By Proposition 8.10, $\mathbf{H}$ may be regarded as a non-commutative field. The group of non-zero quaternions will be denoted by $\mathbf{H}^{\bullet}$.

Proposition 8.11 *For all* $a, b \in \mathbf{H}$, $|a\, b| = |a|\, |b|$.

A quaternion is said to be a *unit quaternion* if $|q| = 1$.

Proposition 8.12 *The set of unit quaternions coincides with the unit sphere* S^3 *in* $\mathbf{R}^4$ *and is a subgroup of the group* $\mathbf{H}^{\bullet}$.

Proposition 8.13 *Let* $q \in \mathbf{H}$ *be such that* $q^2 = -1$. *Then* $q \in S^2$, *the unit sphere in* $\mathbf{R}^3$.

Proof Since q^2 is real and non-positive, $q \in \mathbf{R}^3$ while, since $q^2 = -1$, $|q| = 1$. So $q \in S^2$. □

The *cross product* $a \times b$ of a pair (a, b) of pure quaternions is defined by the formula

$$a \times b = \mathrm{pu}(a\,b).$$

Proposition 8.14 *For all $a, b \in \mathbf{R}^3$*

$$a\,b = -a \cdot b + a \times b, \text{ and } a \times b = \tfrac{1}{2}(a\,b - b\,a) = -(b \times a),$$

while $a \times a = a \cdot (a \times b) = b \cdot (a \times b) = 0$.

If a and b are mutually orthogonal elements of $\mathbf{R}^3$ then a and b anti-commute, that is, $b\,a = -a\,b$, and $a \times b = a\,b$. In particular,

$$\mathrm{i} \cdot (\mathrm{j} \times \mathrm{k}) = \mathrm{i} \cdot (\mathrm{jk}) = \mathrm{i} \cdot \mathrm{i} = -\mathrm{i}^2 = 1.$$

Proposition 8.15 *Let q be any quaternion. Then there exists a non-zero pure quaternion b such that $q\,b$ also is a pure quaternion.*

Proof Let b be any non-zero pure quaternion orthogonal to the pure part of q. Then $q\,b = (\mathrm{re}\,q)b + (\mathrm{pu}\,q) \times b$, being the sum of two pure quaternions, also is pure. □

Corollary 8.16 *Any quaternion q is expressible as the product of a pair of pure quaternions.*

Proof Let b be any non-zero pure quaternion orthogonal to the pure part of q. Then $q\,b = (\mathrm{re}\,q)b + (\mathrm{pu}\,q) \times b$, being the sum of two pure quaternions, also is pure. □

Proposition 8.17 *For any $a, b, c \in \mathbf{H}$,*

$$\mathrm{re}(a\,b\,c) = \mathrm{re}(b\,c\,a) = \mathrm{re}(c\,a\,b).$$

Moreover, for any $a, b, c \in \mathbf{R}^3$,

$$\mathrm{re}(a\,b\,c) = -a \cdot (b \times c) = -\det(a, b, c),$$

where (a, b, c) denotes the matrix with columns a, b and c.

We call the real number $\mathrm{re}(a\,b\,c) = \overline{a} \cdot (b\,c)$ the *scalar triple product* of the quaternions a, b, c, in that order. (In the case that a, b, c are pure it is more usual to take the negative of this as the scalar triple product.)

Automorphisms and anti-automorphisms of H

By Proposition 8.4 the field automorphisms and anti-automorphisms of **H** coincide with the **R**-algebra automorphisms and anti-automorphisms of **H**. In this section we show that they are also closely related to the orthogonal automorphisms of $\mathbf{R}^3$. The relationship one way is given by the next proposition.

Proposition 8.18 *Any automorphism or anti-automorphism u of* **H** *is of the form* $\mathbf{H} \to \mathbf{H};\ a \mapsto \mathrm{re}\, a + t(\mathrm{pu}\, a)$, *where t is an orthogonal automorphism of* $\mathbf{R}^3$.

Proof By Proposition 8.4, Proposition 2.9 and Proposition 8.5, u is a linear map leaving each quaternion fixed and mapping $\mathbf{R}^3$ to itself. Also, for each $x \in \mathbf{R}^3$, $(u(x))^2 = u(x^2) = x^2$, since $x^2 \in \mathbf{R}$, while $|x|^2 = -x^2$. So

$$t : \mathbf{R}^3 \to \mathbf{R}^3;\ x \mapsto u(x)$$

is linear and respects norm. Therefore, by Proposition 5.32, it is an orthogonal automorphism of $\mathbf{R}^3$. □

In the reverse direction we have the following fundamental result.

Proposition 8.19 *Let q be an invertible pure quaternion. Then, for any pure quaternion x, $q\,x\,q^{-1}$ is a pure quaternion, and the map*

$$-\rho_q : \mathbf{R}^3 \to \mathbf{R}^3;\ x \mapsto -q\,x\,q^{-1}$$

is reflection in the plane $(\mathbf{R}\{q\})^{\perp}$.

Proof Since $(q\,x\,q^{-1})^2 = x^2$, which is real and non-positive, $q\,x\,q^{-1}$ is pure, by Proposition 8.5. Also $-\rho_q$ is linear, and $-\rho_q(q) = -q$, while, for any $r \in (\mathbf{R}\{q\})^{\perp}$, $\rho_q(r) = -q\,r\,q^{-1} = r\,q\,q^{-1} = r$. Hence the result. □

Proposition 8.19 is used twice in the proof of Proposition 8.20.

Proposition 8.20 *Each rotation of* $\mathbf{R}^3$ *is of the form* ρ_q *for some non-zero quaternion q, and every such map is a rotation of* $\mathbf{R}^3$.

Proof Since, by Corollary 5.17, any rotation of $\mathbf{R}^3$ is the composite of two plane reflections it follows, by Proposition 8.19, that the rotation can be expressed in the given form. The converse is by Corollary 8.16 and Proposition 8.19. □

In fact, each rotation of $\mathbf{R}^3$ can be so represented by a unit quaternion, unique up to sign. This follows from Proposition 8.21.

Proposition 8.21 *The map $\rho : \mathbf{H}^* \to SO(3)$; $q \mapsto \rho_q$ is a group surjection, with kernel $\mathbf{R}^*$, the restriction of ρ to S^3 also being surjective, with kernel $S^0 = \{\pm 1\}$.*

Proof The map ρ is surjective, by Proposition 8.20, and is a group map since, for all $q, r \in \mathbf{H}^*$, and all $x \in \mathbf{R}^3$,

$$\rho_{qr}(x) = q\,r\,x\,(q\,r)^{-1} = \rho_q\rho_r(x).$$

Moreover, $q \in \ker\rho$ if and only if $q\,x\,q^{-1} = x$, for all $x \in \mathbf{R}^3$, that is if and only if $q\,x = x\,q$, for all $x \in \mathbf{R}^3$. Therefore, by Proposition 8.3, $\ker\rho = \mathbf{R} \cap \mathbf{H}^* = \mathbf{R}^*$.

The restriction of ρ to S^3 also is surjective simply because, for any $\lambda \in \mathbf{R}^*$ and for any $q \in \mathbf{H}^*$, $\rho_{\lambda q} = \rho_q$, and λ may be so chosen that $|\lambda\,q| = 1$. Finally, $\ker(\rho\,|\,S^3) = \ker\rho \cap S^3 = \mathbf{R}^* \cap S^3 = S^0$. □

Proposition 8.22 will be used in Proposition 8.26.

Proposition 8.22 *Any unit quaternion q is expressible in the form $a\,b\,a^{-1}\,b^{-1}$, where a and b are non-zero quaternions.*

Proof By Proposition 8.15 there is, for any unit quaternion q, a non-zero pure quaternion b such that $q\,b$ is a pure quaternion. Since $|\,q\,| = 1$, $|\,q\,b\,| = |\,b\,|$. There is therefore, by Proposition 8.20, a non-zero quaternion a such that $q\,b = a\,b\,a^{-1}$, that is, such that $q = a\,b\,a^{-1}\,b^{-1}$. □

Proposition 8.20 also leads to Proposition 8.23, converse to Proposition 8.18.

Proposition 8.23 *For each $t \in O(3)$, the map*

$$u : \mathbf{H} \to \mathbf{H};\; a \mapsto \operatorname{re} a + t(\operatorname{pu} a)$$

is an automorphism or anti-automorphism of $\mathbf{H}$, u being an automorphism if t is a rotation and an anti-automorphism if t is an anti-rotation of $\mathbf{R}^3$.

Proof For each $t \in SO(3)$, the map u can, by Proposition 8.20, be put in the form

$$\mathbf{H} \to \mathbf{H};\ a \mapsto q\,a\,q^{-1} = \mathrm{re}\,a + q\,(\mathrm{pu}\,a)\,q^{-1},$$

where $q \in \mathbf{H}^{\bullet}$, and such a map is an automorphism of $\mathbf{H}$.

Also, $-1_{\mathbf{R}^3}$ is an anti-rotation of $\mathbf{R}^3$, and if $t = -1_{\mathbf{R}^3}$, u is conjugation, which is an anti-automorphism of $\mathbf{H}$. The remainder of the proposition follows at once, since any anti-automorphism of $\mathbf{H}$ can be expressed as the composite of any particular anti-automorphism, for example conjugation, with some automorphism. □

Corollary 8.24 *An involution of* $\mathbf{H}$ *either is the identity or corresponds to the rotation of* $\mathbf{R}^3$ *through* π *about some axis, that is, reflection in some line through* 0. *Any anti-involution of* $\mathbf{H}$ *is conjugation composed with such an involution, and corresponds either to the reflection of* $\mathbf{R}^3$ *in the origin or to the reflection of* $\mathbf{R}^3$ *in some plane through* 0.

It is convenient to single out one of the non-trivial involutions of $\mathbf{H}$ to be typical of the class. For technical reasons we choose the involution $\mathbf{H} \to \mathbf{H};\ a \mapsto \mathrm{j}\,a\,\mathrm{j}^{-1}$, corresponding to the reflection of $\mathbf{R}^3$ in the line $\mathbf{R}\{\mathrm{j}\}$. This will be called the *main involution* of $\mathbf{H}$ and, for each $a \in \mathbf{H}$, $\widehat{a} = \mathrm{j}\,a\,\mathrm{j}^{-1}$ will be called the *involute* of a. The main involution commutes with conjugation. The composite will be called *reversion* and, for each $a \in \mathbf{H}$, $\widetilde{a} = \widehat{\overline{a}} = \overline{\widehat{a}}$ will be called the *reverse* of a. A reason for this is that $\mathbf{H}$ may be regarded as being generated as a real algebra by i and k, and reversion sends i to i and k to k but sends $\mathrm{i\,k}$ to $\mathrm{k\,i}$, reversing the multiplication. Reversion in a more general setting will be defined later, in Chapter 15.

Proposition 8.25 is required in the proof of Proposition 10.9.

Proposition 8.25 *The map* $\mathbf{H} \to \mathbf{H};\ x \mapsto \widetilde{x}\,x$ *has as image the three-dimensional real linear subspace* $\{y \in \mathbf{H} : \widetilde{y} = y\} = \mathbf{R}\{1, \mathrm{i}, \mathrm{k}\}$.

Proof It is enough to prove that the map $S^3 \to S^3;\ x \mapsto \widetilde{x}\,x = \widehat{x}^{-1}\,x$ has as image the unit sphere in $\mathbf{R}\{1, \mathrm{i}, \mathrm{k}\}$.

So let $y \in \mathbf{H}$ be such that $\overline{y}\,y = 1$ and $\overline{y} = \widehat{y}$. Then

$$1 + y = \overline{y}\,y + 1 = (\widehat{y} + 1)\,y.$$

So, if $y \neq -1$, $y = \widehat{x}^{-1}\,x$, where $x = (1 + y)(|1 + y|)^{-1}$. Finally, $-1 = \widetilde{\mathrm{i}}\,\mathrm{i}$. □

Rotations of $\mathbf{R}^4$

Quaternions may be used to represent rotations of $\mathbf{R}^4$.

Proposition 8.26 *Let q be a unit quaternion, with $\mathbf{H}$ identified with $\mathbf{R}^4$. Then $q_L : \mathbf{R}^4 \to \mathbf{R}^4;\ x \mapsto q\,x$ is a rotation of $\mathbf{R}^4$, as also is the map $q_R : \mathbf{R}^4 \to \mathbf{R}^4;\ x \mapsto x\,q$.*

Proof The map q_L is linear, and preserves norm by Proposition 8.11; so it is orthogonal, by Proposition 5.32. That it is a rotation follows from Proposition 8.22, which states that there exist non-zero quaternions a and b such that $q = a\,b\,a^{-1}\,b^{-1}$, for then $q_L = a_L b_L (a_L)^{-1}(b_L)^{-1}$, implying that $\det_{\mathbf{R}}(q_L) = 1$. Similarly for q_R. □

Proposition 8.27 *The map*

$$\rho : S^3 \times S^3 \to SO(4);\ (q, r) \mapsto q_L\,\bar{r}_R$$

is a group surjection with kernel $\{(1, 1), (-1, -1)\}$.

Proof For any $q, q', r, r' \in S^3$ and any $x \in \mathbf{H}$,

$$\rho(q'q,\, r'r)(x) = (q'q)_L(\overline{r'r})_R x \ = q'q\,x\,\bar{r}\,\overline{r'} = \rho(q', r')\rho(q, r)(x).$$

Therefore, for any $(q, r), (q', r') \in S^3 \times S^3$,

$$\rho((q', r')(q, r)) = \rho(q', r')\rho(q, r);$$

that is, ρ is a group map. That it has the stated kernel follows from the observation that if q and r are unit quaternions such that $q\,x\,\bar{r} = x$, for all $x \in \mathbf{H}$, then, by choosing $x = 1$, $q\,\bar{r} = 1$, from which it follows that $q\,x\,q^{-1} = x$ for all $x \in \mathbf{H}$, or, equivalently, that $q\,x = x\,q$ for all $x \in \mathbf{H}$. This implies, by Proposition 8.3, that $q \in \{1, -1\} = \mathbf{R} \cap S^3$.

To prove that ρ is surjective, let t be any rotation of $\mathbf{R}^4$ and let $s = t(1)$. Then $|s| = 1$ and the map $\mathbf{R}^4 \to \mathbf{R}^4;\ x \mapsto \bar{s}(t(x))$ is a rotation of $\mathbf{R}^4$ leaving 1 and therefore each point of $\mathbf{R}$ fixed. So, by Proposition 8.20, there exists a unit quaternion r such that, for all $x \in \mathbf{R}^4$,

$$\bar{s}(t(x)) = r\,x\,r^{-1}$$

or, equivalently, $t(x) = q\,x\,\bar{r}$, where $q = s\,r$. □

Anti-rotations of $\mathbf{R}^4$ also are easily represented by quaternions, since conjugation is an anti-rotation and since any anti-rotation is the composite of any given anti-rotation and a rotation.

Exercises

8.1 Let q and r be non-zero quaternions such that $|q| = |r|$, and let $\alpha = \frac{1}{2}(q + r)$ and $\beta = \frac{1}{2}(q - r)$. Prove that either α or β is non-zero (cf. Proposition 4.2) and that if α is invertible then $\beta\,\alpha^{-1}$ is a pure quaternion, while if β is invertible then $\alpha\,\beta^{-1}$ is a pure quaternion.

8.2 Let α and β be non-zero quaternions such that $\beta\,\overline{\alpha}$ is pure. Prove that, as vectors of $\mathbf{R}^4$, α and β are mutually orthogonal.

8.3 For any unit quaternion κ and any $\theta \in \mathbf{R}$ let $e^{\kappa\theta}$ denote the quaternion $\cos\theta + \kappa\sin\theta$. Prove that, for any s, $t \in \mathbf{R}$,

$$e^{\mathrm{k}s}e^{\mathrm{i}t}e^{-\mathrm{k}s} = e^{\kappa t}, \text{ where } \kappa = \mathrm{i}\cos 2s - \mathrm{j}\sin 2s.$$

(This identity is the starting point of a recent paper by V.I. Arnol'd (1995), that applies the algebra of quaternions to the geometry of curves on the sphere S^2.)

9
Quaternionic linear spaces

Much of the theory of linear spaces and linear maps over a commutative field, summarised in Chapter 1, extends over $\mathbf{H}$. Because of the non-commutativity of $\mathbf{H}$ it is, however, necessary to distinguish two types of linear space over $\mathbf{H}$, namely *right* linear spaces and *left* linear spaces. There are therefore restrictions on possible types of linear maps over $\mathbf{H}$.

Right and left linear spaces

A *right* linear space over $\mathbf{H}$ consists of an additive group X and a map

$$X \times \mathbf{H} \to X;\ (x, \lambda) \mapsto x\lambda$$

such that the usual distributivity and unity axioms hold and such that, for all $x \in X, \lambda, \lambda' \in \mathbf{H}$,

$$(x\lambda)\lambda' = x(\lambda\lambda').$$

A *left* linear space over $\mathbf{H}$ consists of an additive group X and a map

$$\mathbf{H} \times X \to X;\ (\mu, x) \mapsto \mu x$$

such that the usual distributivity and unity axioms hold and such that, for all $x \in X, \mu, \mu' \in \mathbf{H}$,

$$\mu'(\mu x) = (\mu' \mu)x.$$

The additive group $\mathbf{H}^n$, for any finite n, and in particular $\mathbf{H}$ itself, can be assigned either a right or a left $\mathbf{H}$-linear structure in an obvious way. Unless there is explicit mention to the contrary, it will normally be assumed that the *right* $\mathbf{H}$-linear structure has been chosen. (As we shall see below, a natural notation for $\mathbf{H}^n$ with the obvious left $\mathbf{H}$-linear structure is either $(\mathbf{H}^n)^L$ or $(\mathbf{H}^L)^n$.)

Subspaces of right or left **H**-linear spaces and products of such spaces are defined in the obvious way. Likewise the basis theorems hold, except that care must be taken to put scalar multipliers on the correct side. Where there is a finite basis for an **H**-linear space the number n of elements of the basis is independent of the basis and is by definition the *quaternionic dimension*, $\dim_{\mathbf{H}} X$, of X. *It will be tacitly assumed in all that follows that any* **H**-*linear space encountered is finite-dimensional.*

Linear maps $t : X \to Y$, where X and Y are **H**-linear spaces, may be defined, provided that each of the spaces X and Y is a right linear space or that each is a left linear space. For example, if X and Y are both *right* linear spaces, then t is said to be *linear* or *right linear* if it respects addition and if, for all $x \in X$, $\lambda \in \mathbf{H}$, $t(x\lambda) = (t(x))\lambda$, an analogous definition holding in the *left* case.

The set of linear maps $t : X \to Y$ between either right or left linear spaces X and Y over **H** will be denoted in either case by $L(X, Y)$. However, the usual recipe for $L(X, Y)$ to be a linear space fails. For suppose that we define, for any $t \in L(X, Y)$ and any $\lambda \in \mathbf{H}$, a map $t\lambda : X \to Y$ by the formula $(t\lambda)x = t(x)\lambda$, X and Y being right **H**-linear spaces. Then, for any $t \in L(X, Y)$ and any $x \in X$,

$$t(x)\,\mathrm{k} = (t\,\mathrm{i}\,\mathrm{j})(x) = (t\,\mathrm{i})(x\,\mathrm{j}) = t(x\,\mathrm{j})\mathrm{i} = -t(x)\,\mathrm{k},$$

leading at once to a contradiction if $t \neq 0$, as is possible. Normally $L(X, Y)$ is regarded as a linear space over the *centre* of **H**, namely **R**. In particular, for any right **H**-linear space X, the set $\operatorname{End} X = L(X, X)$ is normally regarded as a *real* algebra.

On the other hand, for any right linear space X over **H**, a left **H**-linear structure can be assigned to $L(X, \mathbf{H})$, by setting $(\mu t)(x) = \mu(t(x))$, for all $t \in L(X, \mathbf{H})$, $x \in \mathbf{H}$ and $\mu \in \mathbf{H}$. This left linear space is called the *linear dual* of X and is also denoted by X^L. The linear dual of a left **H**-linear space is analogously defined. It is a right **H**-linear space.

For any finite-dimensional **H**-linear space X,

$$\dim_{\mathbf{H}} X^L = \dim_{\mathbf{H}} X.$$

Any right **H**-linear map $t : X \to Y$ induces a left linear map $t^L : Y^L \to X^L$ by the formula

$$t^L(\gamma) = \gamma\, t, \text{ for each } \gamma \in Y^L,$$

and if $t \in L(X, Y)$ and $u \in L(W, X)$, W, X and Y all being right **H**-linear spaces, then

$$(t\,u)^L = u^L\, t^L.$$

Quaternionic matrices

Any right $\mathbf{H}$-linear map $t : \mathbf{H}^n \to \mathbf{H}^m$ may be represented in the obvious way by an $m \times n$ matrix with entries in $\mathbf{H}$. In particular, any elements of the right $\mathbf{H}$-linear spaces $\mathbf{H}^n$ and $\mathbf{H}^m$ may be represented by column matrices. Scalar multipliers have, however, to be written on the right and not on the left as has been the custom hitherto. In fact this is really more logical anyway, as the equation $t(x, \lambda) = t(x)\lambda$ may then be regarded as exemplifying the associative law for composition of matrices.

For example, suppose that $t \in \operatorname{End} \mathbf{H}^2$, $x \in \mathbf{H}^2$ and $\lambda \in \mathbf{H}$. Then the statement that $t(x\,\lambda) = t(x)\lambda$ becomes, in matrix notation,

$$\begin{pmatrix} t_{00} & t_{01} \\ t_{10} & t_{11} \end{pmatrix} \begin{pmatrix} x_0\,\lambda \\ x_1\,\lambda \end{pmatrix} = \left(\begin{pmatrix} t_{00} & t_{01} \\ t_{10} & t_{11} \end{pmatrix} \begin{pmatrix} x_0 \\ x_1 \end{pmatrix} \right) \lambda.$$

The left $\mathbf{H}$-linear space $(\mathbf{H}^n)^L$ dual to the right $\mathbf{H}$-linear space $\mathbf{H}^n$ may be identified with the additive group $\mathbf{H}^n$ assigned its left $\mathbf{H}$-linear structure. Elements of this space may be represented by row matrices. A left $\mathbf{H}$-linear map $u : (\mathbf{H}^m)^L \to (\mathbf{H}^n)^L$ is then represented by an $m \times n$ matrix (*sic*) that multiplies the row vector elements of $(\mathbf{H}^m)^L$ on the right.

$\mathbf{H}(n)$ will denote the *real* algebra of $n \times n$ matrices over $\mathbf{H}$.

Any quaternionic linear space X may be regarded as a real linear space, with $\dim_{\mathbf{R}} X = 4 \dim_{\mathbf{H}} X$. Such a space may also be regarded as a complex linear space, once some representative of $\mathbf{C}$ as a subalgebra of $\mathbf{H}$ has been chosen, with $\dim_{\mathbf{C}} X = 2 \dim_{\mathbf{H}} X$. In the following discussion $\mathbf{C}$ *is identified with* $\mathbf{R}\{1, \mathrm{i}\}$ in $\mathbf{H}$, and, for each finite n, $\mathbf{C}^{2n} = \mathbf{C}^n \times \mathbf{C}^n$ is identified with $\mathbf{H}^n$ by the (right) complex linear isomorphism

$$\mathbf{C}^n \times \mathbf{C}^n \to \mathbf{H}^n;\ (u, v) \mapsto u + \mathrm{j}\,v.$$

Proposition 9.1 *Let* $a + \mathrm{j}\,b \in \mathbf{H}(n)$, *where* a *and* $b \in \mathbf{C}(n)$. *Then the corresponding element of* $\mathbf{C}(2n)$ *is* $\begin{pmatrix} a & -\overline{b} \\ b & \overline{a} \end{pmatrix}$.

Proof For any $u, v, u', v' \in \mathbf{C}^n$ and any $a, b \in \mathbf{C}(n)$, the equation $u' + \mathrm{j}\,v' = (a + \mathrm{j}\,b)(u + \mathrm{j}\,b)$ is equivalent to the pair of equations

$$\begin{aligned} u' &= a\,u - \overline{b}\,v, \\ v' &= b\,u + \overline{a}\,v. \end{aligned}$$

□

In particular, when $n = 1$, this becomes an explicit representation of

H as either a complex or a real subalgebra of **C**(2), analogous to the representation of **C** as a subalgebra of **R**(2) given in Corollary 2.2.

Notice that, for any $q = a + \mathrm{j}\, b \in \mathbf{H}$, with $a, b \in \mathbf{C}$,

$$|q|^2 = \overline{q}\, q = \overline{a}\, a + \overline{b}\, b = \det \begin{pmatrix} a & -\overline{b} \\ b & \overline{a} \end{pmatrix}.$$

This remark is a detail in the proof of Proposition 9.3 below.

The lack of commutativity in **H** is most strongly felt when one tries to introduce products, as the following proposition shows.

Proposition 9.2 *Let X, Y and Z be right linear spaces over* **H** *and let $t : X \times Y \to Z$; $(x, y) \mapsto x \cdot y$ be a right bilinear map. Then $t = 0$.*

Proof For any $(x, y) \in X \times Y$,

$$\begin{aligned}(x \cdot y)\,\mathrm{k} &= (x \cdot y)\mathrm{i}\,\mathrm{j} = (x \cdot y\,\mathrm{i})\mathrm{j} = x\,\mathrm{j} \cdot y\,\mathrm{i} = (x\,\mathrm{j} \cdot y)\mathrm{i} \\ &= (x \cdot y)\mathrm{j}\,\mathrm{i} = -(x \cdot y)\mathrm{k}.\end{aligned}$$

Since $\mathrm{k} \neq 0$, $x \cdot y = 0$. So $t = 0$. □

It follows from this, *a fortiori*, that there is no non-trivial n-linear map $X^n \to \mathbf{H}$, for a right **H**-linear space, for any $n > 1$. In particular, there is no direct analogue of the determinant for the algebra of endomorphisms of a quaternionic linear space, in particular the right **H**-linear space $\mathbf{H}^n$. However, any $n \times n$ matrix over **H** is reducible by column operations to a diagonal matrix all of whose entries except one are equal to 1, just as in the real or complex case as described in Chapter 2, though this number is no longer independent of the route taken, as Exercise 9.2 will show. However, any two such numbers have the same absolute value, as follows from the next proposition.

Proposition 9.3 *Let X be a right* **H**-*linear space and let $t : X \to X$ be an* **H**-*linear map. Then* $\det_{\mathbf{C}} t$ *is a non-negative real number.*

Proof Mimic the proof of Proposition 2.3. □

For any **H**-linear endomorphism $t : X \to X$, the (positive) square root of $\det_{\mathbf{C}} t$ is defined to be the *(absolute) determinant* of t, $\det t$. Clearly t is invertible if and only if $\det t \neq 0$.

The group of all invertible right linear maps $t : \mathbf{H}^n \to \mathbf{H}^n$ over **H** or, equivalently, the group of all invertible $n \times n$ matrices over **H** will be denoted by $GL(n;\mathbf{H})$ and the subgroup of all such maps or matrices of determinant 1 by $SL(n;\mathbf{H})$.

Clearly earlier remarks about direct sum decompositions and the matrix algebras ${}^2\mathbf{R}(n)$ and ${}^2\mathbf{C}(n)$ extend to direct decompositions of quaternionic spaces and the (real) matrix algebras ${}^2\mathbf{H}(n)$.

Exercises

9.1 Verify that the matrix $\begin{pmatrix} 1 & \mathrm{i} \\ \mathrm{j} & \mathrm{k} \end{pmatrix}$ is invertible in $\mathbf{H}(2)$, but that the matrix $\begin{pmatrix} 1 & \mathrm{j} \\ \mathrm{i} & \mathrm{k} \end{pmatrix}$ is not.

9.2 Verify that the matrix $\begin{pmatrix} 1 & \mathrm{i} \\ \mathrm{j} & \mathrm{k} \end{pmatrix}$ may be reduced by a pair of elementary column operations not only to $\begin{pmatrix} -2\mathrm{k} & 0 \\ 0 & 1 \end{pmatrix}$ but also to $\begin{pmatrix} 1 & 0 \\ 0 & 2\mathrm{k} \end{pmatrix}$.

9.3 Does a *real* algebra involution of $\mathbf{H}(2)$ necessarily map a matrix of the form $\begin{pmatrix} a & 0 \\ 0 & a \end{pmatrix}$, where $a \in \mathbf{H}$, to one of the same form?

9.4 Verify that the map $\alpha : \mathbf{C}^2 \to \mathbf{H};\ x \mapsto x_0 + \mathrm{j}\, x_1$ is a right $\mathbf{C}$-linear isomorphism, and compute

$$\alpha^{-1}(\widetilde{\alpha(x)}\,\alpha(y)),\ \text{for any } x, y \in \mathbf{C}^2.$$

Let $Q = \{(x, y) \in (\mathbf{C}^2)^2 : x_0 y_0 + x_1 y_1 = 1\}$. Prove that, for any $(a, b) \in \mathbf{H}^\bullet \times \mathbf{C}$,

$$(\alpha^{-1}(\widetilde{a}),\ \alpha^{-1}(a^{-1}(1 + \mathrm{j}\, b))) \in Q$$

and that the map

$$\mathbf{H}^\bullet \times \mathbf{C} \to Q;\ (a, b) \mapsto (\alpha^{-1}(\widetilde{a}),\ \alpha^{-1}(a^{-1}(1 + \mathrm{j}\, b)))$$

is injective.

10
Anti-involutions of $\mathbf{H}(n)$

Proposition 9.2 has shown us that there are no non-trivial quadratic forms on a quaternionic linear space. There are, however, analogues of hermitian forms and these induce adjoint anti-involutions of real quaternionic matrix algebras.

Correlated quaternionic spaces

Let X and Y be right or left quaternionic linear spaces. Then an $\mathbf{H}$-linear map $f : X \to Y$ is said to be *semi-linear* over $\mathbf{H}$ if there is an automorphism or anti-automorphism ψ of $\mathbf{H}$ such that, for all $x \in X$, $\lambda \in \mathbf{H}$, $f(x\lambda) = f(x)\lambda^{\psi}$, $f(x\lambda) = \lambda^{\psi} f(x)$, $f(\lambda x) = \lambda^{\psi} f(x)$ or $f(\lambda x) = f(x)\,\lambda^{\xi}$, as the case may be, ψ being an automorphism of $\mathbf{H}$ if $\mathbf{H}$ operates on X and Y on the same side and an anti-automorphism if $\mathbf{H}$ operates on X and Y on opposite sides. The terms *right*, *right-to-left*, *left* and *left-to-right* semi-linear maps over $\mathbf{H}$ have the obvious meanings.

The semi-linear map t uniquely determines the automorphism or anti-automorphism ψ of $\mathbf{H}$ with respect to which it is semi-linear, unless $t = 0$. On occasion it is convenient to refer directly to ψ, the map t then being said to be $\mathbf{H}^{\psi}$*-linear*.

The following maps are invertible right-semi-linear maps over $\mathbf{H}$:

$\mathbf{H} \to \mathbf{H}$; $x \mapsto, x$
$x \mapsto a\,x$, for any non-zero $a \in \mathbf{H}$,
$x \mapsto x\,b^{-1}$, for any non-zero $b \in \mathbf{H}$,
$x \mapsto a\,x\,b^{-1}$, for any non-zero $a, b \in \mathbf{H}$,
$x \mapsto \widehat{x}\ (= \mathrm{j}\,x\,\mathrm{j}^{-1})$,
and $\mathbf{H}^2 \to \mathbf{H}^2$; $(x, y) \mapsto (a\,x, b\,y)$, for any non-zero $a, b \in \mathbf{H}$,
$(x, y) \mapsto (b\,y, a\,x)$, for any non-zero $a, b \in \mathbf{H}$,

the corresponding automorphisms of $\mathbf{H}$ being, respectively,

$$1_{\mathbf{H}},\ 1_{\mathbf{H}},\ \lambda \mapsto b\lambda b^{-1},\ \lambda \mapsto b\lambda b^{-1},\ \lambda \mapsto \widehat{\lambda} \text{ and } 1_{\mathbf{H}},\ 1_{\mathbf{H}}.$$

By contrast, the map

$$\mathbf{H}^2 \to \mathbf{H}^2;\ (x, y) \mapsto (x a, y b), \text{ with } a, b \in \mathbf{H},$$

is *not* right semi-linear over $\mathbf{H}$, unless $\lambda a = \mu b$, with $\lambda, \mu \in \mathbf{R}$.

The maps

$$\mathbf{H}^2 \to \mathbf{H}^2;\quad (x, y) \mapsto (\overline{x}, \overline{y}),$$
$$(x, y) \mapsto (\overline{y}, \overline{x})$$

are invertible right-to-left $\overline{\mathbf{H}}$-linear maps.

The composite of any two composable semi-linear maps is semi-linear and the inverse of an invertible semi-linear map is semi-linear. An invertible semi-linear map is said to be a *semi-linear isomorphism.*

Proposition 10.1 *Let $f : X \to Y$ be a semi-linear map over $\mathbf{H}$. Then* $\operatorname{im} f$ *is an $\mathbf{H}$-linear subspace of Y and* $\ker f$ *is an $\mathbf{H}$-linear subspace of X.*

Rank and *kernel rank* are defined for semi-linear maps as for linear maps.

Proposition 10.2 *Let $f : X \to Y$ be an $\mathbf{H}^{\psi}$-linear map, where X and Y are finite-dimensional $\mathbf{H}$-linear spaces, and ψ is an automorphism or anti-automorphism of $\mathbf{H}$. Then, for any $\gamma \in Y^L, \psi\gamma f \in X^L$.*

Proof The map $\psi\gamma f$ is certainly $\mathbf{R}$-linear. It remains to consider its interaction with $\mathbf{H}$-multiplication. There are four cases, of which we consider only one, namely the case in which X and Y are each right $\mathbf{H}$-linear. In this case, for each $x \in X$, $\lambda \in \mathbf{H}$,

$$\psi^{-1}\gamma\, t(x\lambda) = \psi^{-1}\gamma(t(x)\,\lambda^{\psi}) = \psi^{-1}((\gamma\, t(x))\lambda^{\psi}) = (\psi^{-1}\gamma\, t(x))\lambda.$$

The proofs in the other three cases are similar. □

The map $f^L : Y^L \to X^L$, defined, for all $\gamma \in Y^L$, by the formula $f^L(\gamma) = \psi\gamma f$, is called the *dual* of f.

Proposition 10.3 *The dual f^L of an $\mathbf{H}^{\psi}$-linear map $f : X \to Y$ is $\mathbf{H}^{\psi^{-1}}$-linear.*

Many properties of the duals of $\mathbf{R}$-linear maps carry over to semi-linear maps over $\mathbf{H}$.

A *correlation* on a finite-dimensional $\mathbf{H}$-linear space X is an $\mathbf{H}$-semi-linear map $\xi : X \to X^L;\ x \mapsto x^\xi$. The map $X \times X \to (a,b) \mapsto a^\xi b = a^\xi(b)$ is the *product* induced by the correlation, and the map $X \to \mathbf{C};\ a \mapsto a^\xi a$ the *form* induced by the correlation. Such a product is $\mathbf{R}$-bilinear, but not, in general, $\mathbf{H}$-bilinear, for although the map

$$X \to \mathbf{H};\ x \mapsto a^\xi x$$

is $\mathbf{H}$-linear, for any $a \in X$, the map

$$X \to \mathbf{H};\ x \mapsto x^\xi b,$$

for any $b \in X$, is, in general, not linear but only (right-to-left) semi-linear over $\mathbf{H}$. As in the parallel $\overline{\mathbf{C}}$ case such a product is said to be *sesqui-linear*.

An $\mathbf{H}^\psi$-correlation $\xi : X \to X^L$ and the induced product on the right $\mathbf{H}$-linear space X are said to be, respectively, *symmetric* or *skew with respect to* ψ or *over* $\mathbf{H}^\psi$ according as, for all a, $b \in X$,

$$b^\xi a = (a^\xi b)^\psi \text{ or } -(a^\xi b)^\psi.$$

A *correlated quaternionic space* is a finite-dimensional $\mathbf{H}$-linear space assigned an $\overline{\mathbf{H}}$-correlation.

As always an invertible correlation is said to be *non-degenerate*.

Semi-linear correlations $\xi, \eta : X \to X^L$ on a right $\mathbf{H}$-linear space X are said to be *equivalent* if, for some invertible $\lambda \in \mathbf{H}, \eta = \lambda\,\xi$. This is clearly an equivalence on any set of semi-linear correlations on X.

There are two quaternionic analogues of Proposition 7.11.

Proposition 10.4 *Let ψ be any anti-involution of $\mathbf{H}$ other than conjugation. Then any skew $\mathbf{H}^\psi$-correlation on a right $\mathbf{H}$-linear space X is equivalent to a symmetric $\overline{\mathbf{H}}$-correlation on X, and conversely.*

Proof We give the proof for the case $\mathbf{H}^\psi = \widetilde{\mathbf{H}}$.

Let ξ be a skew $\widetilde{\mathbf{H}}$-correlation on X. Then $\mathrm{j}\,\xi$ is an $\overline{\mathbf{H}}$-correlation on X, since, for all $x \in X$, $\lambda \in \mathbf{H}$,

$$(\mathrm{j}\,\xi)(x\,\lambda) = \mathrm{j}\,\widetilde{\lambda}\,\xi(x) = \overline{\lambda}\,(\mathrm{j}\,\xi)(x).$$

Moreover, for all $a, b \in X$,

$$b^{j\xi}a = \mathrm{j}\, b^{\xi}a = -\mathrm{j}\,\widetilde{a^{\xi}b} = \overline{a^{\xi}b\,\bar{\mathrm{j}}} = \overline{a^{j\xi}b}.$$

That is, $\mathrm{j}\,\xi$ is symmetric over $\overline{\mathbf{H}}$. □

Proposition 10.5 *Let ψ be as in Proposition 10.4. Then any symmetric $\mathbf{H}^{\psi}$-correlation on a right $\mathbf{H}$-linear space X is equivalent to a skew $\overline{\mathbf{H}}$-correlation on X, and conversely.*

Suppose now that $f : X \to Y$ is a linear map of a right quaternionic space X, with non-degenerate correlation ξ, to a right quaternionic space Y, with correlation η. Since ξ is bijective there is a unique $\mathbf{H}$-linear map $t^{*} : Y \to X$ such that $t^{*} = \xi^{-1}t^{L}\eta$, that is, such that, for any $x \in X$, $y \in Y, t^{*}(y)^{\xi}x = y^{\eta}t(x)$. The map $t^{*} = \xi^{-1}t^{L}\eta$ is called the *adjoint* of t with respect to ξ and η.

The adjoint of a linear endomorphism u of a right quaternionic linear space X with respect to a non-degenerate correlation ξ will be denoted by u^{ξ}. The map u is said to be *self-adjoint* if $u^{\xi} = u$ and *skew-adjoint* if $u^{\xi} = -u$. The real linear subspaces $\{u \in \operatorname{End} X : u^{\xi} = u\}$ and $\{u \in \operatorname{End} X : u^{\xi} = -u\}$ of the real linear space $\operatorname{End} X = L(X, X)$ will be denoted by $\operatorname{End}_{+}(X, \xi)$ and $\operatorname{End}_{-}(X, \xi)$, respectively.

Proposition 10.6 *Let ξ be a non-degenerate correlation on a right quaternionic linear space X. Then the map*

$$\operatorname{End} X \to \operatorname{End} X;\ f \mapsto f^{*}$$

is an anti-involution of End X*, regarded as a real algebra.*

Equivalent semi-linear correlations ξ and η on the right quaternionic space $\mathbf{H}^{n}$ induce the same adjoint anti-automorphism of $\mathbf{H}(n)$.

Symmetric and skew correlations are particular cases of reflexive correlations, a correlation ξ on a quaternionic linear space X being said to be *reflexive* if, for all $a, b \in X$,

$$b^{\xi}a = 0 \Leftrightarrow a^{\xi}b = 0.$$

It is almost the case that any reflexive correlation is equivalent either to a symmetric or to a skew one, a counter-example in the one-dimensional case being provided by the correlation on $\mathbf{H}$ with product

$$\mathbf{H}^{2} \to \mathbf{H}^{2};\ (a, b) \mapsto \bar{a}(1 + \mathrm{j})b.$$

Proposition 10.7 *Any non-zero reflexive $\mathbf{H}^\psi$-correlation ξ on a right $\mathbf{H}$-space X of dimension greater than one is either symmetric or skew.*

Proof This is naturally a further re-run of Proposition 6.1, but one has to remember that ψ is an *anti-automorphism* of $\mathbf{H}$, and that it need not be an anti-involution.

For all $a, b \in X$, $(b^\xi a)^{\psi^{-1}} = 0 \Leftrightarrow b^\xi a = 0 \Leftrightarrow a^\xi b = 0$. That is, for any non-zero $a \in X$, the kernels of the surjective $\mathbf{H}$-*linear* maps $b \mapsto (b^\xi a)^{\psi^{-1}}$ and a^ξ coincide. Therefore, by the quaternionic analogue of Proposition 1.5 there is a non-zero quaternion λ_a such that, for all $b \in X$,

$$(b^\xi a)^{\psi^{-1}} = \lambda_a a^\xi b.$$

To prove that λ_a is independent of a one then considers $c \in X$ linearly independent of a, c existing since $\dim X > 1$. Then, since

$$b^\xi a + b^\xi c = b^\xi(a + c),$$

it follows that

$$\lambda_a a^\xi b + \lambda_c c^\xi b = \lambda_{a+c}(a + c)^\xi b,$$

for all $b \in X$. So

$$a\,\lambda_a^{\psi^{-1}} = \lambda_{a+c}^{\psi^{-1}} = \lambda_c^{\psi^{-1}}.$$

But a and c are linearly independent. So

$$\lambda_a^{\psi^{-1}} = \lambda_{a+c}^{\psi^{-1}} = \lambda_c^{\psi^{-1}},$$

implying that $\lambda_a = \lambda_c$. So, as before, there exists $\lambda \in \mathbf{H}$ such that, for all $a, b \in X, (b^\xi a)^{\psi^{-1}} = \lambda\, a^\xi b$.

Again there are apparently two cases.

However, if $a^\xi a = 0$, for all $a \in X$, it follows that $(b^\xi a)^{\psi^{-1}} = -\lambda b^\xi a$ for all $a, b \in X$. But, for suitable a and b, $b^\xi a = 1$, implying that $\lambda = -1$ and that $\psi = 1_{\mathbf{H}}$. But ψ cannot be the identity, so this case does not arise.

The alternative is that, for some $x \in X$, $x^\xi x \neq 0$, implying that, for some invertible $\mu \in \mathbf{H}$, $(\mu^{-1})^{\psi^{-1}} = \lambda\mu^{-1}$, or, equivalently, $\mu^{-1} = (\mu^{-1})^\psi \lambda^\psi$. Then, for all $a, b \in X$,

$$b^{\mu\xi} a = \mu(\lambda a^\xi b)^\psi = \mu(\mu a^\xi b)^\psi (\mu^{-1})^\psi \lambda^\psi = \mu(\mu a^\xi b)^\psi \mu^{-1} = \mu(a^{\mu\xi} b)^\psi \mu^{-1}.$$

Moreover, for all $\lambda \in \mathbf{H}$, $(b\lambda)^{\mu\xi} a = (\mu\lambda^\psi\mu^{-1}) b^{\mu\xi} a$. The correlation $\mu\xi$, equivalent to ξ, is therefore a symmetric $\mathbf{H}^{\psi'}$-correlation, where, for any $\nu \in \mathbf{H}$,

$$\nu^{\psi'} = \mu\,\nu^\psi \mu^{-1}.$$

That is, the correlation $\mu\xi$, equivalent to ξ, is symmetric. □

There is a converse of Proposition 10.6, analogous to the converses of Propositions 6.2 and 7.12.

Theorem 10.8 *Any anti-involution α of the* real *algebra $\mathbf{H}(n+1)$ is representable as the adjoint anti-involution induced by a non-degenerate reflexive correlation on $\mathbf{H}^{n+1}$.*

Proof This follows the proof of Theorem 6.4 or 7.14. At the end of the first stage of the argument one proves that the map ψ is an anti-automorphism of $\mathbf{H}$, not necessarily an anti-involution. The correlation ξ is defined as before and proved to be $\mathbf{H}^{\psi}$-linear, and injective, and so is a non-degenerate $\mathbf{H}^{\psi}$-linear correlation on $\mathbf{H}^n \times \mathbf{H}$.

The remainder of the proof then goes through without change. □

The next proposition, analogous to Proposition 4.1, is required in the proof of Theorem 10.10.

Proposition 10.9 *(i) Let X be a right $\mathbf{H}$-linear space and ξ a symmetric $\overline{\mathbf{H}}$-correlation and suppose that $x \in X$ is such that $x^{\xi}x \neq 0$. Then there exists $\lambda \in \mathbf{H}$ such that $(x\lambda)^{\xi}(x\lambda) = 1$ or -1.*

(ii) Let X be a right $\mathbf{H}$-linear space and ξ a symmetric $\widetilde{\mathbf{H}}$-correlation and suppose that $x \in X$ is such that $x^{\xi}x \neq 0$. Then there exists $\lambda \in \mathbf{H}$ such that $(x\lambda)^{\xi}(x\lambda) = 1$.

Proof Let ψ be either of the anti-involutions of $\mathbf{H}$. Then, since, for all $\lambda \in \mathbf{H}$, $(x\lambda)^{\xi}(x\lambda) = \lambda^{\psi}(x^{\xi}x)\lambda$, it is enough to prove that, for some $\lambda \in \mathbf{H}$, $(\lambda^{-1})^{\psi}\lambda^{-1} = (\lambda^{\psi})^{-1}\lambda^{-1} = \pm x^{\psi}x$. Now, when ψ is conjugation, $\overline{x^{\xi}x} = x^{\xi}x$, by the symmetry of ξ, and $x^{\xi}x$ is therefore real. So in this case we may take λ^{-1} to be the square root of $|x^{\xi}x|$, proving (i). On the other hand, if $\mathbf{H}^{\psi} = \widetilde{\mathbf{H}}$, $\widetilde{x^{\xi}x} = x^{\xi}x$, and so, by Proposition 8.25, $x^{\xi}x$ belongs to the image of the map $\mathbf{H} \to \mathbf{H}$; $\mu \mapsto \widetilde{\mu}\mu$. Then we may take $\lambda = \mu^{-1}$, proving (ii). □

Theorem 10.10 (The basis theorem for symmetric correlated quaternionic spaces.) *Each finite-dimensional symmetric $\overline{\mathbf{H}}$- or $\widetilde{\mathbf{H}}$-correlated space has an orthonormal basis.*

By Theorem 10.8 this holds also for skew correlated quaternionic spaces since any skew correlation is equivalent to a symmetric one.

Theorem 10.11 (The classification theorem for symmetric correlated quaternionic spaces.) *(i) Let X be a non-degenerate symmetric $\overline{\mathbf{H}}$-correlated space of finite dimension n. Then there exists a unique pair of numbers (p,q), with $p+q=n$, such that X is isomorphic to $\overline{\mathbf{H}}^{p,q}$, this being the right $\mathbf{H}$-linear space $\mathbf{H}^{p+q}$, with the sesqui-linear product*

$$(\mathbf{H}^{p+q})^2 \to \mathbf{H};\ (a,b) \mapsto -\sum_{0\le i<p} \overline{a_i}\, b_i + \sum_{0\le j<q} \overline{a_{p+j}}\, b_{p+j}.$$

(ii) Let X be a non-degenerate symmetric $\widetilde{\mathbf{H}}$-correlated space of finite dimension n. Then X is isomorphic to $\widetilde{\mathbf{H}}^n$, this being the right $\mathbf{H}$-linear space $\mathbf{H}^n$, with the sesqui-linear product

$$(\mathbf{H}^n)^2 \to \mathbf{H};\ (a,b) \mapsto \sum_{0\le i<n} \widetilde{a_i}\, b_i.$$

The pair of numbers (p,q) in (i) is called the *signature* of the correlated space X and of its form. The space and its form are said to be *positive-definite* if $p=0$.

Quaternionic groups

Proposition 10.12 *(i) Let $t = \begin{pmatrix} a & c \\ b & d \end{pmatrix}$ be an endomorphism of the correlated quaternionic space $\overline{\mathbf{H}}^{p,q}$. Then the adjoint of t is*

$$t^{\xi} = \begin{pmatrix} \overline{a}^{\tau} & -\overline{b}^{\tau} \\ -\overline{c}^{\tau} & \overline{d}^{\tau} \end{pmatrix}.$$

(ii) Let t be an endomorphism of the correlated quaternionic space $\widetilde{\mathbf{H}}^n$. Then the adjoint of t is $t^{*} = \widetilde{t}^{\tau}$.

There are quaternionic groups analogous to the orthogonal and symplectic groups over $\mathbf{R}$ or $\mathbf{C}$. The classical notations make no mention of quaternions and so are quite inappropriate for our purposes. It turns out that the natural definitions are as follows: to define the *orthogonal quaternionic group of degree n, $O(n;\mathbf{H})$,* to be the group of automorphisms of the correlated quaternionic space $\widetilde{\mathbf{H}}^n$, and to define the *symplectic quaternionic group of degree $n=p+q$ and signature (p,q), $Sp(p,q;\mathbf{H})$,* to be the group of automorphisms of the correlated quaternionic space $\overline{\mathbf{H}}^{p,q}$. Of course, by Propositions 10.4 and 10.5 either may be regarded as preserving either a symmetric or a skew form. As in the unitary case, but by contrast to the real or complex case, there is no requirement in the symplectic quaternionic case for the degree of the group, the dimension of the correlated space, to be even. The group $Sp(0,n;\mathbf{H})$ is frequently denoted simply by $Sp(n)$, and referred to as the symplectic group of degree n, without even mention of the quaternions $\mathbf{H}$.

Note that $Sp(1) = S^3$, the group of quaternions of absolute value 1.

The rather varied uses of the word 'symplectic' tend to be a bit confusing at first. The word was first used to describe the groups $Sp(2n;\mathbf{R})$ and $Sp(2;\mathbf{C})$ to indicate their connection with the set of null planes in the correlated spaces $\mathbf{R}^{2n}_{\text{sp}}$ and $\mathbf{C}^{2n}_{\text{sp}}$. Such a set of null planes, regarded as a set of projective lines in the associated projective space, is known to projective geometers as a *line complex*. The groups were therefore originally called *complex groups*. This was leading to hopeless confusion when Hermann Weyl (1939) coined the word 'symplectic', derived from the Greek equivalent of the Latin word 'complex'. Whether the situation is any less complicated now is a matter of dispute!

The next proposition continues Proposition 7.20.

Proposition 10.13 *For any finite n, and $p + q = n$, with a rather obvious definition of $Sp(2p, 2q;\mathbf{C})$,*

$$\begin{aligned} Sp(n) &= Sp(2n;\mathbf{C}) \cap GL(n;\mathbf{H}) = Sp(2n;\mathbf{C}) \cap U(n), \\ O(n;\mathbf{H}) &= O(2n;\mathbf{C}) \cap GL(n;\mathbf{H}), \\ Sp(p,q) &= Sp(2p, 2q;\mathbf{C}) \cap U(2p, 2q). \end{aligned}$$

Proof As an example, the equation $Sp(n) = Sp(2n;\mathbf{C}) \cap U(2n)$ follows at once from the observation that, for all $z + \mathrm{j}\,w,\ z' + \mathrm{j}\,w' \in \mathbf{H}$,

$$\overline{(z + \mathrm{j}\,w)}(z' + \mathrm{j}\,w') = (\overline{z}\,z' + \overline{w}\,w') - \mathrm{j}(w\,z' - z\,w') = 0$$

if and only if $\overline{z}\,z' + \overline{w}\,w' = 0$ and $w\,z' - z\,w' = 0$. □

Corollary 10.14 *Each element, either of the group $O(n;\mathbf{H})$ or of the group $Sp(p,q)$, has determinant* 1.

Exercises

10.1 Let $t \in SU(3)$. Show that

$$\overline{t_{02}} = \begin{vmatrix} t_{10} & t_{11} \\ t_{20} & t_{21} \end{vmatrix},\ \overline{t_{12}} = -\begin{vmatrix} t_{00} & t_{01} \\ t_{20} & t_{21} \end{vmatrix},\ \text{and}\ \overline{t_{02}} = \begin{vmatrix} t_{00} & t_{01} \\ t_{10} & t_{11} \end{vmatrix}.$$

10.2 Show that the diagram of maps

$$\begin{array}{ccccc} Sp(1) = SU(2) & \longrightarrow & SU(3) & \longrightarrow & S^5 \\ \downarrow & & \downarrow & & \\ Sp(2) & \longrightarrow & SU(4) & \xrightarrow{\pi} & T \\ \downarrow & & \downarrow & & \\ S^7 & \xrightarrow{1} & S^7 & , & \end{array}$$

where any $z + \mathrm{j}\,w$ in $Sp(1)$ is identified with $\begin{pmatrix} z & -\overline{w} \\ w & \overline{z} \end{pmatrix}$ in $SU(2)$, is commutative, the top row and the two columns being special cases of the left-coset exact pairs defined in Theorem 13.13, and the map π being the surjection, with image T a subset of $SU(4)$, defined, for all $t \in SU(4)$, by the formula $\pi(t) = t\,t^{\sim}$, where

$$t^{\sim} = \begin{pmatrix} t_{11} & -t_{01} & t_{31} & -t_{21} \\ -t_{10} & t_{00} & -t_{30} & t_{20} \\ t_{13} & -t_{03} & t_{33} & -t_{23} \\ -t_{12} & t_{02} & -t_{32} & t_{22} \end{pmatrix}.$$

Hence, by Proposition 3.4, construct a bijection $S^3 \to T$ that makes the square

$$\begin{array}{ccc} SU(3) & \longrightarrow & S^3 \\ \downarrow & & \downarrow \\ SU(4) & \xrightarrow{\pi} & T \end{array}$$

commute and show that this bijection is the restriction to S^3 with target T of an injective real linear map $\gamma : \mathbf{C}^3 \to \mathbf{C}(4)$. (Exercise 10.1 is relevant at one point of the argument. We shall encounter this example again in Proposition 17.3 and in Diagram 24.5.)

11

Tensor products of algebras

The tensor product of algebras is a special case and generalisation of the tensor product of linear spaces that can be defined directly. We have chosen not to develop the theory of tensor products in general, as we have no need of the more general concept.

Tensor products of real algebras

Certain algebras over a *commutative* field $\mathbf{K}$ admit a decomposition somewhat analogous to the direct sum decompositions of a linear space, but involving the multiplicative structure rather than the additive structure.

Suppose that B and C are subalgebras of a finite-dimensional algebra A over $\mathbf{K}$, the algebra being associative and with unit element, such that

(i) for any $b \in B$, $c \in C$, $c\,b = b\,c$,
(ii) A is generated as an algebra by B and C,
(iii) $\dim A = \dim B \dim C$.

Then we say that A is the *tensor product* $B \otimes_{\mathbf{K}} C$ over $\mathbf{K}$, the abbreviation $B \otimes C$ being used when the field $\mathbf{K}$ is not in doubt.

Proposition 11.1 *Let B and C be subalgebras of a finite-dimensional algebra A over $\mathbf{K}$, such that $A = B \times C$, the algebra A being associative and with unit element. Then $B \cap C = \mathbf{K}$ (the field $\mathbf{K}$ being identified with the set of scalar multiples of the unit element $1_{(A)}$).*

It is tempting to suppose that the condition $B \cap C = \mathbf{K}$ can be used as an alternative to condition (iii) in the definition. That this is not the case is shown by the following example.

Example 11.2 *Let* $A = \left\{ \begin{pmatrix} a & b & c \\ 0 & a & 0 \\ 0 & 0 & a \end{pmatrix} \in \mathbf{R}(3) : a, b, c \in \mathbf{R} \right\}$,

$B = \left\{ \begin{pmatrix} a & b & 0 \\ 0 & a & 0 \\ 0 & 0 & a \end{pmatrix} : a, b \in \mathbf{R} \right\}$, *and* $C = \left\{ \begin{pmatrix} a & 0 & c \\ 0 & a & 0 \\ 0 & 0 & a \end{pmatrix} : a, c \in \mathbf{R} \right\}$.

Then A is generated as an algebra by B and C, $B \cap C = \mathbf{R}$, and any element of B commutes with any element of C. But $\dim A = 3$, *while* $\dim B = \dim C = 2$, *so that* $\dim A \neq \dim B \, \dim C$.

Condition (iii) is essential to the proof of the following proposition.

Proposition 11.3 *Let A be a finite-dimensional associative algebra with unit element over a commutative field* $\mathbf{K}$ *and let B and C be subalgebras of A such that $A = B \otimes C$. Also let*

$$\{e_i : 0 \leq i < \dim B\} \text{ and } \{f_j : 0 \leq j < \dim C\}$$

be bases for the linear spaces B and C, respectively. Then the set

$$\{e_i f_j : 0 \leq i < \dim B, 0 \leq j < \dim C\}$$

is a basis for the linear space A.

This may be used in the proof of the next proposition.

Proposition 11.4 *Let A and A' be finite-dimensional associative algebras with unit elements over a field* $\mathbf{K}$ *and let B and C be subalgebras of A, and B' and C' subalgebras of A' such that $A = B \otimes C$ and $A' = B' \otimes C'$. Then, if $B \cong B'$ and if $C \cong C'$, it follows that $A \cong A'$.*

Proposition 11.4 encourages various extensions and abuses of the notation $\otimes$. In particular, if A, B, C, B' and C' are associative algebras with a unit element over a commutative field $\mathbf{K}$ such that

$$A = B \otimes C, \; B' \cong B \text{ and } C' \cong C,$$

then one frequently writes $A \cong B' \otimes C'$, even though there is no unique construction of $B' \otimes C'$. The precise meaning of such a statement will always be clear from the context.

The following propositions involving the tensor product of algebras will be of use in determining the table of Clifford algebras in Chapter 15.

Proposition 11.5 *Let A be an associative algebra with unit element over a commutative field* $\mathbf{K}$ *and let B, C and D be subalgebras of A. Then*

$$A = B \otimes C \;\Leftrightarrow\; A = C \otimes B$$
$$\text{and}\quad A = B \otimes (C \otimes D) \;\Leftrightarrow\; A = (B \otimes C) \otimes D.$$

In the latter case it is usual to write, simply, $A = B \otimes C \otimes D$.

Proposition 11.6 *For any commutative field* $\mathbf{K}$, *and for any finite p, q,*

$$\mathbf{K}(pq) \cong \mathbf{K}(p) \otimes_{\mathbf{K}} \mathbf{K}(q).$$

Proof Let $\mathbf{K}^{pq}$ be identified as a linear space with $\mathbf{K}^{p\times q}$, the linear space of $p \times q$ matrices over $\mathbf{K}$. Then the maps $\mathbf{K}(p) \to \mathbf{K}(pq)$; $a \mapsto a_L$ and $\mathbf{K}(q) \to \mathbf{K}(pq)$; $b \mapsto b^{\tau}_R$ are algebra injections, whose images in $\mathbf{K}(pq)$ satisfy conditions (i)–(iii) for $\otimes$, a_L and b^{τ}_R being defined, for each $a \in \mathbf{K}(p)$ and $b \in \mathbf{K}(q)$, and for each $c \in \mathbf{K}^{p\times q}$, by the formulae

$$a_L(c) = a\,c \text{ and } b^{\tau}_R(c) = c\,b^{\tau}.$$

For example, the commutativity condition (i) follows directly from the associativity of matrix multiplication. □

In particular, for any finite p, q,

$$\mathbf{R}(pq) \cong \mathbf{R}(p) \otimes_{\mathbf{R}} \mathbf{R}(q).$$

In this case we can say slightly more.

Proposition 11.7 *For any finite p, q, let $\mathbf{R}^p$, $\mathbf{R}^q$ and $\mathbf{R}^{pq}$ be regarded as positive-definite quadratic spaces in the standard way, and let $\mathbf{R}^{p\times q}$ be identified with $\mathbf{R}^{pq}$. Then the algebra injections*

$$\mathbf{R}(p) \to \mathbf{R}(pq);\ a \mapsto a_L \ \text{and}\ \mathbf{R}(q) \to \mathbf{R}(pq);\ b \mapsto b^{\tau}_R$$

send the orthogonal elements of $\mathbf{R}(p)$ and $\mathbf{R}(q)$, respectively, to orthogonal elements of $\mathbf{R}(pq)$.

Corollary 11.8 *The product of any finite ordered set of elements belonging either to the copy of $O(p)$ or to the copy of $O(q)$ in $\mathbf{R}(pq)$ is an element of $O(pq)$.*

In what follows, $\mathbf{C}$ and $\mathbf{H}$ will both be regarded as *real* algebras, of dimensions 2 and 4, respectively, and $\otimes = \otimes_{\mathbf{R}}$.

Proposition 11.9

$$\mathbf{R} \otimes \mathbf{R} = \mathbf{R}, \quad \mathbf{C} \otimes \mathbf{R} = \mathbf{C}, \quad \mathbf{H} \otimes \mathbf{R} = \mathbf{H},$$
$$\mathbf{C} \otimes \mathbf{C} \cong {}^2\mathbf{C}, \quad \mathbf{H} \otimes \mathbf{C} \cong \mathbf{C}(2) \quad \mathbf{H} \otimes \mathbf{H} \cong \mathbf{R}(4).$$

Proof The first three of these statements are obvious. To prove that $\mathbf{C} \otimes \mathbf{C} = {}^2\mathbf{C}$ it is enough to remark that ${}^2\mathbf{C}$ is generated as a real algebra by the subalgebras $\left\{\begin{pmatrix} z & 0 \\ 0 & z \end{pmatrix} : z \in \mathbf{C}\right\}$ and $\left\{\begin{pmatrix} z & 0 \\ 0 & \bar{z} \end{pmatrix} : z \in \mathbf{C}\right\}$, each isomorphic to $\mathbf{C}$, conditions (i) to (iii) being easily verified.

To prove that $\mathbf{H} \otimes \mathbf{C} \cong \mathbf{C}(2)$ let $\mathbf{C}^2$ be identified with $\mathbf{H}$ as a right complex linear space by the map $\mathbf{C}^2 \to \mathbf{H};\ (z, w) \mapsto z + \mathrm{j}w$, as before. Then, for any $q \in \mathbf{H}$ and any $c \in \mathbf{C}$, the maps

$$q_L : \mathbf{H} \to \mathbf{H};\ x \mapsto qx \text{ and } c_R : \mathbf{H} \to \mathbf{H};\ x \mapsto xc$$

are complex linear, and the maps

$$\mathbf{H} \to \mathbf{C}(2);\ q \mapsto q_L \text{ and } \mathbf{C} \to \mathbf{C}(2);\ c \mapsto c_R$$

are algebra injections. Conditions (i) and (ii) are obviously satisfied by the images of these injections. To prove (iii) it is enough to remark that the matrices

$$\begin{pmatrix} 1 & 0 \\ 0 & 1 \end{pmatrix}, \begin{pmatrix} \mathrm{i} & 0 \\ 0 & -\mathrm{i} \end{pmatrix}, \begin{pmatrix} 0 & -1 \\ 1 & 0 \end{pmatrix}, \begin{pmatrix} 0 & -\mathrm{i} \\ -\mathrm{i} & 0 \end{pmatrix},$$

$$\begin{pmatrix} \mathrm{i} & 0 \\ 0 & \mathrm{i} \end{pmatrix}, \begin{pmatrix} -1 & 0 \\ 0 & 1 \end{pmatrix}, \begin{pmatrix} 0 & -\mathrm{i} \\ \mathrm{i} & 0 \end{pmatrix}, \begin{pmatrix} 0 & 1 \\ 1 & 0 \end{pmatrix},$$

representing

$$\begin{matrix} 1, & \mathrm{i}_L, & \mathrm{j}_L, & \mathrm{k}_L, \\ \mathrm{i}_R, & \mathrm{i}_L\mathrm{i}_R, & \mathrm{j}_L\mathrm{i}_R, & \mathrm{k}_L\mathrm{i}_R, \end{matrix}$$

respectively, span $\mathbf{C}(2)$ linearly.

The proof that $\mathbf{H} \otimes \mathbf{H} \cong \mathbf{R}(4)$ is similar, the maps

$$q_L : \mathbf{H} \to \mathbf{H};\ x \mapsto qx \text{ and } \tilde{r}_R : \mathbf{H} \to \mathbf{H};\ x \mapsto x\tilde{r}$$

being real linear, for any $q, r \in \mathbf{H}$, and the maps

$$\mathbf{H} \to \mathbf{R}(4);\ q \mapsto q_L \text{ and } \mathbf{H} \to \mathbf{R}(4);\ r \mapsto \tilde{r}_R$$

being algebra injections whose images satisfy conditions (i) — (iii), condition (i), for example, following by the associativity of $\mathbf{H}$. □

In this last case it is worth recalling Proposition 8.26, which states that the image, by either of these injections, of a quaternion of absolute value 1 is an orthogonal element of $\mathbf{R}(4)$.

In fact we have the following analogue to Proposition 8.27.

Proposition 11.10 *For any $q \in S^3$ and any $c \in S^1$ the map*

$$\mathbf{C}^2 \to \mathbf{C}^2;\ x \mapsto q\,x\,c$$

is unitary, $\mathbf{C}^2$ being identified with $\mathbf{H}$ in the usual way. Moreover, any element of $U(2)$ can be so represented, two distinct elements (q, c) and $(q', c') \in S^3 \times S^1$ representing the same unitary map if and only if $(q', c') = -(q, c)$.

Proof The map $x \mapsto q\,x\,c$ is complex linear, for any $(q, c) \in S^3 \times S^1$, since it clearly respects addition, while, for any $\lambda \in \mathbf{C}$, $q(x\,\lambda)c = (q\,x\,c)\lambda$, since $\lambda\,c = c\,\lambda$. To prove that it is unitary it is then enough to show that it respects the hermitian form

$$\mathbf{C}^2 \to \mathbf{R};\ x \mapsto \overline{x}^{\tau} x.$$

However, since, for all $(x_0, x_1) \in \mathbf{C}^2$,

$$\overline{x_0}x_0 + \overline{x_1}x_1 = (\overline{x_0} - \overline{x_1}\,\mathrm{j})(x_0 + \mathrm{j}\,x_1) = |x_0 + \mathrm{j}\,x_1|^2,$$

it is enough to verify instead that the map, regarded as a map from $\mathbf{H}$ to $\mathbf{H}$, preserves the norm on $\mathbf{H}$, and this is obvious.

Conversely, let $t \in U(2)$ and let $r = t(1)$. Then $|r| = 1$ and the map

$$\mathbf{C}^2 \to \mathbf{C}^2;\ x \mapsto \overline{r}\,t(x)$$

is an element of $U(2)$ leaving 1, and therefore every point of $\mathbf{C}$, fixed, and mapping the orthogonal complement in $\mathbf{C}^2$ of $\mathbf{C}$, the complex line $\mathrm{j}\mathbf{C} = \{\mathrm{j}\,z : z \in \mathbf{C}\}$, to itself. It follows that there is an element c of S^1, defined uniquely up to sign, such that, for all $x \in \mathbf{C}^2$, $\overline{r}\,t(x) = \overline{c}\,x\,c$ or, equivalently, $t(x) = q\,x\,c$, where $q = r\,\overline{c}$. Finally, since c is defined up to sign, the pair (q, c) also is defined up to sign.

An alternative proof of the converse goes as follows.

Let $t \in U(2)$. Then $t = c_{\mathbf{R}}u$, where $c^2 = \det t$ and $u \in SU(2)$. Now the matrix of u can readily be shown to be of the form $\begin{pmatrix} a & -\overline{b} \\ b & \overline{a} \end{pmatrix}$, from which it follows that $u = q_L$, where $q = a + \mathrm{j}\,b$. The result follows at once. □

Corollary 11.11 *The following is a commutative diagram of exact sequences of group maps:*

$$
\begin{array}{ccccccccc}
 & & & & \{1\} & & \{1\} & & \\
 & & & & \downarrow & & \downarrow & & \\
 & & \{1\} & \longrightarrow & S^0 & \xrightarrow{1} & S^0 & \longrightarrow & \{1\} \\
 & & \downarrow & & \downarrow \iota & & \downarrow \iota & & \\
\{1\} & \longrightarrow & S^3 & \xrightarrow{\iota} & S^3 \times S^1 & \xrightarrow{\text{proj.}} & S^1 & \longrightarrow & \{1\} \\
 & & \downarrow & & \downarrow f & & \downarrow \text{sq.} & & \\
\{1\} & \longrightarrow & SU(2) & \xrightarrow{\iota} & U(2) & \xrightarrow{\det} & S^1 & \longrightarrow & \{1\} \\
 & & \downarrow & & \downarrow & & \downarrow & & \\
 & & \{1\} & & \{1\} & & \{1\} & , &
\end{array}
$$

the map f being defined by the formula

$$f(q, c) = q_L c_R, \textit{ for all } (q, c) \in S^3 \times S^1,$$

and the map sq. *being just the squaring map $z \mapsto z^2$.*

In particular, $Sp(1) = S^3 \cong SU(2)$.

Now, by Proposition 11.9, $\mathbf{C}(2) \cong \mathbf{H} \otimes \mathbf{C}$, the representative of any $q \in \mathbf{H}$ being q_L and the representative of any $c \in \mathbf{C}$ being c_R. It follows, by Proposition 11.10, that the product of any finite ordered set of elements belonging either to the copy of $Sp(1) = S^3$ or to the copy of $U(1) = S^1$ in $\mathbf{C}(2)$ is an element of $U(2)$.

This result is to be compared with Proposition 11.7 and the remark following the proof of Proposition 11.9. Note also that in the standard inclusion of $\mathbf{C}$ in $\mathbf{R}(2)$ the elements representing the elements of $U(1) = S^1$ are all orthogonal.

For an important application of these remarks see Proposition 15.28.

Complexification

Roughly speaking, *complexification* is the operation of tensoring by $\mathbf{C}$. Thus $\mathbf{C}(2)$ is the complexification of $\mathbf{H}$ and ${}^2\mathbf{C}$ is the complexification of $\mathbf{C}$, but in the applications that we have in mind the object complexified will frequently be a real *superalgebra*, that is a real algebra furnished with an anti-involution, in which case there are *two kinds* of complexification, tensoring by $\mathbf{C}$, that is by $\mathbf{C}$ with the identity as (anti-)involution, and tensoring by $\overline{\mathbf{C}}$, that is by $\mathbf{C}$ with conjugation as (anti-)involution. For $\mathbf{H} \otimes \mathbf{C}$ there are four different outcomes, as follows.

Proposition 11.12 *We consider the isomorphism* $\mathbf{C}(2) \cong \mathbf{H} \otimes \mathbf{C}$, *where the quaternion* $q = w + zj$ *is represented by the matrix*

$$\begin{pmatrix} w & -\overline{z} \\ z & \overline{w} \end{pmatrix}$$

and the complex number c *is represented by the matrix*

$$\begin{pmatrix} c & 0 \\ 0 & c \end{pmatrix}.$$

(i) *Let* $\mathbf{H}$ *be assigned conjugation and* $\mathbf{C}$ *the identity. Then the induced anti-involution of* $\mathbf{C}(2) \cong \mathbf{H} \otimes \mathbf{C}$ *is*

$$\begin{pmatrix} a & c \\ b & d \end{pmatrix} \mapsto \begin{pmatrix} d & -c \\ -b & a \end{pmatrix}.$$

(ii) *Let* $\mathbf{H}$ *be assigned conjugation and* $\mathbf{C}$ *conjugation. Then the induced anti-involution of* $\mathbf{C}(2) \cong \mathbf{H} \otimes \mathbf{C}$ *is conjugate transposition.*

(iii) *Let* $\mathbf{H}$ *be assigned the anti-involution* $q = w + zj \mapsto \widetilde{q} = w - \overline{z}j$, *where* $w, z \in \mathbf{C}$, *and let* $\mathbf{C}$ *be assigned the identity. Then the induced anti-involution of* $\mathbf{C}(2) \cong \mathbf{H} \otimes \mathbf{C}$ *is*

$$\begin{pmatrix} a & c \\ b & d \end{pmatrix} \mapsto \begin{pmatrix} a & b \\ c & d \end{pmatrix}.$$

(iv) *Let* $\mathbf{H}$ *be assigned the anti-involution* $q = w + zj \mapsto \widetilde{q} = w - \overline{z}j$, *where* $w, z \in \mathbf{C}$, *and let* $\mathbf{C}$ *be assigned conjugation. Then the induced anti-involution of* $\mathbf{C}(2) \cong \mathbf{H} \otimes \mathbf{C}$ *is*

$$\begin{pmatrix} a & c \\ b & d \end{pmatrix} \mapsto \begin{pmatrix} \overline{d} & -\overline{c} \\ -\overline{b} & \overline{a} \end{pmatrix}.$$

Proof In each case all one has to verify is that the asserted anti-involution is correct both on the copy of $\mathbf{H}$ in $\mathbf{C}(2)$ and on the copy of $\mathbf{C}$ in $\mathbf{C}(2)$. In every case this is immediate. □

Likewise for $\mathbf{C} \times \mathbf{C} \cong {}^2\mathbf{C}$ there are three different outcomes, as follows.

Proposition 11.13 *We consider the isomorphism* ${}^2\mathbf{C} \cong \mathbf{C} \otimes \mathbf{C}$, *where the complex number* w *in the first factor is represented by the matrix*

$$\begin{pmatrix} w & 0 \\ 0 & \overline{w} \end{pmatrix}$$

and the complex number c in the second factor is represented by the matrix

$$\begin{pmatrix} c & 0 \\ 0 & c \end{pmatrix}.$$

(i) *Let both copies of* **C** *be assigned the identity. Then* $^2\mathbf{C} \cong \mathbf{C} \times \mathbf{C}$ *is assigned the identity.*

(ii) *Let the first copy of* **C** *be assigned the identity and the second copy conjugation. Then* $^2\mathbf{C} \cong \mathbf{C} \times \mathbf{C}$ *is assigned the swap (anti-) involution*

$$\sigma : \begin{pmatrix} a & 0 \\ 0 & d \end{pmatrix} \mapsto \begin{pmatrix} d & 0 \\ 0 & a \end{pmatrix}.$$

(iii) *Let both copies of* **C** *be assigned conjugation. Then* $^2\mathbf{C} \cong \mathbf{C} \times \mathbf{C}$ *is assigned conjugation.*

The recognition of subalgebras

It is an advantage to be able to detect quickly whether or not a subalgebra of a given associative algebra A is isomorphic to one of the algebras

$$\mathbf{R},\ \mathbf{C},\ \mathbf{H},\ ^2\mathbf{R},\ ^2\mathbf{C},\ ^2\mathbf{H},\ \mathbf{R}(2),\ \mathbf{C}(2),\ \text{ or } \mathbf{H}(2),$$

or whether a subalgebra of a given complex associative algebra A is isomorphic to one of the algebras **C**, $^2\mathbf{C}$ or **C**(2). The following proposition is useful in this context.

Proposition 11.14 *Let A be a real associative algebra with unit element* 1. *Then*

(i) 1 *generates* **R**,

(ii) *any two-dimensional subalgebra generated by an element e_0 of A such that $e_0^2 = -1$ is isomorphic to* **C**,

(iii) *any two-dimensional subalgebra generated by an element e_0 of A such that $e_0^2 = 1$ is isomorphic to* $^2\mathbf{R}$,

(iv) *any four-dimensional subalgebra generated by a set $\{e_0, e_1\}$ of mutually anti-commuting elements of A such that $e_0^2 = e_1^2 = -1$ is isomorphic to* **H**,

(v) *any four-dimensional subalgebra generated by a set $\{e_0, e_1\}$ of mutually anti-commuting elements of A such that $e_0^2 = e_1^2 = 1$ is isomorphic to* **R**(2),

(vi) *any eight-dimensional subalgebra generated by a set* $\{e_0, e_1, e_2\}$ *of mutually anti-commuting elements of A such that* $e_0^2 = e_1^2 = e_2^2 = -1$ *is isomorphic to* $^2\mathbf{H}$,

(vii) *any eight-dimensional subalgebra generated by a set* $\{e_0, e_1, e_2\}$ *of mutually anti-commuting elements of A such that* $e_0^2 = e_1^2 = e_2^2 = 1$ *is isomorphic to* $\mathbf{C}(2)$.

Sets of elements meeting the required conditions include

$$\left\{\begin{pmatrix} 0 & 1 \\ 1 & 0 \end{pmatrix}, \begin{pmatrix} 0 & 1 \\ 1 & 0 \end{pmatrix}\right\} \text{ for } \mathbf{R}(2),$$

$$\left\{\begin{pmatrix} \mathrm{i} & 0 \\ 0 & -i \end{pmatrix}, \begin{pmatrix} \mathrm{j} & 0 \\ 0 & -\mathrm{j} \end{pmatrix}, \begin{pmatrix} \mathrm{k} & 0 \\ 0 & -\mathrm{k} \end{pmatrix}\right\}$$

$$\text{or } \left\{\begin{pmatrix} \mathrm{i} & 0 \\ 0 & i \end{pmatrix}, \begin{pmatrix} \mathrm{j} & 0 \\ 0 & -\mathrm{j} \end{pmatrix}, \begin{pmatrix} \mathrm{k} & 0 \\ 0 & \mathrm{k} \end{pmatrix}\right\},$$

$$\text{but not } \left\{\begin{pmatrix} \mathrm{i} & 0 \\ 0 & i \end{pmatrix}, \begin{pmatrix} \mathrm{j} & 0 \\ 0 & \mathrm{j} \end{pmatrix}, \begin{pmatrix} \mathrm{k} & 0 \\ 0 & \mathrm{k} \end{pmatrix}\right\}$$

$$\text{or } \left\{\begin{pmatrix} \mathrm{i} & 0 \\ 0 & -i \end{pmatrix}, \begin{pmatrix} \mathrm{j} & 0 \\ 0 & \mathrm{j} \end{pmatrix}, \begin{pmatrix} \mathrm{k} & 0 \\ 0 & -\mathrm{k} \end{pmatrix}\right\}, \text{ for } {}^2\mathbf{H},$$

$$\text{and } \left\{\begin{pmatrix} 0 & 1 \\ 1 & 0 \end{pmatrix}, \begin{pmatrix} 1 & 0 \\ 0 & -1 \end{pmatrix}, \begin{pmatrix} 0 & -\mathrm{i} \\ \mathrm{i} & 0 \end{pmatrix}\right\}, \text{ for } \mathbf{C}(2).$$

Several of those listed in the following proposition we have had before.

Proposition 11.15 *The subset of matrices of the real algebra* $\mathbf{K}(2)$ *of the form*

(i) $\begin{pmatrix} a & b \\ b & a \end{pmatrix}$ *is a subalgebra isomorphic to* $^2\mathbf{R}$, $^2\mathbf{C}$ *or* $^2\mathbf{H}$,

(ii) $\begin{pmatrix} a & -b \\ b & a \end{pmatrix}$ *is a subalgebra isomorphic to* $\mathbf{C}$, $^2\mathbf{C}$ *or* $\mathbf{C}(2)$,

(iii) $\begin{pmatrix} a & b' \\ b & a' \end{pmatrix}$ *is a subalgebra isomorphic to* $^2\mathbf{R}$, $\mathbf{R}(2)$ *or* $\mathbf{C}(2)$,

(iv) $\begin{pmatrix} a & -b' \\ b & a' \end{pmatrix}$ *is a subalgebra isomorphic to* $\mathbf{C}$, $\mathbf{H}$ *or* $^2\mathbf{H}$,

according as $\mathbf{K} = \mathbf{R}$, $\mathbf{C}$ *or* $\mathbf{H}$, *respectively, where, for any* $a \in \mathbf{K}$, $a' = a$, $\bar{a}$ *or* $\widehat{a}$, *respectively.*

Each of the algebras listed in Proposition 11.15 is induced by a (non-unique) real linear injection. For example, those of the form (iii) may be regarded as being the injections of the appropriate endomorphism algebras induced by the real linear injections

$$\begin{aligned} {}^2\mathbf{R} \to \mathbf{R}^2; &\quad (x, y) \mapsto (x+y, x-y),\\ \mathbf{R}^2 \to \mathbf{C}^2; &\quad (x, y) \mapsto (x+\mathrm{i}\,y, x-\mathrm{i}\,y),\\ \text{and}\quad \mathbf{C}^2 \to \mathbf{H}^2; &\quad (z, w) \mapsto (z+\mathrm{j}\,w, \bar{z}+\mathrm{j}\,\bar{w}). \end{aligned}$$

Real algebras, A, B, C and D, say, frequently occur in a commutative square of algebra injections of the form

$$\begin{array}{ccc} A & \longrightarrow & C \\ \downarrow & & \downarrow \\ B & \longrightarrow & D \end{array},$$

the algebra D being generated by the images of B and C. Examples of such squares, which may easily be constructed using the material of Proposition 11.15, include

$$\begin{array}{ccc} \mathbf{R} & \longrightarrow & {}^2\mathbf{R} \\ \downarrow & & \downarrow \\ \mathbf{C} & \longrightarrow & \mathbf{R}(2) \end{array}, \qquad \begin{array}{ccc} {}^2\mathbf{R} & \longrightarrow & \mathbf{R}(2) \\ \downarrow & & \downarrow \\ \mathbf{R}(2) & \longrightarrow & {}^2\mathbf{R}(2) \end{array},$$

$$\begin{array}{ccc} \mathbf{C} & \longrightarrow & \mathbf{R}(2) \\ \downarrow & & \downarrow \\ \mathbf{H} & \longrightarrow & \mathbf{C}(2) \end{array}, \quad \begin{array}{ccc} \mathbf{C} & \longrightarrow & {}^2\mathbf{C} \\ \downarrow & & \downarrow \\ {}^2\mathbf{C} & \longrightarrow & \mathbf{C}(2) \end{array}, \quad \begin{array}{ccc} {}^2\mathbf{C} & \longrightarrow & \mathbf{C}(2) \\ \downarrow & & \downarrow \\ \mathbf{C}(2) & \longrightarrow & {}^2\mathbf{C}(2) \end{array},$$

$$\begin{array}{ccc} \mathbf{H} & \longrightarrow & \mathbf{C}(2) \\ \downarrow & & \downarrow \\ {}^2\mathbf{H} & \longrightarrow & \mathbf{H}(2) \end{array}, \quad \text{and} \quad \begin{array}{ccc} {}^2\mathbf{H} & \longrightarrow & \mathbf{H}(2) \\ \downarrow & & \downarrow \\ \mathbf{H}(2) & \longrightarrow & {}^2\mathbf{H}(2) \end{array}.$$

Exercise

11.1 Show that there is a group map $SU(2) \to S^3 \times S^3$ making the diagram $\begin{array}{ccc} & & S^3 \times S^3 \\ & \nearrow & \downarrow \\ SU(2) & \to & SO(4) \end{array}$ commute, but that there is no group map $U(2) \to S^3 \times S^3$ that makes $\begin{array}{ccc} & & S^3 \times S^3 \\ & \nearrow & \downarrow \\ U(2) & \to & SO(4) \end{array}$ commute, the vertical map in either case being the group map defined in Proposition 8.26 and the horizontal maps being the standard group injections induced by the usual identification of $\mathbf{C}^2$ with $\mathbf{R}^4$.

12
Anti-involutions of ${}^2\mathbf{K}(n)$

In this chapter we learn to think of the general linear groups $GL(n;\mathbf{K})$ as *unitary groups*! For example an element of $GL(n;\mathbf{C})$ may, according to the context, appear as a matrix of the form

$$\begin{pmatrix} a & 0 \\ 0 & (a^\tau)^{-1} \end{pmatrix} \text{ or } \begin{pmatrix} a & 0 \\ 0 & (\bar{a}^\tau)^{-1} \end{pmatrix},$$

where a is an invertible $n \times n$ matrix over $\mathbf{C}$. The first of these leaves invariant the symmetric sesqui-linear form

$${}^2\mathbf{C}^n \times {}^2\mathbf{C}^n \to {}^2\mathbf{C};\ \left(\begin{pmatrix} x & 0 \\ 0 & y \end{pmatrix}, \begin{pmatrix} x & 0 \\ 0 & y \end{pmatrix}\right) \mapsto \begin{pmatrix} y^\tau x & 0 \\ 0 & x^\tau y \end{pmatrix},$$

with swap playing the part of conjugation, and the second the symmetric sesqui-linear form

$${}^2\mathbf{C}^n \times {}^2\mathbf{C}^n \to {}^2\mathbf{C};\ \left(\begin{pmatrix} x & 0 \\ 0 & y \end{pmatrix}, \begin{pmatrix} x & 0 \\ 0 & y \end{pmatrix}\right) \mapsto \begin{pmatrix} \bar{y}^\tau x & 0 \\ 0 & \bar{x}^\tau y \end{pmatrix},$$

with swap composed with conjugation taking the part of conjugation.

${}^2\mathbf{K}$ *linear spaces and maps*

The *double fields* ${}^2\mathbf{R}$ and ${}^2\mathbf{C}$ were introduced at the very beginning of Chapter 2 and ${}^2\mathbf{H}$ at the end of Chapter 9. These are not fields, for it is not true that every non-zero element has an inverse. Nevertheless much of the standard theory of linear spaces and maps over a field holds also for ${}^2\mathbf{K}$-linear spaces, for $\mathbf{K} = \mathbf{R}$, $\mathbf{C}$ or $\mathbf{H}$, as does the theory of semi-linear correlations.

The part of a ${}^2\mathbf{K}$ linear space is played by a $\mathbf{K}$-linear space A with a prescribed direct sum decomposition $A_0 \oplus A_1$, where A_0 and A_1 are isomorphic linear subspaces of A (so in particular in the case that A is finite-dimensional $\dim A_0 = \dim A_1$) with scalar multiplication defined by

$$((\lambda, \mu), x_0 + x_1) \mapsto (\lambda x_0 + \mu x_1),$$

which may also be written as

$$\left(\begin{pmatrix} \lambda & 0 \\ 0 & \mu \end{pmatrix}, \begin{pmatrix} x_0 & 0 \\ 0 & x_1 \end{pmatrix}\right) \mapsto \begin{pmatrix} \lambda x_0 & 0 \\ 0 & \mu x_1 \end{pmatrix}.$$

${}^2\mathbf{K}$*-linear maps* are defined in the obvious way.

Let $t : X \to Y$ be a ${}^2\mathbf{K}$-linear map. Then t is of the form $\begin{pmatrix} a_0 & 0 \\ 0 & a_1 \end{pmatrix}$, where $a_0 \in L(X_0, Y_0)$ and $a_1 \in L(X_1, Y_1)$. Conversely, any map of this form may be regarded as a ${}^2\mathbf{K}$-linear map.

One has to take care with extending the basis theorems of linear algebra over a field to double fields. We say that an element x of a ${}^2\mathbf{K}$-linear space X is *linearly free* of a subset A of X if, and only if, $(1, 0)x$ is linearly independent of $(1, 0)A$ in the $\mathbf{K}$-linear space $(1, 0)X$ and $(0, 1)x$ is linearly independent of $(0, 1)A$ in the $\mathbf{K}$-linear space $(1, 0)X$. Thus, in ${}^2\mathbf{K}$ itself, $(1, 1)$ is *not* free of $\{(1, 0)\}$. Then a *basis* for X is a free subset of X that spans X.

The following is the basis theorem for ${}^2\mathbf{K}$-linear spaces.

Theorem 12.1 *Let X be a ${}^2\mathbf{K}$-linear space with a basis A. Then $X_0 = (1, 0)X$ and $X_1 = (0, 1)X$ are isomorphic as $\mathbf{K}$-linear spaces, the set $(1, 0)A$ being a basis of the $\mathbf{K}$-linear space X_0 and the set $(0, 1)A$ being a basis of the $\mathbf{K}$-linear space X_1. Moreover any two finite bases for X have the same number of elements.*

Any ${}^2\mathbf{K}$-linear space with a finite basis is isomorphic to the ${}^2\mathbf{K}$-linear space ${}^2\mathbf{K}^n = ({}^2\mathbf{K})^n$, n being the number of elements in the basis.

It should be noted that not every point x of a ${}^2\mathbf{K}$-linear space X spans a ${}^2\mathbf{K}$-line. For this to happen, both $(1, 0)x$ and $(0, 1)x$ must be non-zero. A point that spans a line will be called a *regular* point of X. Similar considerations show that the image and kernel of a ${}^2\mathbf{K}$-linear map need not be ${}^2\mathbf{K}$-linear spaces.

Anti-involutions of ${}^2\mathbf{K}$

For the theory of semi-linear correlations on ${}^2\mathbf{K}$-linear spaces we need to know all possible anti-involutions of ${}^2\mathbf{K}$. We begin by recalling Propositions 2.18 and 2.19, extending these to apply not only to the fields **R** and **C** but also to the non-commutative field **H**, in which case whatever was said earlier about automorphisms applies also to anti-automorphisms.

Proposition 12.2 *Let* $\mathbf{K} = \mathbf{R}$, $\mathbf{C}$ *or* $\mathbf{H}$. *Then any automorphism or anti-automorphism of* ${}^2\mathbf{K}$ *is of the form either*

$$\text{(i)}\ {}^2\mathbf{K} \to {}^2\mathbf{K};\ \begin{pmatrix} \lambda & 0 \\ 0 & \mu \end{pmatrix} \mapsto \begin{pmatrix} \lambda^\chi & 0 \\ 0 & \mu^\phi \end{pmatrix}, \text{ or}$$

$$\text{(ii)}\ {}^2\mathbf{K} \to {}^2\mathbf{K};\ \begin{pmatrix} \lambda & 0 \\ 0 & \mu \end{pmatrix} \mapsto \begin{pmatrix} \mu^\phi & 0 \\ 0 & \lambda^\chi \end{pmatrix},$$

where $\chi, \phi : \mathbf{K} \to \mathbf{K}$ *are both automorphisms or both anti-automorphisms of* $\mathbf{K}$, *any* irreducible *automorphism or anti-automorphism necessarily being of type (ii).*

In case (i) the map is an involution or anti-involution of ${}^2\mathbf{K}$ if and only if χ and ϕ are both involutions or both anti-involutions of **K**,

Case (ii) is of greater interest. Such an automorphism or anti-automorphism sends $\begin{pmatrix} 1 & 0 \\ 0 & -1 \end{pmatrix}$ to $\begin{pmatrix} -1 & 0 \\ 0 & 1 \end{pmatrix} = -\begin{pmatrix} 1 & 0 \\ 0 & -1 \end{pmatrix}$, and is an involution or anti-involution of ${}^2\mathbf{K}$ if and only if it is of the form

$${}^2\mathbf{K} \to {}^2\mathbf{K};\ \begin{pmatrix} \lambda & 0 \\ 0 & \mu \end{pmatrix} \mapsto \begin{pmatrix} \mu^\phi & 0 \\ 0 & \lambda^{\phi^{-1}} \end{pmatrix},$$

where ϕ is an automorphism or anti-automorphism (not necessarily an involution or anti-involution) of **K**.

As in Chapter 2 the *swap* involution

$${}^2\mathbf{K} \to {}^2\mathbf{K};\ \begin{pmatrix} \lambda & 0 \\ 0 & \mu \end{pmatrix} \mapsto \begin{pmatrix} \mu & 0 \\ 0 & \lambda \end{pmatrix}$$

will be denoted by σ, the involution or anti-involution

$${}^2\mathbf{K} \to {}^2\mathbf{K};\ \begin{pmatrix} \lambda & 0 \\ 0 & \mu \end{pmatrix} \mapsto \begin{pmatrix} \mu^\phi & 0 \\ 0 & \lambda^{\phi^{-1}} \end{pmatrix}$$

then being the composite involution or anti-involution

$$(\phi \times \phi^{-1})\sigma = \sigma(\phi^{-1} \times \phi).$$

Proposition 12.3 *For any two automorphisms or two anti-automorphisms* ϕ *and* χ *of* $\mathbf{K}$ *the involutions or anti-involutions* $(\phi \times \phi^{-1})\sigma$ *and* $(\chi \times \chi^{-1})\sigma$ *of* ${}^2\mathbf{K}$ *are similar.*

Proof The proof is just as for Proposition 2.20. □

Semi-linear maps over $^2\mathbf{K}$

Let X and Y be right or left linear spaces over the double field $^2\mathbf{K}$, where $\mathbf{K} = \mathbf{R}$, $\mathbf{C}$ or $\mathbf{H}$. Then a $^2\mathbf{K}$-linear map $f : X \to Y$ is said to be *semi-linear* over $^2\mathbf{K}$ if there is an automorphism or anti-automorphism ψ of $^2\mathbf{K}$ such that, for all $x \in X, \lambda \in {}^2\mathbf{K}$, $f(x\lambda) = f(x)\lambda^\psi$, $f(x\lambda) = \lambda^\psi f(x)$, $f(\lambda x) = \lambda^\psi f(x)$ or $f(\lambda x) = f(x)\lambda^\psi$, as the case may be, ψ being an automorphism of $^2\mathbf{K}$ if $^2\mathbf{K}$ operates on X and Y on the same side and an anti-automorphism if $^2\mathbf{K}$ operates on X and Y on opposite sides. The terms *right*, *right-to-left*, *left* and *left-to-right* semi-linear maps over $^2\mathbf{K}$ have the obvious meanings.

The semi-linear map t uniquely determines the automorphism or anti-automorphism ψ of $^2\mathbf{K}$ with respect to which it is semi-linear, unless $t = 0$. On occasion it is convenient to refer directly to ψ, the map t then being said to be $^2\mathbf{K}^\psi$-linear.

The first of the maps

$$\mathbf{H}^2 \to \mathbf{H}^2;\quad (x, y) \mapsto (\overline{x}, \overline{y}),$$
$$(x, y) \mapsto (\overline{y}, \overline{x})$$

is a right-to-left $^2\overline{\mathbf{H}}$-linear map and the second a right-to-left $^2\overline{\mathbf{H}}^\sigma$-linear map.

Semi-linear maps over $^2\mathbf{K}$ are classified by the following proposition.

Proposition 12.4 *Let X and Y both be $^2\mathbf{K}$-linear spaces. Then any $^2\mathbf{K}^{\chi\times\phi}$-linear map $X \to Y$ is of the form*

$$X_0 \oplus X_1 \to Y_0 \oplus Y_1;\ (x_0, x_1) \mapsto (f(x_0), g(x_1)),$$

where $f : X_0 \to Y_0$ is $\mathbf{K}^\chi$-linear and $g : X_1 \to Y_1$ is $\mathbf{K}^\phi$-linear, while any $^2\mathbf{K}^{\sigma(\chi\times\phi)}$-linear map $X \to Y$ is of the form

$$X_0 \oplus X_1 \to Y_0 \oplus Y_1;\ (x_0, x_1) \mapsto (f(x_0), g(x_1)),$$

where $f : X_0 \to Y_1$ is $\mathbf{K}^\chi$-linear and $g : X_1 \to Y_0$ is $\mathbf{K}^\phi$-linear.

Proof We indicate the proof for a $^2\mathbf{K}^{\sigma(\chi\times\phi)}$-linear map $X \to Y$, assuming, for the sake of definiteness, that X is a right $^2\mathbf{K}$-linear space and Y a left $^2\mathbf{K}$-linear space. Then, for all $a \in X_0$, $b \in X_1$,

$$t(a,0) = t((a,0)(1,0)) = (0,1)t(a,0),$$
$$t(0,b) = t((0,b)(0,1)) = (1,0)t(0,b).$$

So maps $f : X_0 \to Y_1$ and $g : X_1 \to Y_0$ are defined by

$$(0, f(a)) = t(a, 0) \text{ and } (g(b), 0) = t(0, b), \quad \text{for all } (a, b) \in X.$$

It is then a straightforward matter to check that these maps f and g have the required properties.

The proofs in the other cases are similar. □

An $\mathbf{A}^{\psi}$-linear map $t : X \to Y$ is said to be *irreducible* if ψ is irreducible. Otherwise it is said to be *reducible.* If t is irreducible, and if ψ is an involution or anti-involution then, by Proposition 2.17, $\mathbf{A} = \mathbf{K}$ or $^2\mathbf{K}$. Of the two maps in Proposition 12.4 the first is reducible, while the second is irreducible.

As before, the composite of any two composable semi-linear maps is semi-linear and the inverse of an invertible semi-linear map is semi-linear. An invertible semi-linear map is said to be a *semi-linear isomorphism*. *Duals* of $^2\mathbf{K}$-linear spaces and semi-linear maps are then defined just as for $\mathbf{H}$-linear spaces and semi-linear maps. In particular, *correlations* and their induced *sesqui-linear forms* are similarly defined, as are *symmetric* and *skew* correlations. For example, the product

$$^2\mathbf{R} \times {}^2\mathbf{R} \to {}^2\mathbf{R};\ ((a, b), (a', b')) \mapsto (b\,a', a\,b')$$

is a symmetric sesqui-linear product over $^2\mathbf{R}$.

A *correlated* $^2\mathbf{K}$ *space* is a finite-dimensional right $^2\mathbf{K}$-linear space assigned a $^2\mathbf{K}^{\psi}$-correlation, where ψ is some anti-involution of $^2\mathbf{K}$.

Semi-linear correlations $\xi, \eta : X \to X^L$ on a right $^2\mathbf{K}$-linear space X are said to be *equivalent* if, for some invertible $\lambda \in {}^2\mathbf{K}$, $\eta = \lambda\,\xi$. This is clearly an equivalence on any set of semi-linear correlations on X.

There follows an analogue of Propositions 7.11, 10.4 and 10.5.

Proposition 12.5 *Any irreducible skew $^2\mathbf{K}^{\psi}$-correlation on a right $^2\mathbf{K}$-linear space is equivalent to a symmetric correlation on X, and conversely.*

Proof Let ξ be an irreducible skew $^2\mathbf{K}^{\psi}$-correlation on X. Then, for all $a, b \in X$,

$$\begin{aligned} b^{(1,-1)\xi}a &= (1,-1)\,b^{\xi}a = -(a^{\xi}b)^{\psi}\,(1,-1) \\ &= (a^{\xi}b)^{\psi}\,(1,-1)^{\psi}, \quad \text{since } \psi \text{ is a swap anti-involution,} \\ &= (a^{(1,-1)\xi}b)^{\psi}. \end{aligned}$$

That is, the correlation $(1,-1)\xi$ is symmetric.

Similarly, if ξ is symmetric, then $(1,-1)\xi$ is skew. □

The *adjoint* of a ${}^2\mathbf{K}$-linear map $X \to Y$, with respect to correlations ξ on X and η on Y, is then defined just as in the quaternionic case.

Equivalent semi-linear correlations ξ and η on the right quaternionic space $\mathbf{H}^n$ induce the same adjoint anti-automorphism of $\mathbf{H}(n)$.

Proposition 12.6 *Let ξ be a irreducible symmetric or skew correlation on a right ${}^2\mathbf{K}$-linear space X. Then the map*

$$\operatorname{End} X \to \operatorname{End} X;\ t \mapsto t^\xi$$

is a real algebra anti-involution.

As always, symmetric and skew correlations are particular cases of reflexive correlations, a correlation ξ on a ${}^2\mathbf{K}$-linear space X being said to be *reflexive* if, for all $a, b \in X$,

$$b^\xi a = 0 \Leftrightarrow a^\xi b = 0.$$

Proposition 12.7 *Any non-zero irreducible reflexive ${}^2\mathbf{K}^\psi$-correlation ξ on a right ${}^2\mathbf{K}$-space X of dimension greater than one is equivalent either to a symmetric or to a skew one.*

(A counter-example in the one-dimensional case is the correlation on $\mathbf{H}$ with product $\mathbf{H}^2 \to \mathbf{H};\ (a, b) \mapsto \bar{a}(1 + \mathrm{j})b$.)

Proof This is our final re-run of Proposition 6.1, but one has again to remember that ψ is an *anti-automorphism* of ${}^2\mathbf{K}$, and that it need not be an anti-involution.

The proof is basically as before, but care has to be taken, since a non-zero element of ${}^2\mathbf{K}$ or of X is not necessarily regular. One proves that there exists an invertible $\lambda \in {}^2\mathbf{K}$ such that, for all *regular* $a \in X$ and all $b \in X$,

$$(b^\xi a)^{\psi^{-1}} = \lambda a^\xi b.$$

It is then easy to deduce that this formula also holds for all $a \in X$. The remainder of the proof is as before. □

Next we state an analogue of Proposition 6.3.

Proposition 12.8 *Let X be a finite-dimensional ${}^2\mathbf{K}$-linear space, let α be an anti-automorphism of the real algebra* $\operatorname{End} X$ *and let $t \in \operatorname{End} X$, with* $\operatorname{rk} t = 1$. *Then* $\operatorname{rk} t^\alpha = 1$.

Proof This is almost as before. The only difference is that the left ideal generated by an element t of $\operatorname{End} X$, with $\operatorname{rk} t = 1$, is not minimal, but has exactly two minimal left ideals as proper subideals. However, this can only occur if $\operatorname{rk} t = 1$. The details are left to the reader. □

What then follows is our final generalisation of Theorem 6.4.

Theorem 12.9 *Let X be a finite-dimensional $^2\mathbf{K}$-linear space, with* $\dim X > 1$. *Then any anti-involution α of the real algebra* $\operatorname{End} X$ *is representable as the adjoint anti-involution induced by a non-degenerate reflexive correlation on X.*

Proof Entirely as for Theorem 10.8. □

Proposition 12.10 *Let ξ be a non-zero irreducible reflexive correlation on a right $^2\mathbf{K}$-linear space X. Then for some $x \in X$, $x^\xi x$ is invertible.*

Proof Since ξ is an irreducible correlation, $\xi = \sigma(\eta \times \zeta)$, where $\zeta : X_1 \to X_0^L$ and $\eta : X_0 \to X_1^L$ are semi-linear. Since ξ is reflexive,

$$(0, a^\eta b) = (a, 0)^\xi (a, b) = 0 \Leftrightarrow (b^\zeta a, 0) = (0, b)^\xi (a, 0) = 0,$$

that is $a^\eta b \neq 0 \Leftrightarrow b^\zeta a \neq 0$.

Since ξ is non-zero, either η or ζ is non-zero. Suppose that η is non-zero. Then there exists $(a, b) \in X_0 \times X_1$ such that $a^\eta b \neq 0$ and $b^\zeta a \neq 0$, that is such that $(a, b)^\xi (a, b)$ is invertible. □

If ξ is symmetric we can say more, this case being required in the proof of Theorem 12.14 below.

Proposition 12.11 *Let ξ be an irreducible correlation on a right $^2\mathbf{K}$-linear space X, symmetric with respect to the anti-involution $(\phi \times \phi^{-1})\sigma$ of $^2\mathbf{K}$, and suppose that, for some $x \in X$, $x^\xi x$ is invertible. Then there exists $\lambda \in {}^2\mathbf{K}$ such that $(x\lambda)^\xi (x\lambda) = 1$.*

Proof As in the proof of Proposition 12.10, $\xi = \sigma(\eta \times \zeta)$. Now ξ is symmetric with respect to $(\phi \times \phi^{-1})\sigma$, so, for all $(a, b) \in X$,

$$((a, b)^\xi (a, b))^{(\phi \times \phi^{-1})\sigma} = (a, b)^\xi (a, b),$$

that is,

$$((a^\eta b)^\phi, (b^\zeta a)^{\phi^{-1}}) = (b^\zeta a, a^\eta b).$$

In particular, $a^\eta b = 1$ if and only if $b^\zeta = 1$. Now, if $x = (a, b)$ is invertible, $b^\zeta a \neq 0$. Choose $\lambda = ((b^\zeta a)^{-1}, 1)$. □

An invertible correlation is said to be *non-degenerate.*

Proposition 12.12 *Let ξ be a non-degenerate correlation on a finite-dimensional right $\mathbf{K}$-linear space X, and let x be any non-zero element of X. Then there exists $x' \in X$ such that $x^\xi x' = 1$.*

Proposition 12.13 *Let ξ be a non-degenerate irreducible correlation on a finite-dimensional right ${}^2\mathbf{K}$-linear space X, and let x be a regular element of X. Then there exists $x' \in X$ such that $x^\xi x' = 1$ $(= (1, 1))$.*

Theorem 12.14 (Basis theorem.) *Each finite-dimensional irreducible symmetric ${}^2\mathbf{K}^\sigma$-correlated space has an orthonormal basis.*

By Theorem 12.9 this holds also for irreducible skew ${}^2\mathbf{K}^\sigma$-correlated spaces since any skew correlation is equivalent to a symmetric one.

Theorem 12.15 (Classification theorem.) *Let X be a non-degenerate symmetric ${}^2\overline{\mathbf{K}}^\sigma$-correlated space of finite dimension n, where $\mathbf{K} = \mathbf{R}$, $\mathbf{C}$ or $\mathbf{H}$. Then X is isomorphic to $({}^2\overline{\mathbf{K}})^n$, this being the right ${}^2\mathbf{K}$-linear space ${}^2\mathbf{K}^n$, with the sesqui-linear product*

$$({}^2\mathbf{K}^n)^2 \to {}^2\mathbf{K};\ (a, b) \mapsto \sum_{0 \le i < n} \overline{a_i}^\sigma\, b_i.$$

General linear groups

Proposition 12.16 *Let $t = \begin{pmatrix} a & 0 \\ 0 & d \end{pmatrix}$ be an endomorphism of the correlated space $({}^2\overline{\mathbf{K}}^\sigma)^n$, where $\mathbf{K} = \mathbf{R}$, $\mathbf{C}$ or $\mathbf{H}$. Then the adjoint of t is*

$$t^\xi = \begin{pmatrix} \overline{d}^\tau & 0 \\ 0 & \overline{a}^\tau \end{pmatrix}.$$

Corollary 12.17 *The correlated automorphisms of $({}^2\overline{\mathbf{K}}^\sigma)^n$ are the endomorphisms of ${}^2\mathbf{K}^n$ of the form $\begin{pmatrix} a & 0 \\ 0 & (\overline{a}^\tau)^{-1} \end{pmatrix}$, where a is any automorphism of $\mathbf{K}^n$.*

Corollary 12.18 *The group of correlated automorphisms of* $({}^2\overline{K}^{\sigma})^n$ *is isomorphic to the general linear group* $GL(n;\mathbf{K})$.

We define the *determinant* of such a correlated automorphism to be the determinant of $a \in \mathbf{K}(n)$. Accordingly the subgroup of all such correlated automorphisms of determinant equal to 1 is isomorphic to the *special linear group* $SL(n;\mathbf{K})$.

Exercise

12.1 Let tw**K** denote the *twisted square* of **K**, that is, $\mathbf{K} \times \mathbf{K}$ with the product $(\mathbf{K} \times \mathbf{K})^2 \to \mathbf{K} \times \mathbf{K}$; $((\lambda, \mu), (\lambda', \mu')) \mapsto (\lambda\lambda', \mu'\mu)$. Show that, for any finite-dimensional **K**-linear space X, the **K**-linear space $X \times X^L$ may be regarded as a (tw**K**)-linear space by defining scalar multiplication by the formula

$$(x, \omega)(\lambda, \mu) = (x\lambda, \mu\omega),$$

for any $(x, \omega) \in X \times X^L$, $(\lambda, \mu) \in \mathrm{tw}\mathbf{K}$.

Develop the theory of tw**K**-correlated spaces. Show, in particular, that, for any finite-dimensional **K**-linear space X, the map

$$(X \times X^L)^2 \to \mathrm{tw}\mathbf{K},\ ((a, \alpha), (b, \beta)) \mapsto (\alpha(b), \beta(a))$$

is the product of a non-degenerate symmetric $(\mathrm{tw}\mathbf{K})^{\sigma}$-correlation on $X \times X^L$, σ being an *anti*-involution of tw**K**.

13
The classical groups

In this chapter we tie together the results of the last few chapters. The main result, Theorem 13.8, states how any real algebra anti-involution of $\mathbf{A}(n)$ may be regarded as the adjoint involution induced by some appropriate product on the right $\mathbf{A}$-linear space $\mathbf{A}^n$, where $\mathbf{A}$ is equal to $\mathbf{K}$ or ${}^2\mathbf{K}$, and $\mathbf{K} = \mathbf{R}$, $\mathbf{C}$ or $\mathbf{H}$, n being finite. Theorem 13.8 and Theorem 13.7 together classify the anti-involutions of $\mathbf{A}(n)$ into *ten classes*, and associated with these are the ten families of *classical groups*.

The left-coset exact pairs of the last section will play a role in the discussion of Lie groups in Chapter 22.

Equivalent correlated spaces

Let $\mathbf{A} = \mathbf{K}$ or ${}^2\mathbf{K}$, where $\mathbf{K} = \mathbf{R}$, $\mathbf{C}$ or $\mathbf{H}$. Then a *correlated* $\mathbf{A}$*-space* is a finite-dimensional right $\mathbf{A}$-linear space assigned an $\mathbf{A}^\psi$-correlation, where ψ is some anti-involution of ${}^2\mathbf{K}$. One speaks of the $\mathbf{A}^\psi$-correlated space (X, ξ), where ξ is the assigned $\mathbf{A}^\psi$-correlation.

Let (X, ξ) and (Y, η) be correlated $\mathbf{A}$-spaces. Then a *correlated map* $t : (X, \xi) \to (Y, \eta)$ is a (right) $\mathbf{A}^\alpha$-linear map, where α is an automorphism of $\mathbf{A}$, such that, for all $a, b \in X$,

$$t(a)^\eta t(b) = (a^\xi b)^\alpha,$$

an invertible map of this type being a *correlated isomorphism*. If such an isomorphism exists then the correlated spaces are said to be equivalent, $(X, \xi) \cong (Y, \eta)$.

Recall also that two anti-automorphisms of $\mathbf{A}$ are said to be *similar* if there exists an automorphism α of $\mathbf{A}$ such that $\alpha\psi = \chi\alpha$.

The next proposition generalises several propositions of previous chapters.

Proposition 13.1 *Let ξ be an $\mathbf{A}^{\psi}$-correlation on a right* **A***-linear space* X *and let χ be any anti-automorphism of* **A** *similar to ψ. Then there exist a right* **A***-linear space* Y *and an $\mathbf{A}^{\chi}$-correlation η on Y such that* $(Y, \eta) \cong (X, \xi)$.

Proof Since χ and ψ are similar, there exists an automorphism α of A such that $\alpha\psi = \chi\alpha$. Let $Y = X^{\alpha}$, where X^{α} consists of the set X with addition defined as before, but with a new scalar multiplication, namely,

$$X^{\alpha} \times \mathbf{A} \to X^{\alpha};\ (x, \lambda) \mapsto x\lambda^{\alpha^{-1}},$$

and let $\eta : Y \to Y^L$ be defined, for all $a, b \in Y$, by the formula

$$a^{\eta}b = (a^{\xi}b)^{\alpha}.$$

The image of η is genuinely in Y^L since, for any $\mu \in \mathbf{A}$,

$$a^{\eta}(b\,\mu^{\alpha^{-1}}) = (a^{\xi}b\,\mu^{\alpha^{-1}})^{\alpha} = (a^{\xi}b)^{\alpha}\mu.$$

Moreover, for any $\lambda \in \mathbf{A}$,

$$(a\lambda^{\alpha^{-1}})^{\eta}b = ((a\lambda^{\alpha^{-1}})^{\xi}b)^{\alpha} = (\lambda^{\psi\alpha^{-1}}a^{\xi}b)^{\alpha} = \lambda^{\chi}a^{\xi}b,$$

since $\chi = \alpha\psi\alpha^{-1}$. That is, η is $\mathbf{A}^{\chi}$-linear.

Finally, the set identity $(X, \xi) \to (Y, \eta)$ is a correlated isomorphism, since it is a semi-linear isomorphism and, from its very definition,

$$a^{\eta}b = (a^{\xi}b)\alpha, \text{ for all } a, b \in X.$$

□

The *adjoint* t^* of a linear map t between correlated spaces (X, ξ) and (Y, η) is the map $\xi^{-1}t^L\eta : Y \to X$, and is such that, for all $a \in X$, $b \in Y$,

$$b^{\eta}t(a) = t^*(b)^{\xi}a,$$

the adjoint of a linear map $u : X \to X$ with respect to ξ being denoted by u^{ξ}. The map u is said to be *self-adjoint* if $u^{\xi} = u$ and *skew-adjoint* if $u^{\xi} = -u$. The real linear subspaces $\{u \in \operatorname{End} X : u^{\xi} = u\}$ and $\{u \in \operatorname{End} X : u\xi = -u\}$ of the real linear space $\operatorname{End} X = L(X, X)$ will be denoted by $\operatorname{End}_+(X, \xi)$ and $\operatorname{End}_-(X, \xi)$, respectively.

Proposition 13.2 *Let ξ be an irreducible reflexive correlation on a finite-dimensional right* **A***-linear space X of dimension > 1. Then the map* $\operatorname{End} X \to \operatorname{End} X;\ t \mapsto t^{\xi}$ *is a real algebra anti-involution, equivalent correlations inducing the same anti-involution of* $\operatorname{End} X$.

Proposition 13.3 *Let (X, ξ) and (Y, η) be non-degenerate finite-dimensional $\mathbf{A}^{\varphi}$-correlated spaces. Then an* **A**-*linear map $t : (X, \xi) \to (Y, \eta)$ is correlated if and only if $t^{*} t = 1_X$, where t^{*} is the adjoint of t with respect to ξ and η.*

Corollary 13.4 *Let (X, ξ) and (Y, η) be as in Proposition 13.3. Then any correlated map $t : (X, \xi) \to (Y, \eta)$ is injective.*

Corollary 13.5 *Let (X, ξ) be as in Proposition 13.3 and let $t \in \operatorname{End} X$. Then t is a correlated automorphism of (X, ξ) if and only if $t^{\xi} t = 1_X$.*

The final proposition in this section will be used in the construction of charts on quadric Grassmannians in Chapter 14.

Proposition 13.6 *Let (X, ξ) and (Y, η) be as in Proposition 13.3, and suppose, further, that ξ and η are each symmetric or skew. Then, for any $t \in L(X, Y)$, $(t^{*}t)^{\xi} = \pm t^{*}t$, the $+$ sign applying if ξ and η are both symmetric or both skew, and the $-$ sign if one is symmetric and the other skew.*

The ten product types

The following theorem combines many earlier results.

Theorem 13.7 *Let ξ be an irreducible correlation on a right* **A**-*linear space of finite dimension > 1, and therefore equivalent to a symmetric or skew correlation. Then ξ is equivalent to one of the following ten types, these being mutually exclusive.*

0 *a symmetric* **R**-*correlation;*
1 *a symmetric, or equivalently a skew, ${}^2\mathbf{R}^{\sigma}$-correlation;*
2 *a skew* **R**-*correlation;*
3 *a skew* **C**-*correlation;*
4 *a skew $\tilde{\mathbf{H}}$- or equivalently a symmetric $\overline{\mathbf{H}}$-correlation;*
5 *a skew, or equivalently a symmetric, ${}^2\overline{\mathbf{H}}^{\sigma}$-correlation;*
6 *a symmetric $\tilde{\mathbf{H}}$-, or equivalently a skew, $\overline{\mathbf{H}}$-correlation;*
7 *a symmetric* **C**-*correlation;*
8 *a symmetric, or equivalently a skew, $\overline{\mathbf{C}}$-correlation;*
9 *a symmetric, or equivalently a skew, ${}^2\overline{\mathbf{C}}^{\sigma}$-correlation.*

Explicitly this combines Propositions 7.11, 10.4, 10.5, 12.5 and 13.1 and Propositions 6.1. 7.13, 10.7 and 12.7. The logic behind the coding of the ten types listed in the theorem will be explained later.

Theorem 13.8 follows at once from earlier work.

Theorem 13.8 *Let X be a finite-dimensional right $\mathbf{K}$- or ${}^2\mathbf{K}$-linear space. Then any irreducible anti-involution α of the real algebra* End X *is representable as the adjoint anti-involution induced by a symmetric or skew correlation on X.*

By Theorem 13.7 there are ten cases. In each of these there is a family of groups of correlated automorphisms analogous to the orthogonal groups. These are known as the *classical groups.* In a later chapter (Chapter 22) we shall show that they are all Lie groups, groups that are also differentiable manifolds, the group product being smooth. As such each has a real *dimension.* What we then prove is that for the group $G = \{t \in \mathbf{A}(n) : t^\xi t = 1\}$ the dimension is equal to the dimension of $G = \{t \in \mathbf{A}(n) : t^\xi + t = 0\}$ as a real vector space.

First we summarise here the explicit anti-involutions for each type. The notations are all as before, with the additional convention that, if, for some number n, $a \in \mathbf{K}(n)$ and if ψ is any anti-involution of $\mathbf{K}$, then a^ψ denotes the element of $\mathbf{K}(n)$ whose matrix is obtained from the matrix of a by applying the anti-involution ψ to each term of the matrix. The map $a^{\psi\tau}$ is then the transpose of a^ψ.

Table 13.9

X	(X, ξ)	$t \in \operatorname{End} X$	t^ξ
$\mathbf{K}^n$	$(\mathbf{K}^\psi)^n$ $(\mathbf{K}^\psi = \mathbf{C}$ *or* $\widetilde{\mathbf{H}})$	t	$t^{\psi\tau}$
$\mathbf{K}^p \times \mathbf{K}^q$	$\overline{\mathbf{K}}^{p,q}$ $(\overline{\mathbf{K}} = \mathbf{R}, \overline{\mathbf{C}}$ *or* $\overline{\mathbf{H}})$	$\begin{pmatrix} a & c \\ b & d \end{pmatrix}$	$\begin{pmatrix} \overline{a}^\tau & -\overline{b}^\tau \\ -\overline{c}^\tau & \overline{d}^\tau \end{pmatrix}$
$\mathbf{K}^n \times \mathbf{K}^n$	$(\mathbf{K}^\psi)^{2n}_{\text{hb}}$	$\begin{pmatrix} a & c \\ b & d \end{pmatrix}$	$\begin{pmatrix} d^{\psi\tau} & c^{\psi\tau} \\ b^{\psi\tau} & a^{\psi\tau} \end{pmatrix}$
$\mathbf{K}^n \times \mathbf{K}^n$	$(\mathbf{K}^\psi)^{2n}_{\text{sp}}$	$\begin{pmatrix} a & c \\ b & d \end{pmatrix}$	$\begin{pmatrix} d^{\psi\tau} & -c^{\psi\tau} \\ -b^{\psi\tau} & a^{\psi\tau} \end{pmatrix}$
${}^2\mathbf{K}^n$	$({}^2\mathbf{K}^{\psi^\sigma})^n$	$\begin{pmatrix} a & 0 \\ 0 & d \end{pmatrix}$	$\begin{pmatrix} d^{\psi\tau} & 0 \\ 0 & a^{\psi\tau} \end{pmatrix}$

We are now in a position to list the classical groups.

Table 13.10 *The ten families of classical groups, with their real dimensions, are as follows, where* $n = p + q$:

Code	*Group*	*Dimension*
0	$O(p,q;\mathbf{R})$ or $O(p,q)$, with $O(n) = O(0,n)$	$\frac{1}{2}n(n-1)$
1	$GL(n;\mathbf{R})$	n^2
2	$Sp(2n;\mathbf{R})$	$n(2n+1)$
3	$Sp(2n;\mathbf{C})$	$2n(2n+1)$
4	$Sp(p,q;\mathbf{H})$ or $Sp(p,q)$, with $Sp(n) = Sp(0,n)$	$n(2n+1)$
5	$GL(n;\mathbf{H})$	$4n^2$
6	$O(n;\mathbf{H})$	$n(2n-1)$
7	$O(n;\mathbf{C})$	$n(n-1)$
8	$U(p,q)$, with $U(n) = U(0,n)$	n^2
9	$GL(n;\mathbf{C})$	$2n^2$,

the subgroup consisting of all elements of the group of determinant 1 *in each case being as follows:*

Code	*Group*	*Dimension*
0	$SO(p,q;\mathbf{R})$ or $SO(p,q)$, with $SO(n) = SO(0,n)$	$\frac{1}{2}n(n-1)$
1	$SL(n;\mathbf{R})$	n^2-1
2	$Sp(2n;\mathbf{R})$	$n(2n+1)$
3	$Sp(2n;\mathbf{C})$	$2n(2n+1)$
4	$Sp(p,q;\mathbf{H})$ or $Sp(p,q)$, with $Sp(n) = Sp(0,n)$	$n(2n+1)$
5	$SL(n;\mathbf{H})$	$4n^2-1$
6	$SO(n;\mathbf{H})$	$n(2n-1)$
7	$SO(n;\mathbf{C})$	$n(n-1)$
8	$SU(p,q)$, with $SU(n) = SU(0,n)$	n^2-1
9	$SL(n;\mathbf{C})$	$2n^2-2$.

Note that the three families of general linear groups in this context all turn up as *'unitary'* groups. In the simplest case, that of a symmetric ${}^2\mathbf{R}^\sigma$ form, the elements of the group preserving the form actually occur as matrices of the form

$$\begin{pmatrix} a & 0 \\ 0 & (a^\tau)^{-1} \end{pmatrix}$$

where a is any invertible $m \times m$ matrix and a^τ denotes its transpose, the group in this way being isomorphic to the general linear group $GL(m;\mathbf{R})$. In the skew real case or in any of the complex or quaternionic cases the automorphism in the bottom right-hand slot may be different, but that is irrelevant to the nature of the group.

Complexification

There are two distinct complexifications of real algebras furnished with an anti-involution, according to whether the algebra of complex numbers is assigned the identity or conjugation as (anti-)involution. In the tables of the following theorem we use the code numbers for the various types that were introduced in Theorem 13.7.

Theorem 13.11 *Let X be a finite-dimensional right* $\mathbf{K}$- *or* ${}^2\mathbf{K}$-*linear space, where* $\mathbf{K} = \mathbf{R}$, $\mathbf{C}$ *or* $\mathbf{H}$, *and let* α *be an anti-involution of the real algebra* $\operatorname{End} X$ *of type* m, *where* $0 \le m \le 9$. *Then the induced anti-involution of the real algebra* $\operatorname{End} X \otimes \mathbf{C}$ *restricting to* α *on* $\operatorname{End} X$ *and to* the identity *on* $\mathbf{C}$ *is of the type given by the following table:*

0	1	2	3	4	5	6	7	8	9
7	9	3	${}^2 3$	3	9	7	${}^2 7$	9	${}^2 9$,

while the induced anti-involution of the real algebra $\operatorname{End} X \otimes \mathbf{C}$ *restricting to* α *on* $\operatorname{End} X$ *and to* conjugation *on* $\mathbf{C}$ *is of the type given by the following table:*

0	1	2	3	4	5	6	7	8	9
8	9	8	9	8	9	8	9	${}^2 8$	${}^2 9$.

Proof None of these are difficult to verify, if one has the canonical forms of Table 13.9 in mind. Indeed the cases in which $\mathbf{K} = \mathbf{R}$ are immediate. For the cases in which $\mathbf{K} = \mathbf{H}$ the argument is a mild generalisation of Proposition 11.12, while in the cases in which $\mathbf{K} = \mathbf{C}$ the argument is a mild generalisation of Proposition 11.13.

As an example consider the correlated space $\overline{\mathbf{H}}^n$ of type 4, inducing the anti-involution $t \mapsto \bar{t}^\tau$ of $\mathbf{H}(n)$. If each entry $q = z + \mathrm{j}w$ is represented by the 2×2 complex matrix $\begin{pmatrix} z & -\bar{w} \\ w & \bar{z} \end{pmatrix}$, with $\bar{q} = \bar{z} - \mathrm{j}w$ represented by the matrix $\begin{pmatrix} \bar{z} & \bar{w} \\ -w & z \end{pmatrix}$, then the matrix of t may be represented in block form as an element of $\mathbf{C}(n)(2) \cong \mathbf{C}(2n)$, namely $\begin{pmatrix} Z & -\overline{W} \\ W & \overline{Z} \end{pmatrix}$, where Z, $W \in \mathbf{C}(n)$, mapped by the anti-involution to the matrix $\begin{pmatrix} \overline{Z}^\tau & \overline{W}^\tau \\ -W^\tau & Z^\tau \end{pmatrix}$, in accordance with both the anti-involutions of $\mathbf{C}(2n)$, $\begin{pmatrix} a & c \\ b & d \end{pmatrix} \mapsto \begin{pmatrix} d^\tau & -c^\tau \\ -b^\tau & a^\tau \end{pmatrix}$ of type 3 and $\begin{pmatrix} a & c \\ b & d \end{pmatrix} \mapsto \begin{pmatrix} \bar{a}^\tau & \bar{b}^\tau \\ \bar{c}^\tau & \bar{d}^\tau \end{pmatrix}$

of type 8, the first being appropriate if the (anti)-involution on **C** is the identity and the second if it is conjugation.

By contrast, the anti-involution $\mathbf{H}(n) \to \mathbf{H}(n);\ t \mapsto \tilde{t}$, of type 6, transforms to $\begin{pmatrix} Z & -\overline{W} \\ W & \overline{Z} \end{pmatrix} \mapsto \begin{pmatrix} Z^\tau & W^\tau \\ -\overline{W}^\tau & \overline{Z}^\tau \end{pmatrix}$, in accord with both the anti-involutions of $\mathbf{C}(2n)$, $\begin{pmatrix} a & c \\ b & d \end{pmatrix} \mapsto \begin{pmatrix} a^\tau & b^\tau \\ c^\tau & d^\tau \end{pmatrix}$ of type 7 and $\begin{pmatrix} a & c \\ b & d \end{pmatrix} \mapsto \begin{pmatrix} \overline{d}^\tau & -\overline{c}^\tau \\ -\overline{b}^\tau & \overline{a}^\tau \end{pmatrix}$ of type 8, the latter being a different version of type 8 from that above, but type 8 nevertheless, the first again being appropriate if the (anti)-involution on **C** is the identity and the second if it is conjugation. □

For completeness' sake and for future reference we also give the following analogue of Theorem 13.11 involving tensoring by $^2\mathbf{R}$.

Theorem 13.12 *Let X be a finite-dimensional right* **K**- *or* ^{2}**K**-*linear space, where* $\mathbf{K} = \mathbf{R}$, **C** *or* **H**, *and let α be an anti-involution of the real algebra* End X *of type m, where $0 \leq m \leq 9$. Then the induced anti-involution of the real algebra* End $X \otimes {}^2\mathbf{R}$ *restricting to α on* End X *and to* swap *on* $^2\mathbf{R}$ *is of the type given by the following table:*

0	1	2	3	4	5	6	7	8	9
1	^{2}1	1	9	5	^{2}5	5	9	9	^{2}9 .

The reason for separating off cases 8 and 9 from the others in these two theorems will become apparent when we come to applications in a later chapter (Chapter 17).

Quasi-spheres

There are analogues of the unit sphere S^n in $\mathbf{R}^{n+1}$ for all non-degenerate finite-dimensional correlated spaces. Suppose first that (X, ξ) is a symmetric correlated space Then

$$\mathscr{S}(X, \xi) = \{x \in X : x^\xi x = 1\}$$

is defined to be the unit *quasi-sphere* in (X, ξ), with $\mathscr{S}(\mathbf{R}^{n+1}) = S^n$, while $\mathscr{S}(\overline{\mathbf{C}}^{n+1})$ and $\mathscr{S}(\overline{\mathbf{H}}^{n+1})$ are identifiable in an obvious way with S^{2n+1} and S^{4n+3}, respectively, for any n.

Note also that, for $(X, \xi) = {}^2\widetilde{\mathbf{K}}^{n+1}$, with $\widetilde{\mathbf{K}} = \mathbf{R}$, $\mathbf{C}$ or $\widetilde{\mathbf{H}}$,

$$\begin{aligned} \mathscr{S}(X, \xi) &= \{x \in {}^2\,\mathbf{K}^{n+1} : x^{\xi} x = 1\} \\ &= \{x \in {}^2\,\mathbf{K}^{n+1} : (\widetilde{x}_1^{\tau} x_0, \widetilde{x}_0^{\tau} x_1) = (1, 1)\} \\ &= \{x \in {}^2\,\mathbf{K}^{n+1} : \widetilde{x}^{\tau} x_1 = 1\}, \end{aligned}$$

since $\widetilde{x}_1^{\tau} x_0 = 1$ if and only if $\widetilde{x}_0^{\tau} x_1 = 1$.

A slightly different definition is necessary in the essentially skew cases. The appropriate definition is

$$\mathscr{S}(\mathbf{K}_{\rm sp}^{2n}) = \{(x, y) \in (\mathbf{K}_{\rm sp}^{2n})^2 : x \cdot y = 1\},$$

where $\cdot$ denotes the standard product on $\mathbf{K}_{\rm sp}^{2n}$, or, equivalently,

$$\mathscr{S}(\mathbf{K}_{\rm sp}^{2n}) = \left\{ \begin{pmatrix} a & c \\ b & d \end{pmatrix} \in (\mathbf{K}^n)^{2\times 2} : a \cdot d - b \cdot c = 1 \right\},$$

where $\cdot$ denotes the standard scalar product on $\mathbf{K}^n$, $\mathbf{K}$ being either $\mathbf{R}$ or $\mathbf{C}$.

The verification of the following theorem is a straightforward check!

Theorem 13.13 *For any p, q, n let the correlated spaces $\mathbf{R}^{p,q+1}$, $\mathbf{C}^{n+1}$, $\overline{\mathbf{C}}^{p,q+1}$, $\overline{\mathbf{H}}^{n+1}$, $\widetilde{\mathbf{H}}^{p,q+1}$, ${}^2\mathbf{R}^{n+1}$, ${}^2\mathbf{C}^{n+1}$, ${}^2\widetilde{\mathbf{H}}^{n+1}$, $\mathbf{R}_{\rm sp}^{2n+2}$, $\mathbf{C}_{\rm sp}^{2n+2}$ be identified with $\mathbf{R}^{p,q} \times \mathbf{R}$, $\mathbf{C}^n \times \mathbf{C}$, $\overline{\mathbf{C}}^{p,q} \times \overline{\mathbf{C}}$, $\overline{\mathbf{H}}^n \times \overline{\mathbf{H}}$, $\widetilde{\mathbf{H}}^{p,q} \times \widetilde{\mathbf{H}}$, ${}^2\mathbf{R}^n \times {}^2\mathbf{R}$, ${}^2\mathbf{C}^n \times {}^2\mathbf{C}$, ${}^2\widetilde{\mathbf{H}}^n \times {}^2\widetilde{\mathbf{H}}$, $\mathbf{R}_{\rm sp}^{2n} \times \mathbf{R}_{\rm sp}^2$, $\mathbf{C}_{\rm sp}^{2n} \times \mathbf{C}_{\rm sp}^2$, respectively, in the obvious ways. Then the pairs of maps*

$$\begin{array}{rcccll}
O(p, q) & \longrightarrow & O(p, q+1) & \longrightarrow & \mathscr{S}(\mathbf{R}^{p,q+1}), & \\
SO(p, q) & \longrightarrow & SO(p, q+1) & \longrightarrow & \mathscr{S}(\mathbf{R}^{p,q+1}), & p+q>0, \\
O(n; \mathbf{C}) & \longrightarrow & O(n+1, \mathbf{C}) & \longrightarrow & \mathscr{S}(\mathbf{C}^{n+1}), & \\
SO(n; \mathbf{C}) & \longrightarrow & SO(n+1, \mathbf{C} & \longrightarrow & \mathscr{S}(\mathbf{R}^{p,q+1}), & n>0, \\
U(p, q) & \longrightarrow & U(p, q+1) & \longrightarrow & \mathscr{S}(\overline{\mathbf{C}}^{p,q+1}), & \\
SU(p, q) & \longrightarrow & SU(p, q+1) & \longrightarrow & \mathscr{S}(\overline{\mathbf{C}}^{p,q+1}), & p+q>0, \\
O(n; \widetilde{\mathbf{H}}) & \longrightarrow & O(n+1; \widetilde{\mathbf{H}}) & \longrightarrow & \mathscr{S}(\widetilde{\mathbf{H}}^{n+1}), & \\
Sp(p, q) & \longrightarrow & Sp(p, q+1) & \longrightarrow & \mathscr{S}(\overline{\mathbf{H}}^{p,q+1}), & \\
GL(n; \mathbf{R}) & \longrightarrow & GL(n+1; \mathbf{R}) & \longrightarrow & \mathscr{S}({}^2\mathbf{R}^{n+1}), & \\
SL(n; \mathbf{R}) & \longrightarrow & SL(n+1; \mathbf{R}) & \longrightarrow & \mathscr{S}({}^2\mathbf{R}^{n+1}), & n>0, \\
GL(n; \mathbf{C}) & \longrightarrow & GL(n+1; \mathbf{C}) & \longrightarrow & \mathscr{S}({}^2\mathbf{C}^{n+1}), & \\
SL(n; \mathbf{C}) & \longrightarrow & SL(n+1; \mathbf{C}) & \longrightarrow & \mathscr{S}({}^2\mathbf{C}^{n+1}), & \\
GL(n; \mathbf{H}) & \longrightarrow & GL(n+1; \mathbf{H}) & \longrightarrow & \mathscr{S}({}^2\overline{\mathbf{H}}^{n+1}), & \\
SL(n; \mathbf{H}) & \longrightarrow & SL(n+1; \mathbf{H}) & \longrightarrow & \mathscr{S}({}^2\overline{\mathbf{H}}^{n+1}), & \\
Sp(2n; \mathbf{R}) & \longrightarrow & Sp(2n+2; \mathbf{R}) & \longrightarrow & \mathscr{S}(\mathbf{R}_{\rm sp}^{2n+2}), & \\
Sp(2n; \mathbf{C}) & \longrightarrow & Sp(2n+2; \mathbf{C}) & \longrightarrow & \mathscr{S}(\mathbf{C}_{\rm sp}^{2n+2}) &
\end{array}$$

are each left-coset exact (Chapter 3), the first map in each case being the injection $s \mapsto \begin{pmatrix} s & 0 \\ 0 & 1 \end{pmatrix}$ *and the second being, in all but the last two cases, the map* $t \mapsto t(0, 1)$, *the last column of* t, *and, in the last two cases, the map* $t \mapsto (t(0,(1,0)), t(0,(0,1)))$, *the last two columns of* t.

Note, in particular, the left-coset exact pairs of maps

$$\begin{array}{rcccll} O(n) & \longrightarrow & O(n+1) & \longrightarrow & S^n, & \\ SO(n) & \longrightarrow & SO(n+1) & \longrightarrow & S^n, & n>0, \\ U(n) & \longrightarrow & U(n+1) & \longrightarrow & S^{2n+1}, & \\ SU(n) & \longrightarrow & SU(n+1) & \longrightarrow & S^{2n+1}, & n>0, \\ Sp(n) & \longrightarrow & Sp(n+1) & \longrightarrow & S^{4n+3}. & \end{array}$$

For applications of Theorem 13.13 see Corollary 22.38 and Corollary 22.40.

Witt decompositions

The Witt construction of Proposition 5.25 generalises to each of the ten classes of correlated space as follows.

Proposition 13.14 *Let* (X, ξ) *be a non-degenerate irreducible finite-dimensional symmetric or skew* $\mathbf{A}^\psi$*-correlated space, and suppose that* W *is a one-dimensional null subspace of* X. *Then there exists another one-dimensional null subspace* W' *distinct from* W *such that the plane spanned by* W *and* W' *is, respectively, a hyperbolic or symplectic* $\mathbf{A}^\psi$*-plane, that is, isomorphic to* $(\mathbf{A}^\psi)^2_{\text{hb}}$ *or to* $(\mathbf{A}^\psi)^2_{\text{sp}}$.

Proof In the argument which follows, the upper of two alternative signs refers to the symmetric case and the lower to the skew case.

Let w be a regular element of W. Since X is non-degenerate there exists, by Proposition 12.12 or 12.13, an element $x \in X$ such that $w^\xi x = 1$. Then, for any $\lambda \in \mathbf{A}$,

$$(x + w\lambda)^\xi (x + w\lambda) = x^\xi x + \lambda^\psi \pm \lambda,$$

this being 0 if $\lambda = \mp\frac{1}{2}x^\xi x$, since $x^\xi x = \pm(x^\xi x)^\psi$. Let $w' = x \mp \frac{1}{2} w\, x^\xi x$. Then $w^\xi w' = 1$, $w'^\xi w = \pm 1$ and $w'^\xi w' = 0$. Now let $W' = \mathbf{A}\{w'\}$ be a ${}^2\mathbf{K}$-line in the ${}^2\mathbf{K}$ case, since $w'^\xi w = \pm 1 (= \pm(1,1))$. Then the plane spanned by W and W' is a hyperbolic or symplectic $\mathbf{A}^\psi$-plane. □

Corollary 13.15 *Let W be a null subspace of a non-degenerate irreducible finite-dimensional symmetric or skew $\mathbf{A}^{\varphi}$-correlated space X. Then there exists a null subspace W' of X such that $X = W \oplus W' \oplus (W \oplus W')^{\perp}$.*

Such a decomposition of X will be called a *Witt decomposition* of X with respect to the null subspace W.

Corollary 13.16 *Let (X, ξ) be a non-degenerate irreducible symmetric or skew finite-dimensional $\mathbf{A}^{\varphi}$-correlated space. Then there is a unique number k such that X is isomorphic either to $(\mathbf{A}^{\varphi})^{2k}_{\mathrm{hb}} \times Y$ in the symmetric case, or to $(\mathbf{A}^{\varphi})^{2k}_{\mathrm{sp}} \times Y$ in the skew case, where in either case Y is a subspace of X admitting no non-zero null subspace.*

The number k is the *(Witt) index* of the correlated space (X, ξ). It is the dimension of the null space of greatest dimension in (X, ξ).

Proposition 13.17 *The index of a non-degenerate finite-dimensional $\mathbf{A}^{\varphi}$-correlated space (X, ξ) is at most half the dimension of X.*

Proposition 13.18 *The correlated spaces $\mathbf{R}^{n,n+k}$, $\mathbf{R}^{2n}_{sp}$, $\mathbf{C}^{2n}$, $\mathbf{C}^{2n+1}$, $\overline{\mathbf{C}}^{n,n+k}$, $\mathbf{C}^{2n}_{sp}$, $\widetilde{\mathbf{H}}^{2n}$, $\widetilde{\mathbf{H}}^{2n+1}$, $\overline{\mathbf{H}}^{n,n+k}$, $({}^2\mathbf{K}^{\varphi})^{2n}$ and $({}^2\mathbf{K}^{\varphi})^{2n+1}$ all have index n for any non-negative integers n and k.*

14
Quadric Grassmannians

The central objects of study in this chapter are the quadric Grassmannians of finite-dimensional correlated spaces. Particular topics include parabolic charts on a quadric Grassmannian and various coset space representations of quadric Grassmannians.

There is no attempt to be exhaustive. The purpose of the chapter is to provide a fund of illustrative examples.

All linear spaces will be finite-dimensional linear spaces over $\mathbf{A} = \mathbf{K}$ or ${}^2\mathbf{K}$, where $\mathbf{K} = \mathbf{R}$, $\mathbf{C}$ or $\mathbf{H}$. On a first reading one should assume that $\mathbf{A} = \mathbf{R}$ or $\mathbf{C}$ and that the anti-involution ψ is the identity, ignoring references to the more complicated cases.

We start by considering ordinary Grassmannians of linear spaces.

Grassmannians

Let X be a right $\mathbf{A}$-linear space of dimension n. Then, for any k, the set $\mathscr{G}_k(X)$ of linear subspaces of X of dimension k over $\mathbf{A}$ is the *Grassmannian* of *linear k-planes* in X. In the case that $\mathbf{A}$ is the field of real numbers $\mathbf{R}$ there are also the Grassmannians $\mathscr{G}_k^+(X)$ of oriented linear k-planes in X.

An important example is the Grassmannian $\mathscr{G}_1(X)$ of lines in X through 0, also called the *projective space* of the linear space X. The projective space $G_1(\mathbf{K}^{n+1})$ is also denoted by $\mathbf{K}P^n$, and said to be *n-dimensional*. A zero-dimensional projective space is called a *projective point*, a one-dimensional projective space is called a *projective line* and a two-dimensional projective space is called a *projective plane*. Each point of the projective space $\mathscr{G}_1(X)$ of a linear space X is a line through 0 in X, this line being determined by any one of its points x other than 0. The line, or projective point, $\mathbf{K}\{x\}$ will also be denoted by $[x]$. For example, $[x, y]$ denotes a point of $\mathbf{K}P^2$, namely the line through 0 and

(x, y). (Confusion here with the closed intervals of $\mathbf{R}$, which are similarly denoted, is most unlikely in practice.)

Let X be a finite-dimensional $\mathbf{K}$-linear space and let $x \in X\backslash\{0\}$. Then we denote by $\mathbf{K}\{x\}$ the linear subspace of X dimension 1 spanned by x. The map

$$X\backslash\{0\} \to \mathcal{G}_1(X); \; x \mapsto [x]$$

is called the *Hopf map* over the projective space $\mathcal{G}_1(X)$.

Various subsets of the Grassmannian $\mathcal{G}_k(X)$ may be regarded in a natural way as linear spaces.

Proposition 14.1 *For any linear subspace W of X of dimension k, any complementary linear subspace Y of X of codimension k and any linear map $t : W \to Y$,* graph t *is an element of $\mathcal{G}_k(X)$.*

The injective map $L(W, Y) \to \mathcal{G}_k(X)$; $t \mapsto$ graph t will be called a *standard chart* on $\mathcal{G}_k(X)$.

A *standard atlas* on $\mathcal{G}_k(X)$ is a set of standard charts on $\mathcal{G}_k(X)$ such that every point of $\mathcal{G}_k(X)$ is in the image of at least one chart, the *dimension* of $\mathcal{G}_k(X)$ being defined to be the dimension over $\mathbf{A}$ of any of the linear spaces $L(W, Y)$, namely $k(n-k)$, where $n = \dim X$.

Example 14.2 *The maps*

$$\mathbf{K} \to \mathbf{K}P^1; \; x \mapsto [x, 1] \text{ and } y \mapsto [1, y]$$

form a standard atlas for $\mathbf{K}P^1$.

That is $\mathbf{K}P^1$ may be thought of simply as the union of two copies of the field $\mathbf{K}$ glued together by the map $\mathbf{K} \rightarrowtail \mathbf{K}$; $x \mapsto x^{-1}$. Only one point of the second copy of $\mathbf{K}$ fails to correspond to a point of the first. This point $[1, 0]$ is often denoted by ∞ and called the *point at infinity* on the projective line. Every other point $[x, y]$ of $\mathbf{K}P^1$ is represented by a unique point $x y^{-1}$ in the first copy of $\mathbf{K}$. When we are using this representation of $\mathbf{K}P^1$ we shall simply write $\mathbf{K} \cup \{\infty\}$ in place of $\mathbf{K}P^1$.

Example 14.3 *The maps*

$$\mathbf{K}^2 \to \mathbf{K}P^2; \; (x, y) \mapsto [x, y, 1], \; (x, z) \mapsto [x, 1, z] \text{ and } (y, z) \mapsto [1, y, z]$$

form a standard atlas for $\mathbf{K}P^2$.

As was the case with $\mathbf{K}P^1$ it is often convenient in working with $\mathbf{K}P^2$ to regard the first of these charts as standard and to regard all the points not lying in its image as *lying at infinity*. Observe that the set of points then lying at infinity is a projective line, namely the projective space of the plane $\{(x, y, z) \in \mathbf{K}^3 : z = 0\}$ in $\mathbf{K}^3$.

Quadric Grassmannians

Now let ξ be an irreducible symmetric or skew correlation on the right $\mathbf{A}$-linear space X. The kth quadric Grassmannian of the correlated space is, by definition, the subset $\mathcal{N}_k(X, \xi)$ of $\mathcal{G}_k(X)$ consisting of the k-dimensional null subspaces of (X, ξ).

Proposition 14.4 *Let ξ be such a correlation on X and let η be any correlation equivalent to ξ. Then, for each k, $\mathcal{N}_k(X,\eta) = \mathcal{N}_k(X,\xi)$. In particular, $\mathcal{N}_k(X,-\xi) = \mathcal{N}_k(X,\xi)$, for each k.*

The inverse image of $\mathcal{N}_k(X, \xi)$ by any one of the standard charts of $\mathcal{G}_k(X)$ will be called an *affine form* of $\mathcal{N}_k(X, \xi)$ or simply an *affine* quadric Grassmannian. We shall mainly be concerned with the case when ξ is non-degenerate. In this case the dimension of a null subspace is at most half the dimension of X.

When (X, ξ) is a non-degenerate neutral space, necessarily of even dimension, null subspaces of half the dimension of X exist. Such subspaces will be termed *semi-neutral* (or, when $(X, \xi) \cong \mathbf{R}^{2n}_{sp}$ or $\mathbf{C}^{2n}_{sp}$, *Lagrangian* (Arnol'd (1974)) subspaces) and the set of semi-neutral subspaces of (X, ξ) will be called a *semi-neutral* quadric Grassmannian.

There are null lines in (X, ξ) unless (X, ξ) is positive- or negative-definite. The subset $\mathcal{N}_1(X, \xi)$ of the projective space $\mathcal{G}_1(X)$ is called the *projective quadric* of (X, ξ).

A line W in X is null with respect to the correlation ξ on X, or, equivalently, is a point of the projective quadric, if and only if, for every $x \in W$, $x^\xi x = 0$. This equation is frequently referred to as the *equation of the quadric* $\mathcal{N}_1(X, \xi)$.

Just as the elements of $\mathcal{G}_k(X)$ may, when $k \geq 1$, be interpreted as $(k-1)$-dimensional projective subspaces of the projective space $\mathcal{G}_1(X)$ rather than as k-dimensional linear subspaces of X, so the elements of $\mathcal{N}_k(X, \xi)$ may, when $k \geq 1$, be interpreted as $(k-1)$-dimensional projective spaces lying on the projective quadric $\mathcal{N}_1(X, \xi)$ rather than as k-dimensional null subspaces of (X, ξ). We shall refer to this as the projective interpretation of the quadric Grassmannians.

When (X, ξ) is isomorphic either to $\mathbf{R}^{2n}_{\text{sp}}$ or to $\mathbf{C}^{2n}_{\text{sp}}$, every line in (X, ξ) is null. In these cases, therefore, the first interesting quadric Grassmannian is not $\mathcal{N}_1(X, \xi)$ but $\mathcal{N}_2(X, \xi)$, the set of null planes in (X, ξ). This set is usually called the *(projective) line complex* of (X, ξ), the terminology reflecting the projective rather than the linear interpretation of $\mathcal{N}_2(X, \xi)$. See also the remarks preceding Proposition 10.13.

Grassmannians as coset spaces

Proposition 14.5 *Let $\mathbf{R}^n$ be identified with $\mathbf{R}^k \times \mathbf{R}^{n-k}$. Then the map $f : O(\mathbf{R}^n) \to \mathcal{G}_k(\mathbf{R}^n)$; $t \mapsto t(\mathbf{R}^k \times \{0\})$ is surjective, its fibres being the left cosets in $O(\mathbf{R}^n)$ of the subgroup $O(\mathbf{R}^k) \times O(\mathbf{R}^{n-k})$.*

Proof With $\mathbf{R}^n$ identified with $\mathbf{R}^k \times \mathbf{R}^{n-k}$, any linear map $t : \mathbf{R}^n \to \mathbf{R}^n$ is of the form $\begin{pmatrix} a & c \\ b & d \end{pmatrix}$, where $a \in L(\mathbf{R}^k, \mathbf{R}^k)$, $b \in L(\mathbf{R}^k, \mathbf{R}^{n-k})$, $c \in L(\mathbf{R}^{n-k}, \mathbf{R}^k)$ and $d \in L(\mathbf{R}^{n-k}, \mathbf{R}^{n-k})$. Since the first k columns of the matrix span $t(\mathbf{R}^k \times \{0\})$, $t(\mathbf{R}^k \times \{0\}) = \mathbf{R}^k \times \{0\}$ if and only if $b = 0$. However, if t is orthogonal with $b = 0$, then c also is zero, since any two columns of the matrix are mutually orthogonal. The subgroup $O(\mathbf{R}^k) \times O(\mathbf{R}^{n-k})$, consisting of all $\begin{pmatrix} a & 0 \\ 0 & d \end{pmatrix} \in O(\mathbf{R}^k \times \mathbf{R}^{n-k})$, is therefore the fibre of f over $\mathbf{R}^k \times \{0\}$.

The map f is surjective by Proposition 5.2. This follows since any element of $\mathcal{G}_k(\mathbf{R}^n)$ is a non-degenerate subspace of $\mathbf{R}^n$ and so has an orthonormal basis that extends to an orthonormal basis for the whole of $\mathbf{R}^n$.

Finally, if t and $u \in O(\mathbf{R}^n)$ are such that $t(\mathbf{R}^k \times \{0\}) = u(\mathbf{R}^k \times \{0\})$, then $(u^{-1}t)(\mathbf{R}^k \times \{0\}) = \mathbf{R}^k \times \{0\}$, from which it follows that the fibres of the map f are left cosets in $O(\mathbf{R}^n)$ of the subgroup $O(\mathbf{R}^k) \times O(\mathbf{R}^{n-k})$. □

Accordingly the pair of maps $O(k) \times O(n-k) \xrightarrow{\iota} O(n) \xrightarrow{f} \mathcal{G}_k(\mathbf{R}^n)$ is left-coset exact, and the induced bijection

$$O(n)/(O(k) \times O(n-k)) \to \mathcal{G}_k(\mathbf{R}^n)$$

is a coset space representation of $\mathcal{G}_k(\mathbf{R}^n)$.

Proposition 14.6 *For each finite n, k, with k < n, there are coset space representations*

$$\begin{aligned} U(n)/(U(k)\times U(n-k)) &\longrightarrow \mathscr{G}_k(\mathbf{C}^n),\\ Sp(n)/(Sp(k)\times Sp(n-k)) &\longrightarrow \mathscr{G}_k(\mathbf{H}^n),\\ SO(n)/(SO(k)\times SO(n-k)) &\longrightarrow \mathscr{G}_k^+(\mathbf{R}^n),\end{aligned}$$

analogous to the coset space representation

$$O(n)/(O(k)\times O(n-k)) \longrightarrow \mathscr{G}_k(\mathbf{R}^n)$$

constructed in Proposition 14.5.

Charts on quadric Grassmannians

Let (X, ξ) be any non-degenerate irreducible symmetric or skew $\mathbf{A}^\psi$-correlated space, and consider the quadric Grassmannian $\mathscr{N}_k(X, \xi)$.

By Corollary 13.15 there is, for any $W \in \mathscr{N}_k(X, \xi)$, a Witt decomposition $W\oplus W'\oplus Z$ of X, where $W' \in \mathscr{N}_k(X, \xi)$ and $Z = (W\oplus W')^\perp$. There are, moreover, linear isomorphisms $\mathbf{A}^k \to W$ and $\mathbf{A}^k \to W'$ such that the product on X induced by ξ is given with respect to these isomorphisms by the formula

$$(a, b, c)^\xi(a', b', c') = b^\eta\, a' \pm a^\eta\, b' + c^\zeta\, c',$$

where η is the (symmetric) correlation on $(\mathbf{A}^\psi)^k$ and ζ is the correlation induced on Z by ξ, and where X has been identified with $W \times W' \times Z$ to simplify notations.

Both here and in the subsequent discussion, where there is a choice of sign the upper sign applies when ξ (and therefore ζ) is symmetric and the lower sign when ξ (and ζ) is skew.

Now let $Y = W' \oplus Z \cong W \times Z$ and consider the standard chart on $\mathscr{G}_k(X)$,

$$L(W, Y) \to \mathscr{G}_k(X);\ (s, t) \mapsto \operatorname{graph}(s, t).$$

The inverse image by this chart of the quadric Grassmannian $\mathscr{N}_k(X, \xi)$ is given by the following proposition.

Proposition 14.7 *Let* $(s, t) \in L(W, Y)$, *the notations and sign conventions being those just introduced. Then*

$$\operatorname{graph}(s, t) \in \mathscr{N}_k(X, \xi) \Leftrightarrow s \pm s^\eta + t^* t = 0,$$

where t^* *is the adjoint of* T *with respect to the correlations* η *on* $\mathbf{A}^k$ *and* ζ *on* Z.

In particular, when $Z = \{0\}$, that is, when $\mathcal{N}_k(X, \xi)$ is semi-neutral, the counterimage of $\mathcal{N}_k(X, \xi)$ by the chart

$$L(W, Y) \to \mathcal{G}_k(X);\ (s, t) \mapsto \operatorname{graph}(s, t)$$

is a real linear subspace of its source.

Proof For all $a, b \in W$,

$$\begin{aligned}(a, s(a), t(a))\xi(b, s(b), t(b)) &= s(a)^\eta b \pm a^\eta s(b) + t(a)^\zeta t(b) \\ &= (s(a) \pm s^\eta(a) + t^* t(a))^\eta b\end{aligned}$$

by the definition of the adjoint preceding Proposition 13.2. Therefore

$$\operatorname{graph}(s, t) \in \mathcal{N}_k(X, \xi) \Leftrightarrow s \pm s^\eta + t^* t = 0.$$

The second part of the proposition follows from the remark that $\operatorname{End}_+(A^k, \eta)$ and $\operatorname{End}_-(\mathbf{A}^k, \eta)$ are real linear subspaces of $\operatorname{End}(A^k)$, while $t = 0$, when $Z = \{0\}$. □

Proposition 14.8 *Let the notations be as above. Then the map*

$$f : \operatorname{End}_\mp(A^k, \eta) \times L(\mathbf{A}^k, Z) \to L(\mathbf{A}^k, Y);\ (s, t) \mapsto (s - \tfrac{1}{2}t^* t, t)$$

is injective, with image the affine form of $\mathcal{N}_k(X, \xi)$ in $L(\mathbf{A}^k, Y)$ $(= L(W, Y))$.

Proof That the map f is injective is obvious. That its image is as stated follows from the fact that, for any $t \in L(\mathbf{A}^k, Z)$, $(t^* t)^\eta = \pm t^* t$, by Proposition 13.6. Therefore, for any (s, t),

$$(s - \tfrac{1}{2}t^* t) \pm (s - \tfrac{1}{2}t^* t)^\eta + t^* t = 0.$$

That is, the image of f is a subset of $\mathcal{N}_k(X, \xi)$, by Proposition 14.7. Conversely, if $\operatorname{graph}(s', t') \in \mathcal{N}_k(X, \xi)$, let $s = s' + \tfrac{1}{2}t'^* t'$ and let $t = t'$. Then $s \pm s^\eta = 0$; so $s \in \operatorname{End}_\mp(\mathbf{A}^k, \eta)$ and $t \in L(\mathbf{A}^k, Z)$, while $s' = s - \tfrac{1}{2}t^* t$ and $t' = t$.

That is, the affine form of $\mathcal{N}_k(X, \xi)$ is a subset of $\operatorname{im} f$. The image of f and the affine quadric Grassmannian coincide. □

The composite of the map f with the inclusion of $\operatorname{im} f$ in $\mathcal{N}_k(X, \xi)$ will be called a *parabolic chart* on $\mathcal{N}_k(X, \xi)$ at W. A set of parabolic charts on $\mathcal{N}_k(X, \xi)$, one at each point of $\mathcal{N}_k(X, \xi)$, will be called a *parabolic atlas* for $\mathcal{N}_k(X, \xi)$.

Quadric Grassmannians as coset spaces

Coset space representations analogous to those of Proposition 14.6 exist for each of the quadric Grassmannians.

We begin by considering a particular case, the semi-neutral Grassmannian $\mathcal{N}_n(\mathbf{C}^{2n}_{\mathrm{hb}})$ of the neutral **C**-correlated space $\mathbf{C}^{2n}_{\mathrm{hb}}$.

Proposition 14.9 *There exists a bijection*

$$O(2n)/U(n) \longrightarrow \mathcal{N}_n(\mathbf{C}^{2n}_{\mathrm{hb}}),$$

where $O(2n)/U(n)$ denotes the set of left cosets in $O(2n)$ of the standard image of $U(n)$ in $O(2n)$.

Proof The bijection is constructed as follows.

The linear space underlying the correlated space $\mathbf{C}^{2n}_{\mathrm{hb}}$ is $\mathbf{C}^n \times \mathbf{C}^n$, and this same linear space also underlies the positive-definite correlated space $\overline{\mathbf{C}}^n \times \overline{\mathbf{C}}^n$. Any linear map $t : \mathbf{C}^n \times \mathbf{C}^n \to \mathbf{C}^n \times \mathbf{C}^n$ that respects both correlations is of the form $\begin{pmatrix} a & \overline{b} \\ b & \overline{a} \end{pmatrix}$, with $a^\tau b + b^\tau a = 0$ and $\overline{a}^\tau a = \overline{b}^\tau b = 1$, for, by Table 13.9, the respective adjoints of any such map $t = \begin{pmatrix} a & c \\ b & d \end{pmatrix}$ are $\begin{pmatrix} d^\tau & c^\tau \\ b^\tau & a^\tau \end{pmatrix}$ and $\begin{pmatrix} \overline{a}^\tau & \overline{c}^\tau \\ \overline{b}^\tau & \overline{d}^\tau \end{pmatrix}$, and these are equal if and only if $d = \overline{a}$ and $c = \overline{b}$. By Proposition 11.15 such a map may be identified with an element of $O(2n)$ or, when $b = 0$, with an element of $U(n)$, the injection $U(n) \to O(2n)$ being the standard one.

Suppose that W is any null subspace of $\mathbf{C}^{2n}_{\mathrm{hb}}$ of dimension n. A positive-definite orthonormal basis may be chosen for W as a subspace of $\overline{\mathbf{C}}^n \times \overline{\mathbf{C}}^n$. Suppose this is done, and the basis elements arranged in some order to form the columns of a $2n \times n$ matrix $\begin{pmatrix} a \\ b \end{pmatrix}$. Then W is the image of the null subspace $\mathbf{C}^n \times \{0\}$ by the map $\begin{pmatrix} a & \overline{b} \\ b & \overline{a} \end{pmatrix}$. Moreover, $a^\tau b + b^\tau a = 0$, since W is null for the hyperbolic correlation, while $\overline{a}^\tau a + \overline{b}^\tau b = 1$, since the basis chosen for W is orthonormal with respect to the positive-definite correlation.

Now let f be the map

$$O(2n) \longrightarrow \mathcal{N}_n(\mathbf{C}^{2n}_{\mathrm{hb}});\ \begin{pmatrix} a & \overline{b} \\ b & \overline{a} \end{pmatrix} \mapsto \operatorname{im}\begin{pmatrix} a \\ b \end{pmatrix}.$$

The map is clearly surjective; so none of the fibres is null. Secondly, $f^{-1}(\mathbf{C}^n \times \{0\}) = U(n)$. Finally, by an argument similar to that used in the proof of Proposition 14.5, the remaining fibres of f are the left cosets in $O(2n)$ of the subgroup $U(n)$. □

Theorem 14.10 *Let* $(X, \xi) = (\mathbf{A}^{\psi})^{2n}_{\mathrm{hb}}$ *or* $(\mathbf{A}^{\psi})^{2n}_{\mathrm{sp}}$, *for some* n, *where* ψ *is irreducible. Then in each of the ten standard cases there is a coset space representation on the semi-neutral Grassmannian* $\mathcal{N}_n(X, \xi)$, *as follows:*

$$\begin{aligned}
(O(n) \times O(n))/O(n) &\longrightarrow \mathcal{N}_n(\mathbf{R}^{2n}_{\mathrm{hb}}),\\
U(n)/O(n) &\longrightarrow \mathcal{N}_n(\mathbf{R}^{2n}_{\mathrm{sp}}),\\
O(2n)/U(n) &\longrightarrow \mathcal{N}_n(\mathbf{C}^{2n}_{\mathrm{hb}}),\\
(U(n) \times U(n))/U(n) &\longrightarrow \mathcal{N}_n(\overline{\mathbf{C}}^{2n}_{\mathrm{hb}}) = \mathcal{N}_n(\overline{\mathbf{C}}^{2n}_{\mathrm{sp}}),\\
Sp(n)/U(n) &\longrightarrow \mathcal{N}_n(\mathbf{C}^{2n}_{\mathrm{sp}}),\\
U(2n)/Sp(n) &\longrightarrow \mathcal{N}_n(\widetilde{\mathbf{H}}^{2n}_{\mathrm{hb}}) = \mathcal{N}_n(\overline{\mathbf{H}}^{2n}_{\mathrm{sp}}),\\
(Sp(n) \times Sp(n))/Sp(n) &\longrightarrow \mathcal{N}_n(\widetilde{\mathbf{H}}^{2n}_{\mathrm{sp}}) = \mathcal{N}_n(\overline{H}^{2n}_{\mathrm{hb}}),\\
O(2n)/(O(n) \times O(n)) &\longrightarrow \mathcal{N}_n({}^2\mathbf{R}^{2n}_{\mathrm{hb}}),\\
U(2n)/(U(n) \times U(n)) &\longrightarrow \mathcal{N}_n({}^2\overline{\mathbf{C}}^{2n}_{\mathrm{hb}}),\\
Sp(2n)/(Sp(n) \times Sp(n)) &\longrightarrow \mathcal{N}_n({}^2\overline{\mathbf{H}}^{2n}_{\mathrm{hb}}).
\end{aligned}$$

Proof The third of these is the case considered in Proposition 14.9. The details in each of the other cases follow the details of this case, but using the appropriate part of Proposition 11.15. □

Cayley charts

The first of the cases listed in Theorem 14.10 merits further discussion in view of the following remark.

Proposition 14.11 *Let* f *be the map*

$$O(n) \times O(n) \longrightarrow O(n);\ (a, b) \mapsto a b^{-1}.$$

Then the inverse image of the identity element of $O(n)$ *is the image of* $O(n)$ *by the injective group map*

$$O(n) \longrightarrow O(n) \times O(n);\ a \mapsto (a, a)$$

and the induced map

$$(O(n) \times O(n))/O(n) \longrightarrow O(n)$$

is bijective.

That is $O(n)$ may be represented as the semi-neutral Grassmannian $\mathcal{N}_n(\mathbf{R}^{2n}_{\mathrm{hb}})$. The charts on $O(n)$ corresponding to the parabolic charts on $\mathcal{N}_n(\mathbf{R}^{2n}_{\mathrm{hb}})$ will be called the *Cayley charts* on $O(n)$. The following is an independent account of this case.

Let $(X, \xi) \cong \mathbf{R}^{n,n} \cong \mathbf{R}^{2n}_{\mathrm{hb}}$, and consider the quadric $\mathcal{N}_1(X, \xi)$. Its equation may be taken to be either

$$x^\tau x = y^\tau y, \quad \text{where } (x, y) \in \mathbf{R}^n \times \mathbf{R}^n,$$

or

$$u^\tau v = 0, \quad \text{where } (u, v) \in \mathbf{R}^n \times \mathbf{R}^n,$$

according to the isomorphism chosen, the two models being related, for example, by the equations

$$u = x + y, \quad v = -x + y.$$

Now any n-dimensional subspace of $\mathbf{R}^n \times \mathbf{R}^n$ may be represented as the image of an injective linear map

$$(a, b) = \begin{pmatrix} a \\ b \end{pmatrix} : \mathbf{R}^n \mapsto \mathbf{R}^n \times \mathbf{R}^n.$$

Proposition 14.12 *The linear space $\mathcal{N}_m(a, b)$, where (a, b) is an injective element of $L(\mathbf{R}^n, \mathbf{R}^n \times \mathbf{R}^n)$, is a null subspace of $\mathbf{R}^{n,n}$ if and only if a and b are bijective and $b\,a^{-1} \in O(n)$.*

Proof $\Rightarrow$: Let $\operatorname{im}(a, b) \in \mathcal{N}_n(\mathbf{R}^{n,n})$, let $w \in \mathbf{R}^n$ be such that $x = a(w) = 0$ and let $y = b(w)$. Since (x, y) belongs to an null subspace of $\mathbf{R}^{n,n}$, $x^\tau x = y^\tau y$, but $x = 0$, so that $y = 0$. Since (a, b) is injective, it follows that $w = 0$ and therefore that a is injective. So a is bijective. Similarly, b is bijective. Since a is bijective, a^{-1} exists; so, for any $(x, y) = (a(w)\, b(w))$, $y = b\,a^{-1}(x)$. But $y^\tau y = x^\tau x$. So $b\,a^{-1} \in O(n)$.

$\Leftarrow$: Suppose that a and b are bijective; then, as above, for any $(x, y) \in \mathcal{N}_m(a, b)$, $y = b\,a^{-1}(x)$. If also $b\,a^{-1} \in O(n)$, then $y^\tau y = x^\tau x$. □

Corollary 14.13 *Any n-dimensional null subspace of $\mathbf{R}^{n,n}$ has an equation of the form $y = t(x)$, where $t \in O(n)$, and any n-plane with such an equation is null.*

Proposition 14.14 *Any element of $\mathcal{N}_n(\mathbf{R}^{n,n})$ may be represented as the image of a linear map $(a, b) : \mathbf{R}^n \to \mathbf{R}^n \times \mathbf{R}^n$ with a and b each orthogonal.*

This leads at once to the coset space representation for $\mathcal{N}_n(\mathbf{R}^{n,n})$ whose existence is asserted in Theorem 14.10.

Note that $\mathcal{N}_n(\mathbf{R}^{n,n}) = \{\operatorname{graph} t : t \in O(n)\}$ divides into two disjoint classes, according as t preserves or reverses orientation.

So far we have considered the projective quadric $\mathcal{N}_1(\mathbf{R}^{n,n})$. We now consider the quadric $\mathcal{N}_1(\mathbf{R}^{2n}_{\text{hb}})$. Let $s \in \operatorname{End}(\mathbf{R}^n)$ be such that $\operatorname{graph} s \in \mathcal{N}_n(\mathbf{R}^{2n}_{\text{hb}})$. Then, for all $u, u' \in \mathbf{R}^n$,

$$s(u')^{\tau}u + u'^{\tau}s(u) = 0,$$

implying that $s+s^{\tau} = 0$, that is, that $s \in \operatorname{End}_{-}(\mathbf{R}^n)$, this being a particular case of Proposition 14.7.

Now $\operatorname{graph} s = \operatorname{m}(1, s)$. We can transfer to $\mathcal{N}_1(\mathbf{R}^{n,n})$ by the map $\frac{1}{\sqrt{2}}\begin{pmatrix} 1 & -1 \\ 1 & 1 \end{pmatrix} : \mathbf{R}^{2n}_{\text{hb}} \to \mathbf{R}^{n,n}$. Then the image of $\operatorname{graph} s$ in $\mathbf{R}^{n,n}$, namely

$$\operatorname{im} \frac{1}{\sqrt{2}}\begin{pmatrix} 1 & -1 \\ 1 & 1 \end{pmatrix}\begin{pmatrix} 1 \\ s \end{pmatrix} = \operatorname{im}\begin{pmatrix} 1-s \\ 1+s \end{pmatrix},$$

is an element of $\mathcal{N}_n(\mathbf{R}^{n,n})$. So, by Proposition 14.12, or by Exercise 2.1, $1-s$ is invertible. By Proposition 14.12 again, or by Exercise 2.2, the product $(1+s)(1-s) = 1-s^2 = (1-s)(1+s)$. Moreover, since

$$(1-s) = 1+s^{\tau} = (1+s)^{\tau}, \quad (1+s)(1-s)^{-1} \in SO(n).$$

The following proposition sums this all up.

Proposition 14.15 *For any $s \in \operatorname{End}_{-}(\mathbf{R}^n)$, the endomorphism $1-s$ is invertible, and $(1+s)(1-s)^{-1} \in SO(n)$. Moreover, the map*

$$\operatorname{End}_{-}(\mathbf{R}^n) \to SO(n);\ s \mapsto (1+s)(1-s)^{-1}$$

is injective.

The map of Proposition 14.15 is the Cayley chart on $SO(n)$ (or $O(n)$) at ${}^n1$. For $n \geq 2$ it is not surjective even on $SO(n)$. For example, when $n = 2$ the rotation $-{}^21$ does not lie in its image.

The direct analogue of Proposition 14.15, with $\mathbf{R}^{p,q}$ in place of $\mathbf{R}^n$ and $SO(p,q)$ in place of $SO(n)$, is not true when both p and q are non-zero; for $\begin{pmatrix} 0 & 1 \\ 1 & 0 \end{pmatrix} \in \operatorname{End}_{-}(\mathbf{R}^{1,1})$, but $\begin{pmatrix} 1 & -1 \\ -1 & 1 \end{pmatrix}$ is not invertible. There is, however, the following partial analogue.

Proposition 14.16 *For any $s \in \operatorname{End}_-(\mathbf{R}^{p,q})$ for which $1 - s$ is invertible, $(1+s)(1-s)^{-1} \in SO(p,q)$. Moreover, the map*

$$\operatorname{End}_-(\mathbf{R}^{p,q}) \rightarrowtail SO(p,q);\ s \mapsto (1+s)(1-s)^{-1}$$

is injective.

The map given in Proposition 14.16 is, by definition, the *Cayley chart* on $SO(p,q)$ or $O(p,q)$) at ${}^n1$.

An entirely analogous discussion to that given above for the orthogonal group $O(n)$ can be given also for both the unitary group $U(n)$ and the symplectic group $Sp(n)$.

It was remarked above that the semi-neutral Grassmannian $\mathcal{N}_n(\mathbf{R}^{2n}_{\text{hb}})$ divides into two parts, the parts corresponding to the orientations of $\mathbf{R}^n$. The semi-neutral Grassmannian $\mathcal{N}_n(\mathbf{C}^{2n}_{\text{hb}})$ divides similarly into two parts, the parts corresponding, in the coset space representation

$$O(2n)/U(n) \rightarrow \mathcal{N}_n(\mathbf{C}^{2n}_{\text{hb}}),$$

to the two orientations of $\mathbf{R}^{2n}$. (By Corollary 2.4, any element of $U(n)$ preserves the orientation of $\mathbf{R}^{2n}$.)

Further coset space representations

Coset space representations analogous to those listed above for the semi-neutral quadric Grassmannians exist for all the quadric Grassmannians. The results are summarised in the following theorem.

Theorem 14.17 *Let (X, ξ) be a non-degenerate irreducible symmetric or skew $\mathbf{A}^\psi$-correlated space of dimension n. Then, for each k, in each of the ten standard cases, there is a coset space decomposition of the quadric Grassmannian $\mathcal{N}_k(X\ \xi)$ as follows:*

$$\begin{aligned}
(O(p) \times O(q))/(O(k) \times O(p-k) \times O(q-k)) &\longrightarrow \mathcal{N}_k(\mathbf{R}^{p,q}),\\
U(n)/(O(k) \times U(n-k)) &\longrightarrow \mathcal{N}_k(\mathbf{R}^{2n}_{\text{sp}}),\\
O(n)/(U(k) \times O(n-2k)) &\longrightarrow \mathcal{N}_k(\mathbf{C}^n),\\
(U(p) \times U(q))/(U(k) \times U(p-k) \times U(q-k)) &\longrightarrow \mathcal{N}_k(\overline{\mathbf{C}}^{p,q}),\\
Sp(n)/(U(k) \times Sp(n-k)) &\longrightarrow \mathcal{N}_k(\mathbf{C}^{2n}_{\text{sp}}),\\
U(n)/(Sp(k) \times U(n-2k)) &\longrightarrow \mathcal{N}_k(\overline{\mathbf{H}}^n),\\
(Sp(p) \times Sp(q))/(Sp(k) \times Sp(p-k) \times Sp(q-k)) &\longrightarrow \mathcal{N}_k(\overline{\mathbf{H}}^{p,q}),\\
O(n)/(O(k) \times O(k) \times O(n-2k)) &\longrightarrow \mathcal{N}_k({}^2\mathbf{R}^n),\\
U(n)/(U(k) \times U(k) \times U(n-2k)) &\longrightarrow \mathcal{N}_k({}^2\overline{\mathbf{C}}^n),\\
Sp(n)/(Sp(k) \times Sp(k) \times Sp(n-2k)) &\longrightarrow \mathcal{N}_k({}^2\overline{\mathbf{H}}^n).
\end{aligned}$$

The resourceful reader will be able to supply the proof!

Certain of the cases where $k = 1$ are of especial interest, and we conclude by considering several of these.

Consider first the real projective quadric $\mathcal{N}_1(\mathbf{R}^{p,q})$, where $p \geq 1$ and $q \geq 1$.

Proposition 14.18 *The map*

$$S^{p-1} \times S^{q-1} \longrightarrow \mathcal{N}_1(\mathbf{R}^{p,q});\ (x, y) \mapsto \mathbf{R}\{(x, y)\}$$

is surjective, the fibre over $\mathbf{R}\{(x, y)\}$ *being the set* $\{(x, y), (-x, -y)\}$.

That is, there is a bijection

$$(S^{p-1} \times S^{q-1})/S^0 \longrightarrow \mathcal{N}_1(\mathbf{R}^{p,q}),$$

where the action of S^0 on $S^{p-1} \times S^{q-1}$ is defined by the formula

$$(x, y)(-1) = (-x, -y),$$

for all $(x, y) \in S^{p-1} \times S^{q-1}$.

This result is in accord with the representation of $\mathcal{N}_1(\mathbf{R}^{p,q})$ given in Theorem 14.17, in view of the familiar coset space representations

$$O(p)/O(p-1) \longrightarrow S^{p-1} \text{ and } O(q)/O(q-1) \longrightarrow S^{q-1}$$

of Theorem 13.13.

The complex projective quadric $\mathcal{N}_1(\mathbf{C}^n)$ handles rather differently.

Lemma 14.19 *For any finite n let* $z = x + \mathrm{i}y \in \mathbf{C}^n$, *where* $x, y \in \mathbf{R}^n$. *Then* $z^{(2)} = 0$ *if and only if* $x^{(2)} = y^{(2)}$ *and* $x \cdot y = 0$.

Now let $\mathbf{R}(x, y)$ denote the oriented plane spanned by any orthonormal pair (x, y) of elements of $\mathbf{R}^n$.

Proposition 14.20 *For any orthonormal pair* (x, y) *of elements of* $\mathbf{R}^n$, $\mathbf{C}(x + \mathrm{i}\,y) \in \mathcal{N}_1(\mathbf{C}^n)$ *and the map*

$$\mathcal{G}_2^+(\mathbf{R}^n) \to \mathcal{N}_1(\mathbf{C}^n);\ \mathbf{R}(x, y) \mapsto \mathbf{C}\{x + \mathrm{i}y\}$$

is well-defined and bijective.

The coset space representation

$$SO(n)/(SO(2) \times SO(n-2)) \to \mathcal{G}_2^+(\mathbf{R}^n)$$

given in Proposition 14.6 is in accord with the coset space representation

$$O(n)/(U(1) \times O(n-2)) \to \mathcal{N}_1(\mathbf{C}^n)$$

given in Theorem 14.17, since $SO(2) \cong S^1 \cong U(1)$.

Now consider $\mathcal{N}_1(\mathbf{R}^{2n}_{\text{sp}})$. In this case every line is null; so $\mathcal{N}_1(\mathbf{R}^{2n}_{\text{sp}})$ coincides with $\mathcal{G}_1(\mathbf{R}^{2n})$, for which we already have a coset space representation $O(2n)/(O(1) \times O(2n-1))$, equivalent, by Theorem 13.13, to S^{2n-1}/S^0, where the action of -1 on S^{2n-1} is the antipodal map. By Theorem 14.17 there is also a representation $U(n)/(O(1) \times U(n-1))$. This also is equivalent to S^{2n-1}/S^0 by the standard representation (Theorem 13.13 again)

$$U(n)/U(n-1) \to S^{2n-1}.$$

Finally, the same holds for $\mathcal{N}_1(\mathbf{C}^{2n}_{\text{sp}})$, which coincides with $\mathcal{G}_1(\mathbf{C}^{2n})$, for which we already have a representation $U(n)/(U(1) \times U(2n-1))$, equivalent to S^{4n-1}/S^1. Here the action of S^1 is right multiplication, S^{4n-1} being identified with the quasi-sphere $\mathcal{S}(\overline{\mathbf{C}}^{2n})$ in $\mathbf{C}^{2n}$. Theorem 14.17 provides the alternative representation $Sp(n)/(U(1) \times Sp(n-1))$, also equivalent to S^{4n-1}/S^1 via the standard representation (Theorem 13.13 yet again)

$$Sp(n)/Sp(n-1) \to S^{4n-1}.$$

Exercises

14.1 Show that the fibres of the restriction of the Hopf map

$$\mathbf{C}^2 \to \mathbf{C}P^1;\ (z_0, z_1) \mapsto [z_0, z_1]$$

to the sphere $S^3 = \{(z_0, z_1) \in \mathbf{C}^2 : \overline{z_0}z_0 + \overline{z_1}z_1 = 1\}$ are circles, any two of which link. (Cf. Exercise 5.8 and Hopf (1931) and (1935)).

14.2 Show that the fibres of the restriction of the Hopf map

$$\mathbf{H}^2 \to \mathbf{H}P^1;\ (q_0, q_1) \mapsto [q_0, q_1]$$

to the sphere $S^7 = \{(q_0, q_1) \in \mathbf{H}^2 : \overline{q_0}q_0 + \overline{q_1}q_1 = 1\}$ are 3-spheres, any two of which link.

14.3 Prove that the map $\mathbf{R}P^3 \to SO(3)$, induced by the map $\rho : \mathbf{H}^\bullet \to SO(3);\ q \mapsto \rho_q$ of Proposition 8.21, is bijective. In this representation of $SO(3)$ by $\mathbf{R}P^3$ how are the rotations of $\mathbf{R}^3$ about a specified axis through 0 represented?

15
Clifford algebras

We saw in Chapter 8 how well-adapted the algebra of quaternions is to the study of the groups $SO(3)$ and $SO(4)$. In either case the centre of interest is a real quadratic space X, in the one case $\mathbf{R}^3$ and in the other case $\mathbf{R}^4$, and in either case the real associative algebra $\mathbf{H}$ contains both $\mathbf{R}$ and X as linear subspaces, there being an anti-involution, namely conjugation, of the algebra, such that, for all $x \in X$,

$$\bar{x}\,x = x^{(2)}.$$

In the former case, when $\mathbf{R}^3$ is identified with the subspace of pure quaternions, this formula can also be written in the simpler form

$$x^2 = -x^{(2)}.$$

In an analogous, but more elementary way, the real algebra of complex numbers $\mathbf{C}$ may be used in the study of the group $SO(2)$.

Our aim is to put these rather special cases into a wider context. To keep the algebra simple, the emphasis is laid at first on generalising the second of the two displayed formulae. It is shown that, for any finite-dimensional real quadratic space X, there is a real associative algebra, A say, with unit element 1, containing isomorphic copies of $\mathbf{R}$ and X as linear subspaces in such a way that, for all $x \in X$, $x^2 = -x^{(2)}$.

If the algebra A is also generated as a ring by the copies of $\mathbf{R}$ and X or, equivalently, as a real algebra by $\{1\}$ and X, then A is said to be a *(real) Clifford algebra* for X (Clifford's term (1876) was *geometric algebra*). It is shown that such an algebra can be chosen so that there is also on A an algebra anti-involution

$$A \to A;\ a \mapsto a^-$$

such that, for all $x \in X$, $x^- = -x$.

To simplify notations in the above definitions, $\mathbf{R}$ and X have been identified with their copies in A. More strictly, there are linear injections $\alpha : \mathbf{R} \to A$ and $\iota : X \to A$ such that, for all $x \in X$,

$$(\iota(x))^2 = -\alpha(x^{(2)}),$$

the unit element in A being $\alpha(1)$.

The minus sign in the formula $x^2 = -x^{(2)}$ can be a bit of a nuisance at times. One could get rid of it at the outset, simply by replacing the quadratic space X by its negative. However, it turns up anyway in applications, and so we keep it in.

Proposition 15.1 *Let A be a Clifford algebra for a real quadratic space X and let W be a linear subspace of X. Then the subalgebra of A generated by W is a Clifford algebra for W.*

By Proposition 5.29 and Proposition 15.1 the existence of a Clifford algebra for an arbitrary n-dimensional quadratic space X is implied by the existence of a Clifford algebra for the neutral non-degenerate space $\mathbf{R}^{n,n}$. Such an algebra is constructed below in Corollary 15.18. (An alternative construction of a Clifford algebra for a real quadratic space X depends on the prior construction of the (infinite-dimensional) *tensor algebra* of X, regarded as a linear space. The Clifford algebra is then defined as a quotient algebra of the tensor algebra, this definition being that adopted by Chevalley (1954). For details see, for example, Atiyah, Bott and Shapiro (1964).

Clifford algebras for low-dimensional spaces

Examples of Clifford algebras are easily given for low-dimensional non-degenerate quadratic spaces.

For example, $\mathbf{R}$ itself is a Clifford algebra both for $\mathbf{R}^{0,0}$ and for $\mathbf{R}^{1,0}$, $\mathbf{C}$, regarded as a real algebra, is a Clifford algebra for $\mathbf{R}^{0,1}$, and $\mathbf{H}$, regarded as a real algebra, is a Clifford algebra for both $\mathbf{R}^{0,2}$ and $\mathbf{R}^{0,3}$, it being usual, in the former case, to identify $\mathbf{R}^{0,2}$ with the linear image in $\mathbf{H}$ of $\{\mathrm{i}, \mathrm{k}\}$, while, in the latter case, $\mathbf{R}^{0,3}$ has necessarily to be identified with the linear image of $\{\mathrm{i}, \mathrm{j}, \mathrm{k}\}$, the space of pure quaternions. Moreover, it follows easily from Exercises 4.4 and 4.8 that $\mathbf{R}(2)$ is a Clifford algebra for each of the spaces $\mathbf{R}^{2,0}$, $\mathbf{R}^{1,1}$ and $\mathbf{R}^{2,1}$.

It is provocative to arrange these examples in a table as follows.

Table 15.2 *Clifford algebras for* $\mathbf{R}^{p,q}$, *for low values of p and q*

$-p+q$	-4	-3	-2	-1	0	1	2	3	4
$p+q$									
0					$\mathbf{R}$				
1				$\mathbf{R}$		$\mathbf{C}$			
2			$\mathbf{R}(2)$		$\mathbf{R}(2)$		$\mathbf{H}$		
3		?		$\mathbf{R}(2)$		?		$\mathbf{H}$	
4	?		?		?		?		?

A complete table of Clifford algebras for the non-degenerate quadratic spaces $\mathbf{R}^{p,q}$ is given later in this chapter as Table 15.27. As can be seen from that table, one can always choose as Clifford algebra for such a space the space of endomorphisms of some finite-dimensional linear space over $\mathbf{R}$, $\mathbf{C}$, $\mathbf{H}$, ${}^2\mathbf{R}$ or ${}^2\mathbf{H}$, the endomorphism space being regarded as a real algebra.

In Chapter 16 we examine in some detail how a Clifford algebra A for a real quadratic space X may be used in the study of the group of orthogonal automorphisms of X. Here we only make two preliminary remarks.

Proposition 15.3 *Let a, $b \in X$. Then, in A,*

$$a \cdot b = -\frac{1}{2}(a\,b + b\,a).$$

In particular, a and b are mutually orthogonal if and only if a and b anti-commute.

Proof

$$\begin{aligned} 2a \cdot b &= a \cdot a + b \cdot b - (a-b) \cdot (a-b) \\ &= -a^2 - b^2 + (a-b)^2 \\ &= -a\,b - b\,a. \end{aligned}$$

□

Proposition 15.4 *Let $a \in X$. Then a is invertible in A if and only if it is invertible with respect to the scalar product, when $a^{-1} = -a^{(-1)}$.*

Proof $\Rightarrow$: Let $b = a^{-1}$, in A. Then $a^{(2)}\, b = -a^2\, b = -a$, implying that $a^{(2)} \neq 0$ and that $b = a^{(-1)}$.

$\Leftarrow$: Let $b = a^{(-1)} = -(a^{(2)})^{-1}\, a$. Then $b\, a = -(a^{(2)})^{-1} a^2 = 1$. Similarly, $a\, b = 1$. That is, $b = a^{-1}$. □

Notice that the inverse in A of an element of X is also an element of X.

Orthonormal subsets

One of the characteristic properties of a Clifford algebra may be re-expressed in terms of an orthonormal basis as follows.

Proposition 15.5 *Let X be a finite-dimensional real quadratic space with an orthonormal basis $\{e_i : 0 \leq i < n\}$, where $n = \dim X$, and let A be a real associative algebra with unit element* 1 *containing* **R** *and X as linear subspaces. Then $x^2 = -x^{(2)}$, for all $x \in X$, if and only if*

$$e_i^2 = -e_i^{(2)}, \textit{ for all } i,$$

$$e_i\, e_j + e_j\, e_i = 0 \textit{ for all distinct } i \textit{ and } j.$$

This prompts the following definition.

An *orthonormal subset* of a real associative algebra A with unit element 1 is a linearly independent subset S of mutually anti-commuting elements of A, the square a^2 of any element $a \in S$ being 0, 1 or -1.

Proposition 15.6 *Let S be a subset of mutually anti-commuting elements of the algebra A such that the square a^2 of any element $a \in S$ is* 1 *or* -1. *Then S is an orthonormal subset in A*

All that has to be verified is the linear independence of S.

An orthonormal subset S each of whose elements is invertible, as in Proposition 15.6, is said to be *non-degenerate.* If p of the elements of S have square $+1$ and if the remaining q have square -1, then S is said to be of *type* (p, q).

Proposition 15.7 *Let X be the linear image of an orthonormal subset S of a real associative algebra A. Then there is a unique quadratic form on X such that, for all $a \in S, a^{(2)} = -a^2$, and, if S is of type (p, q), X with this structure is isomorphic to* $\mathbf{R}^{p,q}$. *If S also generates A, then A is a Clifford algebra for the quadratic space X.*

The dimension of a Clifford algebra

There is an obvious upper bound to the linear dimension of a Clifford algebra for a finite-dimensional real quadratic space.

It is convenient first of all to introduce the following notation. Suppose that $(e_i : 0 \leq 1 < n)$ is an n-tuple of elements of an associative algebra A. Then, for each naturally ordered subset I of $\mathbf{n} = \{0, 1, ..., n-1\}$, e_I will denote the product $\prod_{i \in I} e_i$, with $e_\emptyset = 1$, where $\emptyset$ denotes the empty set. In particular, $e_\mathbf{n} = \prod_{i \in \mathbf{n}} e_i$.

Proposition 15.8 *Let A be a real associative algebra with unit element 1 (identified with $1 \in \mathbf{R}$) and suppose that $(e_i : i \in \mathbf{n})$ is an n-tuple of elements of A generating A such that, for any $i, j \in \mathbf{n}$,*

$$e_i e_j + e_j e_i \in \mathbf{R}.$$

Then the set $\{e_I : I \subset \mathbf{n}\}$ spans A linearly.

Corollary 15.9 *Let A be a Clifford algebra for an n-dimensional quadratic space X. Then* $\dim A \leq 2^n$.

The following theorem gives the complete set of possible values for $\dim A$, when X is non-degenerate.

Theorem 15.10 *Let A be a Clifford algebra for an n-dimensional non-degenerate real quadratic space X of signature (p, q). Then* $\dim A = 2^n$ *or* 2^{n-1}, *the lower value being a possibility only if $p - q - 1$ is divisible by 4, in which case n is odd and $e_\mathbf{n} = +1$ or -1 for any ordered orthonormal basis $(e_i : i \in \mathbf{n})$ for X.*

Proof Let $(e_i : i \in \mathbf{n})$ be an ordered orthonormal basis for X. Then, for each $I \subset \mathbf{n}$, e_I is invertible in A and so is non-zero.

To prove that the set $\{e_I : I \subset \mathbf{n}\}$ is linearly independent, it is enough to prove that if there are real numbers λ_I, for each $I \subset \mathbf{n}$, such that $\sum_{I \subset \mathbf{n}} \lambda_I(e_I) = 0$, then, for each $J \subset \mathbf{n}$, $\lambda_J = 0$. Since, for any $J \subset \mathbf{n}$,

$$\sum_{I \subset \mathbf{n}} \lambda_I(e_I) = 0 \Leftrightarrow \sum_{I \subset \mathbf{n}} \lambda_I(e_I)(e_J)^{-1} = 0,$$

thus making λ_J the coefficient of $e_\emptyset$, it is enough to prove that

$$\sum_{I \subset \mathbf{n}} \lambda_I(e_I) = 0 \Rightarrow \lambda_\emptyset = 0.$$

Suppose, therefore, that $\sum_{I \subset \mathbf{n}} \lambda_I(e_I) = 0$. We assert that this implies either that $\lambda_\emptyset = 0$, or, if n is odd, that $\lambda_\emptyset + \lambda_{\mathbf{n}}(e_{\mathbf{n}}) = 0$. This is because, for each $i \in \mathbf{n}$ and each $I \subset \mathbf{n}$, e_i either commutes or anti-commutes with e_I. So

$$\sum_{I \subset \mathbf{n}} \lambda_I(e_I) = 0 \Rightarrow \sum_{I \subset \mathbf{n}} \lambda_I e_i(e_I)e_i^{-1} = \sum_{I \subset \mathbf{n}} \zeta_{i,i}\lambda_I(e_I) = 0$$

where $\zeta_{I,i} = 1$ or -1 according as e_i commutes or anti-commutes with e_I. It follows that $\sum_I \lambda_I(e_I) = 0$, with the summation over all I such that e_I commutes with *each* e_i. Now there are at most only two such subsets of $\mathbf{n}$, namely $\emptyset$, since $e_\emptyset = 1$, and, when n is odd, $\mathbf{n}$ itself. This proves the assertion.

From this it follows that the subset $\{e_I : I \subset \mathbf{n}, \#I \text{ even}\}$ is linearly independent in A for all n and that the subset $\{e_I : I \subset \mathbf{n}\}$ is linearly independent for all even n. For n odd, either $\{e_I : I \subset \mathbf{n}\}$ is linearly independent or $e_{\mathbf{n}}$ is real.

To explore this last possibility further let $n = p + q = 2k + 1$. Then $(e_{\mathbf{n}})^2 = (e_{2k+1})^2 = (-1)^{k(2k+1)+q}$. But, since $e_{\mathbf{n}}$ is real, $(e_{\mathbf{n}})^2$ is positive. Therefore $(e_{\mathbf{n}})^2 = 1$, implying that $k(2k+1)+q$ is divisible by 2, that is, $4k^2 + p + 3q - 1$, or, equivalently, $p - q - 1$, is divisible by 4. Conversely, if $p - q - 1$ is divisible by 4, n is odd.

Finally, if $e_{\mathbf{n}} = \pm 1$, n being odd, then, for each $I \subset \mathbf{n}$ with $\#i$ odd, $e_I = \pm e_{\mathbf{n}\setminus I}$. Since the subset $\{e_I : I \subset \mathbf{n}, \#I \text{ even}\}$ is linearly independent in A, it follows in this case that $\dim A = 2^{n-1}$. □

The lower value of the dimension of a Clifford algebra of a non-degenerate finite-dimensional real quadratic space does occur; for, as has already been noted, $\mathbf{R}$ is a Clifford algebra for $\mathbf{R}^{1,0}$ and $\mathbf{H}$ is a Clifford algebra for $\mathbf{R}^{0,3}$.

Corollary 15.11 indicates how Theorem 15.10 is used in practice.

Corollary 15.11 *Let A be a real associative algebra with an orthonormal subset $\{e_i : i \in \mathbf{n}\}$ of type (p, q), where $p+q = n$. Then, if $\dim A = 2^{n-1}$, A is a Clifford algebra for $\mathbf{R}^{p,q}$ while, if $\dim A = 2^n$ and if $e_{\mathbf{n}} \neq \pm 1$, then A is again a Clifford algebra for $\mathbf{R}^{p,q}$, it being necessary to check that $e_{\mathbf{n}} \neq \pm 1$ only when $p - q - 1$ is divisible by 4.*

For example, $\mathbf{R}(2)$ is now seen to be a Clifford algebra for $\mathbf{R}^{2,0}$ simply because $\dim \mathbf{R}(2) = 2^2$ and because the set $\left\{\begin{pmatrix} 1 & 0 \\ 0 & 1 \end{pmatrix}, \begin{pmatrix} 1 & 0 \\ 0 & -1 \end{pmatrix}\right\}$ is an orthonormal subset of $\mathbf{R}(2)$ of type $(2,0)$.

Proposition 15.12 *The real algebra ${}^2\mathbf{R}$ is a Clifford algebra for $\mathbf{R}^{1,0}$.*

Universal Clifford algebras

The special role played by a Clifford algebra of dimension 2^n, for an n-dimensional real quadratic space X, is brought out by the following theorem.

Theorem 15.13 *Let A be a Clifford algebra for an n-dimensional real quadratic space X, with $\dim A = 2^n$, let B be a Clifford algebra for a real quadratic space Y, and suppose that $t : X \to Y$ is an orthogonal map. Then there is a unique algebra map $t_A : A \to B$ sending $1_{(A)}$ to $1_{(B)}$ and a unique algebra-reversing map $(t_A)^\sim : A \to B$ sending $1_{(A)}$ to $1_{(B)}$ such that the diagrams*

$$\begin{array}{ccc} X & \xrightarrow{t} & Y \\ \downarrow{\scriptstyle \iota} & & \downarrow{\scriptstyle \iota} \\ A & \xrightarrow{t_A} & B \end{array} \quad \textit{and} \quad \begin{array}{ccc} X & \xrightarrow{t} & Y \\ \downarrow{\scriptstyle \iota} & & \downarrow{\scriptstyle \iota} \\ A & \xrightarrow{t_A^\sim} & B \end{array}$$

commute.

Proof We construct t_A, the construction of $(t_A)^\sim$ being similar.

Let $(e_i : i \in \mathbf{n})$ be an ordered orthonormal basis for X. Then, if t_A exists, $t_A(e_I) = \prod_{i\in I} t(e_i)$, for each non-null $I \subset \mathbf{n}$, while $t_A(1_{(A)}) = 1_{(B)}$, by hypothesis. Conversely, since the set $\{e_I : I \subset \mathbf{n}\}$ is a basis for A, there is a *unique linear* map $t_A : A \to B$ such that, for each $I \subset \mathbf{n}$, $t_A(e_I) = \prod_{i\in I} t(e_i)$.

In particular, since, for each $i \in \mathbf{n}$, $t_A(e_I) = t(e_i)$, the diagram $\begin{array}{ccc} X & \xrightarrow{t} & Y \\ \downarrow{\scriptstyle \iota} & & \downarrow{\scriptstyle \iota} \\ A & \xrightarrow{t_A} & B \end{array}$ commutes. It only remains to check that t_A respects products, and for this it is enough to check that, for any $I, J \subset \mathbf{n}$,

$$t_A((e_I)(e_J)) = t_A(e_I)t_A(e_J).$$

The verification is straightforward, if slightly tedious, and depends on the fact that, since t is orthogonal, $t(e_i))^2 = e_i^2$, for any $i \in \mathbf{n}$, and $t(e_j)t(e_i) = -t(e_i)t(e_j)$, for any distinct $i, j \in \mathbf{n}$. The final details are left as an exercise. □

Theorem 15.13 is amplified and extended in Theorem 15.32 in the particular case that $Y = X$ and $B = A$. Immediate corollaries include the following.

Corollary 15.14 *Let A and B be 2^n-dimensional Clifford algebras for an n-dimensional real quadratic space X. Then $A \cong B$.*

Corollary 15.15 *Any Clifford algebra B for an n-dimensional quadratic space X is isomorphic to some quotient of any given 2^n-dimensional Clifford algebra A for X.*

Proof What has to be verified is that the map $(1_X)_A : A \to B$ is a *surjective* algebra map. This verification is left as an exercise. □

A 2^n-dimensional real Clifford algebra for an n-dimensional quadratic space X is said to be a *universal* real Clifford algebra for X. Since any two universal Clifford algebras for X are isomorphic, and since the isomorphism between them is essentially unique, one often speaks loosely of *the* universal Clifford algebra for X. The existence of such an algebra for any X has, of course, still to be proved.

We shall denote the universal real Clifford algebra for the quadratic space $\mathbf{R}^{p,q}$ by $\mathbf{R}_{p,q}$.

The construction of universal Clifford algebras

Corollary 15.11 may now be applied to the construction of universal Clifford algebras for each non-degenerate quadratic space $\mathbf{R}^{p,q}$. The following elementary proposition is used frequently.

Proposition 15.16 *Let a and b be elements of an associative algebra A with unit element 1. Then, if a and b commute, $(ab)^2 = a^2b^2$, so that, in particular,*

$$\begin{aligned} a^2 = b^2 = -1 &\Rightarrow (ab)^2 = 1, \\ a^2 = -1 \text{ and } b^2 = 1 &\Rightarrow (ab)^2 = -1, \\ a^2 = b^2 = 1 &\Rightarrow (ab)^2 = 1, \end{aligned}$$

while, if a and b anti-commute, $(ab)^2 = -a^2b^2$, and

$$\begin{aligned} a^2 = b^2 = -1 &\Rightarrow (ab)^2 = -1, \\ a^2 = -1 \text{ and } b^2 = 1 &\Rightarrow (ab)^2 = 1, \\ a^2 = b^2 = 1 &\Rightarrow (ab)^2 = -1. \end{aligned}$$

The first stage in the construction is to show how to construct the universal Clifford algebra $\mathbf{R}_{p+1,q+1}$ for $\mathbf{R}^{p+1,q+1}$, given $\mathbf{R}_{p,q}$, the universal Clifford algebra for $\mathbf{R}^{p,q}$. This leads directly to the existence theorem.

Proposition 15.17 *Let X be an* **A**-*linear space, where* $\mathbf{A} = \mathbf{K}$ *or* ${}^2\mathbf{K}$ *and* $\mathbf{K} = \mathbf{R}, \mathbf{C}$ *or* $\mathbf{H}$, *and let S be an orthonormal subset of* $\operatorname{End} X$ *of type* (p, q), *generating* $\operatorname{End} X$ *as a real algebra. Then the set of matrices*

$$\left\{\begin{pmatrix} a & 0 \\ 0 & -a \end{pmatrix} : a \in S\right\} \cup \left\{\begin{pmatrix} 0 & 1 \\ 1 & 0 \end{pmatrix}, \begin{pmatrix} 0 & -1 \\ 1 & 0 \end{pmatrix}\right\}$$

is an orthonormal subset of $\operatorname{End} X^2$ *of type* $(p+1, q+1)$, *generating* $\operatorname{End} X^2 \otimes_{\mathbf{R}} \mathbf{R}(2)$ *as a real algebra.*

Corollary 15.18 *For each n, the endomorphism algebra* $\mathbf{R}(2^n)$ *is a universal Clifford algebra for the neutral non-degenerate space* $\mathbf{R}^{n,n}$. *That is,* $\mathbf{R}_{n,n} \cong \mathbf{R}(2^n)$.

Proof By induction. The basis is that **R** is a universal Clifford algebra for $\mathbf{R}^{0,0}$, and the step is Proposition 15.17. □

We return to Proposition 15.17 in Chapter 18.

Theorem 15.19 (Existence theorem.) *Every finite-dimensional quadratic space has a universal Clifford algebra.*

Proof This follows at once from the remarks following Proposition 15.1, from Proposition 15.8 and from Corollary 15.18. □

One might conjecture that, for any finite p, q, $\mathbf{R}_{q,p}$ is isomorphic to $\mathbf{R}_{p,q}$. This is not quite the case, the true state of affairs being given by the corollary to the next proposition.

Proposition 15.20 *Let S be an orthonormal subset of type* $(p+1, q)$, *generating a real associative algebra A. Then, for any* $a \in S$ *with* $a^2 = 1$, *the set*

$$\{b\,a : b \in S \backslash \{a\}\} \cup \{a\}$$

is an orthonormal subset of type $(q+1, p)$ *generating A.*

Corollary 15.21 *The universal Clifford algebras* $\mathbf{R}_{p+1,q}$ *and* $\mathbf{R}_{q+1,p}$ *are isomorphic.*

Proposition 15.22 *The universal Clifford algebra* $\mathbf{R}_{0,q}$ *is isomorphic to* $\mathbf{R}$, $\mathbf{C}$, $\mathbf{H}$, ${}^2\mathbf{H}$, *or* $\mathbf{H}(2)$, *according as* $q = 0, 1, 2, 3$ *or* 4.

Proof By Corollary 15.11 it is enough, in each case, to exhibit an orthonormal subset of the appropriate type with the product of its members, in any order, not equal to 1 or -1, for each algebra has the correct real dimension, namely 2^n. Appropriate orthonormal subsets are

$$\emptyset \text{ for } \mathbf{R},\ \{\mathrm{i}\} \text{ for } \mathbf{C},\ \{\mathrm{i}, \mathrm{k}\} \text{ for } \mathbf{H},$$

$$\left\{ \begin{pmatrix} \mathrm{i} & 0 \\ 0 & -\mathrm{i} \end{pmatrix}, \begin{pmatrix} \mathrm{j} & 0 \\ 0 & -\mathrm{j} \end{pmatrix}, \begin{pmatrix} \mathrm{k} & 0 \\ 0 & -\mathrm{k} \end{pmatrix} \right\} \text{ for } {}^2\mathbf{H},$$

and $$\left\{ \begin{pmatrix} \mathrm{i} & 0 \\ 0 & -\mathrm{i} \end{pmatrix}, \begin{pmatrix} \mathrm{j} & 0 \\ 0 & -\mathrm{j} \end{pmatrix}, \begin{pmatrix} \mathrm{k} & 0 \\ 0 & -\mathrm{k} \end{pmatrix}, \begin{pmatrix} 0 & -1 \\ 1 & 0 \end{pmatrix} \right\} \text{ for } \mathbf{H}(2). \qquad \square$$

Proposition 15.23 $\mathbf{R}_{0,1} \cong \mathbf{C}$, $\mathbf{R}_{3,0} \cong \mathbf{R}_{1,2} \cong \mathbf{C} \otimes \mathbf{R}(2) \cong \mathbf{C}(2)$, *while* $\mathbf{R}_{3,1} \cong \mathbf{R}_{2,2} \cong \mathbf{R}(4)$.

This completes the construction of the algebras $\mathbf{R}_{p,q}$, for $p + q \leq 4$.

Finally, here is a more sophisticated result, leading to the 'periodicity theorem'.

Proposition 15.24 *Let* $S = \{e_i : i \in \mathbf{4}\}$ *be an orthonormal subset of type* $(0, 4)$ *of an associative algebra* A *with unit element* 1 *and let* R *be an orthonormal subset of type* (p, q) *of* A *such that each element of* S *anti-commutes with every element of* R. *Then there exists an orthonormal subset* R' *of type* (p, q) *such that each element of* S *commutes with every element of* R'. *Conversely, the existence of* R' *implies the existence of* R.

Proof Let $a = e_4 = e_0 e_1 e_2 e_3$ and let $R' = \{a\,b : b \in \mathbf{R}\}$. Since a commutes with every element of R and anti-commutes with every element of S and since $a^2 = 1$, it follows at once that R' is of the required form. The converse is similarly proved. $\square$

Corollary 15.25 *For all* p, q,

$$\mathbf{R}_{p,q+4} \cong \mathbf{R}_{p,q} \otimes \mathbf{R}_{0,4} \cong \mathbf{H}(2).$$

For example, by Proposition 11.9,

$$\begin{array}{lllll} & \mathbf{R}_{0,5} & \cong & \mathbf{C} \otimes \mathbf{H}(2) & \cong & \mathbf{C}(4), \\ & \mathbf{R}_{0,6} & \cong & \mathbf{H} \otimes \mathbf{H}(2) & \cong & \mathbf{R}(8), \\ & \mathbf{R}_{0,7} & \cong & {}^2\mathbf{H} \otimes \mathbf{H}(2) & \cong & {}^2\mathbf{R}(8), \\ \text{and} & \mathbf{R}_{0,8} & \cong & \mathbf{H}(2) \otimes \mathbf{H}(2) & \cong & \mathbf{R}(16). \end{array}$$

Corollary 15.26 (The periodicity theorem.) *For all finite p, q,*

$$\mathbf{R}_{p,q+8} \cong \mathbf{R}_{p,q} \otimes \mathbf{R}(16).$$

By putting together Propositions 15.22, 15.12, 15.17, and 15.20, and these last two corollaries, we can construct any $\mathbf{R}_{p,q}$. Table 15.27 shows them all, for $0 \leq p, q < 7$.

Table 15.27

$p \downarrow$, $q \rightarrow$							
$\mathbf{R}$	$\mathbf{C}$	$\mathbf{H}$	$^2\mathbf{H}$	$\mathbf{H}(2)$	$\mathbf{C}(4)$	$\mathbf{R}(8)$	$^2\mathbf{R}(8)$
$^2\mathbf{R}$	$\mathbf{R}(2)$	$\mathbf{C}(2)$	$\mathbf{H}(2)$	$^2\mathbf{H}(2)$	$\mathbf{H}(4)$	$\mathbf{C}(8)$	$\mathbf{R}(16)$
$\mathbf{R}(2)$	$^2\mathbf{R}(2)$	$\mathbf{R}(4)$	$\mathbf{C}(4)$	$\mathbf{H}(4)$	$^2\mathbf{H}(4)$	$\mathbf{H}(8)$	$\mathbf{C}(16)$
$\mathbf{C}(2)$	$\mathbf{R}(4)$	$^2\mathbf{R}(4)$	$\mathbf{R}(8)$	$\mathbf{C}(8)$	$\mathbf{H}(8)$	$^2\mathbf{H}(8)$	$\mathbf{H}(16)$
$\mathbf{H}(2)$	$\mathbf{C}(4)$	$\mathbf{R}(8)$	$^2\mathbf{R}(8)$	$\mathbf{R}(16)$	$\mathbf{C}(16)$	$\mathbf{H}(16)$	$^2\mathbf{H}(16)$
$^2\mathbf{H}(2)$	$\mathbf{H}(4)$	$\mathbf{C}(8)$	$\mathbf{R}(16)$	$^2\mathbf{R}(16)$	$\mathbf{R}(32)$	$\mathbf{C}(32)$	$\mathbf{H}(32)$
$\mathbf{H}(4)$	$^2\mathbf{H}(4)$	$\mathbf{H}(8)$	$\mathbf{C}(16)$	$\mathbf{R}(32)$	$^2\mathbf{R}(32)$	$\mathbf{R}(64)$	$\mathbf{C}(64)$
$\mathbf{C}(8)$	$\mathbf{H}(8)$	$^2\mathbf{H}(8)$	$\mathbf{H}(16)$	$\mathbf{C}(32)$	$\mathbf{R}(64)$	$^2\mathbf{R}(64)$	$\mathbf{R}(128)$

The pattern from the top left to the bottom right of the table is derived from Proposition 15.17, while the symmetry about the line with equation $-p + q = -1$ is derived from Proposition 15.20. Squares like those in Table 15.27 have already made a brief appearance at the end of Chapter 11. There are clearly (non-unique) algebra injections $\mathbf{R}_{p,q} \to \mathbf{R}_{p+1,q}$ and $\mathbf{R}_{p,q} \to \mathbf{R}_{p,q+1}$, for any p, q such that the squares commute.

Table 15.27 exhibits each of the universal Clifford algebras $\mathbf{R}_{p,q}$ as the real algebra of endomorphisms of a right $\mathbf{A}$-linear space V of the form $\mathbf{A}^m$, where $\mathbf{A} = \mathbf{R}$, $\mathbf{C}$, $\mathbf{H}$, $^2\mathbf{R}$ or $^2\mathbf{H}$. This space is called the *(real) spinor space* or *space of (real) spinors* of the orthogonal space $\mathbf{R}^{p,q}$. It is identifiable with a minimal left ideal of the algebra, and to that extent is non-unique. However it can be proved that any two minimal left ideals are equivalent, so that the slight ambiguity in the definition in practice is unimportant.

Proposition 15.28 *Let* $\mathbf{R}_{p,q} = \mathbf{A}(m)$, *according to Table 15.27 or its extension by Corollary 15.26. Then the representative in* $\mathbf{A}(m)$ *of any element of the standard orthonormal basis for* $\mathbf{R}^{p+q}$ *is orthogonal with respect to the standard positive-definite correlation on* $\mathbf{A}^m$.

Proof This follows from Proposition 11.9 and Corollary 11.11, and its truth, readily checked, for small values of p and q. □

When $\mathbf{K}$ is a double field (${}^2\mathbf{R}$ or ${}^2\mathbf{H}$), the $\mathbf{K}$-linear spaces $V(1,0)$ and $V(0,1)$ are called the *(real) half-spinor spaces* or *spaces of (real) half-spinors*, the endomorphism algebra of either being a non-universal Clifford algebra of the appropriate orthogonal space.

Complex Clifford algebras

The real field may be replaced throughout the above discussion by the field of complex numbers $\mathbf{C}$, and indeed by any commutative field, though the case of characteristic 2 needs careful handling – we ignore such matters here! The notation $\mathbf{C}_n$ will denote the *universal complex Clifford algebra* for $\mathbf{C}^n$ unique up to isomorphism.

Proposition 15.29 *For any finite p, q with $n = p + q$, $\mathbf{C}_n \cong \mathbf{R}_{p,q} \otimes_{\mathbf{R}} \mathbf{C}$, $\cong$ denoting a real algebra isomorphism.*

Corollary 15.30 *For any finite k, $\mathbf{C}_{2k} \cong \mathbf{C}(2^k)$ and $\mathbf{C}_{2k+1} \cong {}^2\mathbf{C}(2^k)$.*

The *complex spinor* and *half-spinor spaces* are defined analogously to their real counterparts.

Superfields

A further generalisation of the concept of a Clifford algebra involves the concept of a superfield. A *superfield*, $\mathbf{L}^\alpha$, with fixed field $\mathbf{K}$, consists of a commutative algebra $\mathbf{L}$ with unit element 1 over a commutative field $\mathbf{K}$ and an involution α of $\mathbf{L}$, whose set of fixed points is the set of scalar multiples of 1, identified as usual with $\mathbf{K}$. (The algebra $\mathbf{L}$ need not be a field.) Examples include $\mathbf{R}$, $\overline{\mathbf{C}}$ and ${}^2\mathbf{R}^\sigma$, each with fixed field $\mathbf{R}$, and $\mathbf{C}$ and ${}^2\mathbf{C}^\sigma$, each with fixed field $\mathbf{C}$, where in the latter two examples σ denotes the swap involution.

Let X be a finite-dimensional quadratic space over a commutative field $\mathbf{K}$. Let $\mathbf{L}^\alpha$ be a superfield with fixed field $\mathbf{K}$ and let A be an associative $\mathbf{L}$-algebra with unit element, the algebra $\mathbf{L}$ being identified with the subalgebra generated by $\mathbf{L}$ and the unit element. Then A is said to be an *$\mathbf{L}^\alpha$-Clifford algebra* for X if it contains X as a $\mathbf{K}$-linear subspace in such a way that, for all $x \in X$, $x^{(2)} = -x^{(2)}$, provided also that A is generated as a ring by $\mathbf{L}$ and X or, equivalently, as an $\mathbf{L}$-algebra by 1 and X.

All that has been said before about real Clifford algebras generalises to $\mathbf{L}^\alpha$-Clifford algebras also. The notations $\overline{\mathbf{C}}_{p,q}$ and ${}^2\mathbf{R}^\sigma_{p,q}$ will denote the universal $\overline{\mathbf{C}}$- and ${}^2\mathbf{R}^\sigma$-Clifford algebras for $\mathbf{R}^{p,q}$, and the notation ${}^2\mathbf{C}^\sigma_n$ the universal ${}^2\mathbf{C}$-Clifford algebra for $\mathbf{C}^n$, for any finite p, q, n.

Proposition 15.31 *Let $\mathbf{L}^\alpha$ be a superfield with fixed field $\mathbf{K}$, let X be a $\mathbf{K}$-quadratic space and let A and B be universal $\mathbf{K}$- and $\mathbf{L}^\alpha$-Clifford algebras, respectively, for X. Then, as $\mathbf{K}$-algebras, $B \cong A \otimes_{\mathbf{K}} \mathbf{L}$.*

Note that, as complex algebras, $\overline{\mathbf{C}}_{p,q}$ and $\mathbf{C}_n$ are isomorphic, for any finite n, p, q with $n = p + q$. The detailed construction of the tables of $\mathbf{L}^\alpha$-Clifford algebras is left to the reader.

Involutions and anti-involutions

Certain involutions and anti-involutions of Clifford algebras play major roles in the description of their structure. These arise as corollaries of the following theorem which amplifies and extends Theorem 15.13 in various ways, in the particular case that $Y = X$ and $B = A$.

Theorem 15.32 *Let A be a universal $\mathbf{L}^\alpha$-Clifford algebra for a finite-dimensional $\mathbf{K}$-orthogonal space X, $\mathbf{L}^\alpha$ being a superfield with involution α and fixed commutative field $\mathbf{K}$. Then, for any orthogonal automorphism $t : X \to X$, there is a unique $\mathbf{L}$-algebra automorphism $t_A : A \to A$, sending any $\lambda \in \mathbf{L}$ to λ, and a unique $\mathbf{K}$-algebra anti-automorphism $t_A^{\sim} : A \to A$, sending any λ to λ^α, such that the diagrams*

$$\begin{array}{ccc} X & \xrightarrow{t} & X \\ \downarrow{\scriptstyle \iota} & & \downarrow{\scriptstyle \iota} \\ A & \xrightarrow{t_A} & A \end{array} \qquad \text{and} \qquad \begin{array}{ccc} X & \xrightarrow{t} & X \\ \downarrow{\scriptstyle \iota} & & \downarrow{\scriptstyle \iota} \\ A & \xrightarrow{t_A^{\sim}} & A \end{array}$$

commute. Moreover, $(1_X)_A = 1_A$ and, for any $t, u \in O(X)$,

$$(u\,t)_A = u_A\, t_A = u_A^{\sim} t_A^{\sim}.$$

If t is an orthogonal involution of X, then t_A is an algebra involution of A and $t_A^{\sim}$ is an algebra anti-involution of A.

The involution of A induced by the orthogonal involution -1_X will be denoted by $a \mapsto \widehat{a}$ and called the *main involution* or the *grade involution* of A. The anti-involutions of A induced by the orthogonal involutions 1_X and -1_X will be denoted by $a \mapsto a^{\sim}$ and $a \mapsto a^{-}$ and called, respectively, *reversion* and *conjugation*, $a^{\sim}$ being called the *reverse* of a and a^{-} the *conjugate* of a. (The reason for preferring $a^{\sim}$ to $\tilde{a}$ and a^{-} to $\bar{a}$ will become apparent later.)

Of the two anti-involutions conjugation is the more important. Reversion takes its name from the fact that the reverse of a product of a finite number of elements of X is just their product in the reverse order.

For example, consider $a = 1 + e_0 + e_1e_2 + e_0e_1e_2 \in \mathbf{R}_{0,3}$.

Then $\widehat{a} = 1 - e_0 + e_1e_2 - e_0e_1e_2,$

$a^{\sim} = 1 + e_0 + e_2e_1 + e_2e_1e_0 = 1 + e_0 - e_1e_2 - e_0e_1e_2,$

while $a^{-} = 1 - e_0 + e_2e_1 - e_2e_1e_0 = 1 - e_0 - e_1e_2 + e_0e_1e_2.$

Proposition 15.33 *Let A be a universal $\mathbf{L}^{\alpha}$-Clifford algebra for a finite-dimensional $\mathbf{K}$-quadratic space X, $\mathbf{L}^{\alpha}$ being a superfield with commutative fixed field $\mathbf{K}$. Then, for any $a \in A$, $a^{-} = (\widehat{a})^{\sim} = \widehat{a^{\sim}}$.*

Proof Each of the three anti-involutions is the unique anti-involution of A induced by -1_X. □

Even Clifford algebras

Let A be a universal $\mathbf{L}^{\alpha}$-Clifford algebra for a finite-dimensional $\mathbf{K}$-quadratic space X. By Proposition 1.6 the main involution induces a direct sum decomposition $A^0 \oplus A^1$ of A, where

$$A^0 = \{a \in A : \widehat{a} = a\} \text{ and } A^1 = \{a \in A : \widehat{a} = -a\}.$$

Clearly A^0 is an $\mathbf{L}$-subalgebra of A. This subalgebra is called the *even Clifford algebra* for X. It is unique up to isomorphism. Any element $a \in A$ may be uniquely expressed as the sum of its *even part* $a^0 \in A^0$ and its *odd part* $a^1 \in A^1$. For $a = 1 + e_0 + e_1e_2 + e_0e_1e_2$, $a^0 = 1 + e_1e_2$ and $a^1 = e_0 + e_0e_1e_2$.

The even Clifford algebras for the non-degenerate real or complex finite-dimensional quadratic spaces are determined by the next proposition.

Proposition 15.34 *Let A be a universal $\mathbf{L}^{\alpha}$-Clifford algebra for a finite-dimensional $\mathbf{K}$-quadratic space X, $\mathbf{L}^{\alpha}$ being a superfield with fixed field $\mathbf{K}$, and let S be an orthonormal basis for X of type (p, q). Then, for any $a \in S$, the set $\{ab : b \in S \backslash \{a\}\}$ is an orthonormal subset of A^0 generating A^0, and of type $(p, q-1)$ or $(q, p-1)$, according as $a^2 = -1$ or 1. In either case, moreover, the induced isomorphism of A^0 with the universal*

L^α-Clifford algebra of a $(p+q-1)$-dimensional quadratic space respects conjugation, but not reversion.

Proof The first part is clear, by Proposition 15.16. For the last part it is enough to consider generators and to remark that if a and b are anti-commuting elements of an algebra sent to $-a$ and $-b$, respectively, by an anti-involution of the algebra, then, again by Proposition 15.16, ab is sent to $-ab$. On the other hand, if a and b are sent to a and b, respectively, by the anti-involution, then ab is sent to $-ab$ and not to ab. □

Corollary 15.35 *For any finite p, q, n,*

$$\begin{aligned} \mathbf{R}^0_{p,q+1} &\cong \mathbf{R}_{p,q}, & \mathbf{R}^0_{p+1,q} &\cong \mathbf{R}_{q,p} \\ \overline{\mathbf{C}}^0_{p,q+1} &\cong \overline{\mathbf{C}}_{p,q}, & \overline{\mathbf{C}}^0_{p+1,q} &\cong \overline{\mathbf{C}}_{q,p} \\ ({}^2\mathbf{R}^\sigma)^0_{p,q+1} &\cong ({}^2\mathbf{R}^\sigma)_{p,q}, & ({}^2\mathbf{R}^\sigma)^0_{p+1,q} &\cong ({}^2\mathbf{R}^\sigma)_{q,p}, \\ \mathbf{C}^0_{n+1} &\cong \mathbf{C}_n \quad \text{and} & ({}^2\mathbf{C}^\sigma)^0_{n+1} &\cong ({}^2\mathbf{C}^\sigma)_n. \end{aligned}$$

It follows from Corollary 15.35 in particular, that the table of the even Clifford algebras $\mathbf{R}^0_{p,q}$, with $p+q>0$, is, apart, from relabelling, the same as Table 15.27, except that there is an additional line of entries down the left-hand side matching the existing line of entries across the top row. The symmetry about the main diagonal in the table of even Clifford algebras expresses the fact that the even Clifford algebras of a finite-dimensional non-degenerate quadratic space and of its negative are isomorphic.

So far we have considered only universal Clifford algebras. The usefulness of the non-universal Clifford algebras is limited by the following proposition.

Proposition 15.36 *Let A be a non-universal Clifford algebra for a non-degenerate finite-dimensional quadratic space X. Then either 1_X or -1_X induces an anti-involution of A, but not both.*

If 1_X induces an anti-involution of A, we say that A is a Clifford algebra for X *with reversion*, and, if -1_X induces an anti-involution of A, we say that A is a Clifford algebra for X *with conjugation*.

Proposition 15.37 *The non-universal Clifford algebras for the quadratic spaces $\mathbf{R}^{0,4k+3}$ have conjugation, but not reversion, while those for the quadratic spaces $\mathbf{R}^{4k+1,0}$ have reversion, but not conjugation.*

Exercises

15.1 Let A be a Clifford algebra for a finite-dimensional null real, or complex, quadratic space X such that, for some ordered basis $(e_i : i \in \mathbf{n})$ for X, $\prod_{i \in \mathbf{n}} e_i \neq 0$. Prove that A is a universal Clifford algebra for X. (Try first the case that $n = \dim X = 3$.)

(The universal Clifford algebra, $\bigwedge X$, for a finite-dimensional linear space X, regarded as a null quadratic space by having assigned to it the zero quadratic form, is called the *exterior* or *Grassmann* algebra for X, Grassmann's term being the *extensive* algebra for X (Grassmann (1844)). The square of any element of X in $\bigwedge X$ is 0 and any two elements of X anti-commute.)

15.2 Let X be a real or complex n-dimensional linear space and let a be an element of $\bigwedge X$ expressible in terms of some basis $\{e_i : i \in \mathbf{n}\}$ for X as a linear combination of k-fold products of the e_i's for some positive integer k. Show that if $\{f_i : i \in \mathbf{n}\}$ is any other basis for X then a is a linear combination of k-fold products of the f_i's. Show by an example that the analogous proposition is false for an element of a universal Clifford algebra of a non-degenerate real or complex quadratic space.

15.3 Let X be as in Exercise 15.2. Verify that the set of elements of $\bigwedge X$ expressible in terms of a basis $\{e_i : i \in \mathbf{n}\}$ for X as a linear combination of k-fold products of the e_i's is a linear space of dimension $\binom{n}{k}$ where $\binom{n}{k}$ is the coefficient of x^k in the polynomial $(1 + x)^n$. (This linear space, which is defined by Exercise 15.2 independently of the choice of basis for X, is denoted by $\bigwedge^k X$.)

15.4 Let X be as in Exercise 15.2, let $(a_i : i \in \mathbf{n})$ be a k-tuple of elements of X, let $(e_i : i \in \mathbf{n})$ be an ordered basis for X and let $t : X \to X$ be the linear map sending e_i to a_i, for all $i \in \mathbf{n}$. Prove that in $\bigwedge X$

$$\prod_{i \in \mathbf{n}} a_i = (\det t) \prod_{i \in \mathbf{n}} e_i.$$

15.5 Let X be as in Exercise 15.2 and let $(a_i : i \in \mathbf{k})$ be a k-tuple of elements of X. Prove that the elements of $(a_i : i \in \mathbf{k})$ are linearly independent in X if and only if, in $\bigwedge X$, $\prod_{i \in \mathbf{k}} a_k \neq 0$.

15.6 Let X be as in Exercise 15.2 and let $(a_i : i \in \mathbf{k})$ and $(b_i : i \in \mathbf{k})$ be ordered sets of k linearly independent elements of X. Prove that the k-dimensional linear subspaces of X spanned by these coincide if and only if $\prod_{i\in\mathbf{k}} b_i$ is a non-zero scalar multiple of $\prod_{i\in\mathbf{k}} a_i$.

(Consider first the case $k = 2$. In this case

$$a_0a_1 = b_0b_1 \Rightarrow a_0a_1b_0 = 0,$$

from which it follows that b_0 is linearly dependent on a_0 and a_1. It should now be easy to complete the argument, not only in this case, but in the general case also.)

15.7 Construct an injective map $\mathscr{G}_k(X) \to \mathscr{G}_1(\bigwedge^k X)$, where X is as in Exercise 15.2.

(Use Exercise 15.6. This is the link between Grassmannians and Grassmann algebras.)

16
Spin groups

We turn to applications of Clifford algebras to groups of quadratic automorphisms and rotation groups in particular. The letter X will denote a finite-dimensional real quadratic space, whose elements will be termed *vectors*, and A will normally denote a universal real Clifford algebra for X – we shall make some remarks at the end about the case where A is non-universal. For each $x \in X$, $x^{(2)} = x^{-}x = \widehat{x}\,x = -x^2$. Also, since A is universal, $\mathbf{R} \cap X = \{0\}$. The subspace $\mathbf{R} \oplus X$ of A will be denoted by Y, the space of *paravectors* of X, and the letter y will generally be reserved as notation for a paravector. The space Y will be assigned the quadratic form

$$Y \to \mathbf{R};\ y \mapsto y^{-}y.$$

It is then the orthogonal direct sum of the quadratic spaces $\mathbf{R}$ and X. If $X \cong \mathbf{R}^{p,q}$, then $Y \cong \mathbf{R}^{p,q+1}$.

Clifford groups

Our first proposition singles out a certain subset of a universal Clifford algebra A of a finite-dimensional real quadratic space X that turns out to be a subgroup of A.

Proposition 16.1 *Let g be an invertible element of A such that, for each $x \in X$, $g\,x\,\widehat{g}^{-1} \in X$. Then the map*

$$\rho_{X,g} : x \mapsto g\,x\,\widehat{g}^{-1}$$

is an orthogonal automorphism of X.

Proof For each $x \in X$,

$$(\rho_{X,g}(x))^{(2)} = (\widehat{gx\widehat{g}^{-1}})g\,x\,\widehat{g}^{-1} = \widehat{g}\,\widehat{x}\,g^{-1}\,g\,x\,\widehat{g}^{-1} = \widehat{x}\,x = x^{(2)},$$

since $\widehat{x}\,x \in \mathbf{R}$. So $\rho_{X,g}$ is an orthogonal map. Moreover, it is injective, since $g\,x\,\widehat{g}^{-1} = 0 \Rightarrow x = 0$ (this does not follow from the orthogonality of $\rho_{X,g}$ if X is degenerate). Finally, since X is finite-dimensional, $\rho_{X,g}$ must also be surjective. □

The element g will be said to *induce* or *represent* the orthogonal transformation $\rho_{X,g}$ and the set of all such elements g will be denoted by $\Gamma(X)$ or simply by Γ.

Proposition 16.2 *The subset Γ is a subgroup of A.*

Proof The closure of Γ under multiplication is obvious. That Γ is also closed with respect to inversion follows from the remark that, for any $g \in \Gamma$, the inverse of $\rho_{X,g}$ is $\rho_{X,g^{-1}}$. Of course $1_{(A)} \in \Gamma$. So Γ is a group. □

The group Γ is called the *Clifford group* (or *Lipschitz group* – see the historical remarks following Theorem 18.9) for X in the Clifford algebra A. Since the universal algebra A is uniquely defined up to isomorphism, Γ is also uniquely defined up to isomorphism.

There are similar propositions concerning the action of A on the space of paravectors $Y = \mathbf{R} \oplus X$.

Proposition 16.3 *Let g be an invertible element of A such that, for each $y \in Y$, $g\,y\,\widehat{g}^{-1} \in Y$. Then the map*

$$\rho_{Y,g} : y \mapsto g\,y\,\widehat{g}^{-1}$$

is an orthogonal automorphism of Y.

Proposition 16.4 *The subset $\Omega = \{g \in A : y \in Y \Rightarrow g\,y\,\widehat{g}^{-1} \in Y\}$ is a subgroup of A.*

From now on we suppose that X is non-degenerate, and prove that in this case every orthogonal automorphism of X is represented by an element of Γ. Recall that, by Theorem 5.15, every orthogonal automorphism of X is the composite of a finite number of hyperplane reflections.

Proposition 16.5 *Let a be an invertible element of X. Then $a \in \Gamma$, and the map $\rho_{X,a}$ is reflection in the hyperplane $(\mathbf{R}\{a\})^{\perp}$.*

Proof By Proposition 5.1, $X = \mathbf{R}\{a\} \otimes (\mathbf{R}\{a\})^\perp$, so any point of X is of the form $\lambda a + b$, where $\lambda \in \mathbf{R}$ and $b \cdot a = 0$. By Proposition 15.3, $b\,a = -a\,b$. Therefore, since $\widehat{a} = -a$,

$$\rho_{X,a}(\lambda a + b) = -a(\lambda a + b)a^{-1} = -\lambda a + b.$$

Hence the result. □

The next proposition characterises the field of real numbers within the algebra A.

Proposition 16.6 *Let $a \in A$ be such that $a x = x \widehat{a}$, for all $x \in X$, A being a universal Clifford algebra for the finite-dimensional non-degenerate real quadratic space X. Then $a \in \mathbf{R}$.*

Proof Let $a = a^0 + a^1$, where $a^0 \in A^0$ and $a^1 \in A^1$. Then, since $a x = x \widehat{a}$,

$$a^0 x = x a^0 \text{ and } a^1 x = -x a^1,$$

for all $x \in X$, in particular for each element e_i of some orthonormal basis $\{e_i : 0 \le i < n\}$ for X.

Now, by an argument used in the proof of Theorem 15.10, a^0 commutes with each e_i if and only if $a^0 \in \mathbf{R}$, and by a similar argument a^1 anti-commutes with each e_i if and only if $a^1 = 0$. So $a \in \mathbf{R}$. □

Theorem 16.7 *With X as above, the map*

$$\rho_X : \Gamma(X) \to O(X);\ g \mapsto \rho_{X,g}$$

is a surjective group map with kernel $\mathbf{R}^\bullet$.

Proof To prove that ρ_X is a group map, let $g, g' \in \Gamma$. Then, for all $x \in X$,

$$\begin{aligned}\rho_{X,gg'}(x) &= g g' x \left(\widehat{gg'}\right)^{-1} \\ &= g g' x \widehat{g'}^{-1} \widehat{g}^{-1} \\ &= \rho_{X,g}\, \rho_{X,g'}(x).\end{aligned}$$

So $\rho_{X,gg'} = \rho_{X,g}\, \rho_{X,g'}$, which is what had to be proved.

The surjectivity of ρ_X is an immediate corollary of Theorem 5.15 and Proposition 16.5.

Finally, suppose that $\rho_{X,g} = \rho_{X,g'}$, for $g, g' \in \Gamma$. Then, for all $x \in X$, $g x \widehat{g}^{-1} = g' x \widehat{g'}^{-1}$, implying that $(g^{-1}g')x = x(\widehat{g^{-1}g'})$, and therefore that

$g^{-1}g' \in \mathbf{R}$, by Proposition 16.6. Moreover, $g^{-1}g'$ is invertible and is therefore non-zero. Hence the result. □

We isolate part of this as a corollary.

Corollary 16.8 *Any element g of the Clifford group $\Gamma(X)$ of a non-degenerate finite-dimensional real quadratic space X is representable as the product of a finite number of elements of X.*

An element g of $\Gamma(X)$ represents a *rotation* of X if and only if g is the product of an *even* number of elements of X. The set of such elements will be denoted by $\Gamma^0 = \Gamma^0(X)$. An element g of Γ represents an *anti-rotation* of X if and only if g is the product of an *odd* number of elements of X. The set of such elements will be denoted by $\Gamma^1 = \Gamma^1(X)$. Clearly, $\Gamma^0 = \Gamma \cap A^0$ is a subgroup of Γ, while $\Gamma^1 = \Gamma \cap A^1$.

Since, for any $a \in A^0$, $\widehat{a} = a$, the rotation induced by an element g of Γ^0 is of the form

$$X \to X;\ x \mapsto g x g^{-1}.$$

Similarly, since, for any $a \in A^1$, $\widehat{a} = -a$, the rotation induced by an element g of Γ^1 is of the form

$$X \to X;\ x \mapsto -g x g^{-1}.$$

An analogous discussion to that just given for the group Γ can be given for the subgroup Ω of the universal Clifford algebra A for X consisting of those invertible elements g of A such that, for all $y \in Y$, $g y \widehat{g}^{-1} \in Y$. However, the properties of this group are deducible directly from the preceding discussion, by virtue of the following proposition.

The notations are as follows. As before, X denotes an n-dimensional non-degenerate real quadratic space, of signature (p, q), say. This can be considered as the subspace of $\mathbf{R}^{p,q+1}$ consisting of those elements of $\mathbf{R}^{p,q+1}$ whose last co-ordinate, labelled the nth, is zero. The subalgebra $\mathbf{R}_{p,q}$ of $\mathbf{R}_{p,q+1}$ generated by X is a universal Clifford algebra for X, as also is the even Clifford algebra $\mathbf{R}^0_{p,q+1}$, by the linear injection

$$X \to \mathbf{R}^0_{p,q+1};\ x \mapsto x\, e_n.$$

(See Proposition 15.34.) The linear space $Y = \mathbf{R} \oplus X$ is assigned the quadratic form $y \mapsto y^- y$.

Proposition 16.9 *Let $\theta : \mathbf{R}_{p,q} \to \mathbf{R}^0_{p,q+1}$ be the isomorphism of universal*

Clifford algebras induced, according to Theorem 15.13, by 1_X. *Then*

(i) *the map*

$$u : Y \to \mathbf{R}^{p,q+1};\ y \mapsto \theta(y)\,e_n^{-1}$$

is an orthogonal isomorphism,

(ii) *for any* $g \in \Omega$, $\theta(g) \in \Gamma^0(p, q+1)$ *and the diagram*

$$\begin{array}{ccc} Y & \xrightarrow{\rho'_g} & Y \\ \downarrow u & & \downarrow u \\ \mathbf{R}^{p,q+1} & \xrightarrow{\rho_{\theta(g)}} & \mathbf{R}^{p,q+1} \end{array}$$

commutes,

(iii) *the map* $\Omega \to \Gamma^0(p, q+1);\ g \mapsto \theta(g)$ *is a group isomorphism.*

Proof

(i) Since θ respects conjugation, and since $e_n^{-} e_n = 1$,

$$(\theta(y)\,e_n^{-1}) = y^{-} y, \text{ for any } y \in Y.$$

(ii) First observe that, for any $g \in \mathbf{R}_{p,q}$, $\theta(g)\,e_n = e_n\,\theta(\widehat{g})$, for the isomorphism θ and the isomorphism $g \mapsto e_n\,\theta(\widehat{g})\,e_n^{-1}$ agree on X. Now let $g \in \Omega$. Then, for any $u(y) \in \mathbf{R}^{p,q+1}$, where $y \in Y$,

$$\begin{aligned} \theta(g)(\theta(y)\,e_n^{-1})\theta(g)^{-1} &= \theta(g)\,\theta(y)\,\theta(\widehat{g})^{-1}\,e_n^{-1} \\ &= \theta(g\,y\,\widehat{g}^{-1})e_n^{-1} = u\,\rho'_g(y). \end{aligned}$$

So $\theta(g) \in \Gamma^0(p, q+1)$, and the diagram commutes.

(iii) The map is clearly a group map, since θ is an algebra isomorphism. One proves that it is invertible by showing, by essentially the same argument as in (ii), that, for any $h \in \Gamma^0(p, q+1)$, $\theta^{-1}(h) \in \Omega$.

□

Corollary 16.10 *The orthogonal transformations of* Y *represented by the elements of* Ω *are the rotations of* Y.

Since conjugation, restricted to Y, is an anti-rotation of Y, the anti-rotations of Y also are representable by elements of Ω in a simple manner.

It remains to make a few remarks about the non-universal case. We suppose that A is a non-universal Clifford algebra for X (you might have in mind $\mathbf{H}$ as a non-universal Clifford algebra for $\mathbf{R}^{0,3}$). Since the main involution is not now defined, we cannot proceed exactly as before. However in the case that we have just been discussing, $\widehat{g} = g$ or $-g$, for any $g \in \Gamma$, according as $g \in \Gamma^0$ or Γ^1. What is then true is the following.

Proposition 16.11 *Let g be an invertible element of the non-universal Clifford algebra A for X such that, for all $x \in X$, $g x g^{-1} \in X$. Then the map $X \to X;\ g x g^{-1}$ is a rotation of X, while the map $X \to X;\ x \mapsto -g x g^{-1}$ is an anti-rotation of X.*

In this case $\Gamma = \Gamma^0 = \Gamma^1$.

The discussion involving $Y = \mathbf{R} \oplus X$ requires that conjugation be defined, but if this is met by the non-universal Clifford algebra A, then A may be used also to describe the rotations of Y. The restriction to Y of conjugation is, as before, an anti-rotation of Y.

Pin groups and Spin groups

The Clifford group $\Gamma(X)$ of a quadratic space X is larger than is necessary if our interest is in representing orthogonal transformations of X. Use of a *quadratic norm* N on the Clifford algebra A leads to the definition of subgroups of Γ that are less redundant for this purpose. This quadratic norm $N : A \to A$ is defined by the formula

$$N(a) = a^{-} a, \text{ for any } a \in A,$$

A denoting, as before the universal Clifford algebra of the non-degenerate finite-dimensional real quadratic space X.

Proposition 16.12

(i) *For any $g \in \Gamma$, $N(g) \in \mathbf{R}$,*
(ii) $N(1) = 1$,
(iii) *for any g, $g' \in \Gamma$, $N(g g') = N(g) N(g')$,*
(iv) *for any $g \in \Gamma, N(g) \neq 0$ and $N(g^{-1}) = (N(g))^{-1}$,*
(v) *for any $g \in \Gamma$, there exists a unique positive real number λ such that $|N(\lambda g)| = 1$, namely $\lambda = \sqrt{(|N(g)|)^{-1}}$.*

Proof That $N(1) = 1$ is obvious. All the other statements follow directly from the observation that, by Theorem 16.7, any $g \in \Gamma$ is expressible (not necessarily uniquely) in the form

$$\prod_{i \in \mathbf{k}} x_i = x_0\, x_1 \dots x_{k-2}\, x_{k-1},$$

where, for all $i \in \mathbf{k}, x_i \in X$, k being finite; for it follows that

$$g^- = \prod_{i \in \mathbf{k}} x_{k-1-i}^- = x_{k-1}^- x_{k-2}^- \ldots x_1^- x_0^-,$$

and that

$$N(g) = g^- g = \prod_{i \in \mathbf{k}} N(x_i),$$

where, for each $i \in k$, $N(x_i) = -x_i^2 \in \mathbf{R}$. □

For X and $\Gamma = \Gamma(X)$ as above we now define

$$Pin\, X = \{g \in \Gamma : |N(g)| = 1\} \text{ and } Spin\, X = \{g \in \Gamma^0 : |N(g)| = 1\}.$$

Proposition 16.13 *As subgroups of Γ and Γ^0 respectively the groups $Pin\, X$ and $Spin\, X$ are normal subgroups, the quotient groups $\Gamma/Pin\, X$ and $\Gamma^0/Spin\, X$ each being isomorphic to $\mathbf{R}^+ = \{\lambda \in \mathbf{R} : \lambda > 0\}$.*

The next proposition asserts the important fact that the *Pin* and *Spin* groups *doubly cover* the relevant orthogonal and special orthogonal groups.

Proposition 16.14 *Let X be a non-degenerate quadratic space of positive finite dimension. Then the maps*

$$Pin\, X \to O(X);\ g \mapsto \rho_{X,g} \text{ and } Spin\, X \to SO(x);\ g \mapsto \rho_{X,g}$$

are surjective, the kernel in each case being isomorphic to S^0.

When $X = \mathbf{R}^{p,q}$ the standard notations for $\Gamma(X)$, $\Gamma^0(X)$, $Pin\, X$ and $Spin\, X$ will be $\Gamma(p, q)$, $\Gamma^0(p, q)$, $Pin(p, q)$ and $Spin(p, q)$. Since $\mathbf{R}^0_{q,p} \cong \mathbf{R}^0_{p,q}$, $\Gamma^0(q\, p) \cong \Gamma^0(p, q)$ and $Spin(q, p) \cong Spin(p, q)$. Finally, $\Gamma^0(0, n)$ is often abbreviated to $\Gamma^0(n)$ and $Spin(0, n)$ to $Spin(n)$.

That Proposition 16.12(i) is a genuine restriction on g is illustrated by the element $1 + e_4 \in \mathbf{R}_{0,4}$, since $N(1 + e_4) = 2(1 + e_4) \notin \mathbf{R}$. That the same proposition does not, in general, provide a sufficient condition for g to belong to Γ is illustrated by the element $1 + e_6 \in \mathbf{R}_{0,6}$, for, since $N(1 + e_6) = (1 - e_6)(1 + e_6) = 2$, the element is invertible, but, by explicit computation of the element $(1 + e_6)\, e_0\, (1 + e_6)^{-1}$, it can be shown that it does not belong to Γ. However, the condition is sufficient when $p+q \leq 5$, as the following proposition shows.

Proposition 16.15 *Let A be a universal Clifford algebra for a real non-degenerate quadratic space X with* $\dim X \leq 5$. *Then*

$$Spin\, X = \{g \in A^0 : N(g) = \pm 1\}.$$

Proof The proof is given in full for the hardest case, namely when $\dim X = 5$. The proofs in the other cases may be obtained from this one simply by deleting the irrelevant parts of the argument.

From the definition there is inclusion one way ($\subset$). What has to be proved, therefore, is that, for all $g \in A^0$ such that $N(g) = \pm 1$,

$$x \in X \Rightarrow g\,x\,g^{-1} \in X.$$

Let $\{e_i : i \in \mathbf{5}\}$ be an orthonormal basis for X. Then, since $X \subset A^1$ and $g \in A^0$, $x' = g\,x\,g^{-1} \in A^1$, for any $x \in X$. So there are real numbers a_i, b_{jkl}, c such that

$$x' = \sum_{i \in \mathbf{5}} a_i\, e_i + \sum_{0 \leq j < k < l < 5} b_{jkl} e_j\, e_k\, e_l + c\, e_{\mathbf{5}}.$$

Now $(x')^{-} = (g\,x\,g^{-1})^{-} = -x'$, since $g^{-1} = \pm g^{-}$, while $(e_i)^{-} = -e_i$, $(e_j\, e_k\, e_l)^{-} = e_j\, e_k\, e_l$, and $e_{\mathbf{5}}{}^{-} = -e_{\mathbf{5}}$. So, for $0 \leq j < k < l < 5, b_{jkl} = 0$. That is

$$x' = x'' + c\, e_{\mathbf{5}}, \text{ for some } x' \in X$$

The argument ends at this point if $n < 5$. Otherwise it remains to prove that $c = 0$. Now $x'^2 = x^2 \in \mathbf{R}$. Also $\prod e_{\mathbf{5}}$ commutes with each e_i and so with x''. So

$$x''^2 + 2c\,x''(e_{\mathbf{5}}) + c^2(e_{\mathbf{5}})^2 \in \mathbf{R}.$$

Since x''^2 and $c^2(e_{\mathbf{5}})^2 \in \mathbf{R}$, and $e_{\mathbf{5}} \notin \mathbf{R}$, either $c = 0$ or $x'' = 0$. Whichever is the correct alternative it is the same for every x, for, if there were an element of each kind, their sum would provide a contradiction. Since the map

$$X \to A;\ x \mapsto g\,x\,g^{-1}$$

is injective, it follows that $c = 0$. Therefore $g\,x\,g^{-1} \in X$, for each $x \in X$. □

To use Proposition 16.15 we need to know the form that conjugation takes on the Clifford algebra. This is the subject of the next chapter.

17
Conjugation

This chapter is concerned with classifying the conjugation and reversion anti-involutions of universal Clifford algebras of non-degenerate real or complex quadratic spaces. Now, according to Table 15.27 and Corollary 15.30 any such Clifford algebra is representable as an endomorphism algebra $\mathbf{A}(m)$, for some number m, where $\mathbf{A} = \mathbf{K}$ or ${}^2\mathbf{K}$, where $\mathbf{K} = \mathbf{R}$, $\mathbf{C}$ or $\mathbf{H}$. Moreover any correlation on the spinor space $\mathbf{A}^m$ induces an anti-involution of the Clifford algebra, namely the appropriate adjoint anti-involution, and conversely any anti-involution of A is so induced by a symmetric or skew correlation on the spinor space. So the problem reduces to determining in each case which anti-involution it is out of a list which we essentially already have. The job of identification is made easier by the fact that an anti-involution of an algebra is uniquely determined, by Proposition 2.8, by its restriction to any subset that generates the algebra.

The algebras $\mathbf{R}_{0,n}$

For the Clifford algebras $\mathbf{R}_{0,n}$ the determination of the conjugation anti-involution is made easy by Proposition 17.1.

Proposition 17.1 *Conjugation on the Clifford algebra* $\mathbf{R}_{0,n} = \mathbf{A}(m)$ *is the adjoint anti-involution induced by the standard positive-definite correlation on the spinor space* $\mathbf{A}^m$.

Proof By Proposition 15.28, $\overline{e_i}^{\tau}\, e_i = 1$, for any element e_i of the standard orthonormal basis for $\mathbf{R}^{0,n}$, here identified with its image in the Clifford algebra $\mathbf{A}(m)$. Also by the definition of conjugacy on $\mathbf{R}_{0,n}$, $e_i^- = -e_i$. But

$e_i^2 = -1$. So, for all $i \in \mathbf{n}$, $e_i^- = \overline{e_i}^\tau$, from which the result follows at once, by Proposition 2.8. □

This indicates, incidentally, why we wrote a^-, and not $\overline{a}$, for the conjugate of an element a of a Clifford algebra A, the reason for writing $a^\sim$ and not $\widetilde{a}$, for the reverse of a, being similar. The notation $\widehat{a}$ is less harmful in practice, for, in the context of Proposition 17.1 at least, $\widehat{a}$ in either of its senses coincides with $\widehat{a}$ in its other sense.

As an immediate application of Proposition 17.1 we have information on the groups $Spin(n)$, for small n.

Proposition 17.2

$Spin(1) \cong O(1) \cong S^0$, $\quad Spin(2) \cong U(1) \cong S^1$,
$Spin(3) \cong Sp(1) \cong S^3$, $\quad Spin(4) \cong Sp(1) \times Sp(1) \cong S^3 \times S^3$,
$Spin(5) \cong Sp(2)$, $\quad Spin(6)$ *is a subgroup of* $U(4)$.

Proof Apply Propositions 16.15 and 17.1. □

In the case of $Spin(n)$, for $n = 1, 2, 3, 4$, what this proposition does is to put into a wider relationship with each other various results which we have had before. It may be helpful to look at these cases in turn.

$\mathbf{R}^2$: The universal Clifford algebra $\mathbf{R}_{0,2}$ is $\mathbf{H}$, while the universal Clifford algebra $\mathbf{R}_{2,0}$ is $\mathbf{R}(2)$, the even Clifford algebras $\mathbf{R}^0_{0,2}$ and $\mathbf{R}^0_{2,0}$ each being isomorphic to $\mathbf{C}$.

Suppose that we use $\mathbf{R}_{0,2} = \mathbf{H}$ to describe the rotations of $\mathbf{R}^2$, $\mathbf{R}^2$ being identified with $\mathbf{R}\{\mathrm{j}, \mathrm{k}\}$ and $\mathbf{R}^0_{0,2} = \mathbf{C}$ being identified with $\mathbf{R}\{1, \mathrm{i}\}$. Then the rotation of $\mathbf{R}^2$ represented by $g \in Spin(2) = U(1)$ is the map

$$x \mapsto g x g^{-1} = g x \widetilde{g},$$

that is the map

$$\begin{aligned}(x_0 + x_1\mathrm{i})\mathrm{j} = (x_0\mathrm{j} + x_1\mathrm{k}) &\mapsto (a + b\mathrm{i})^2(x_0 + x_1\mathrm{i})\mathrm{j} \\ &= (a + b\mathrm{i})(x_0\mathrm{j} + x_1\mathrm{k})(a - b\mathrm{i}),\end{aligned}$$

where $x = x_0\mathrm{j} + x_1\mathrm{k}$ and $g = a + b\mathrm{i}$.

On the other hand, by Corollary 16.10, we may use $\mathbf{C}$ directly, $\mathbf{R}^2$ being identified with $\mathbf{C}$. Then the rotation of $\mathbf{R}^2$ represented by g is the map

$$y \mapsto g y \widehat{g}^{-1} = g y g = g^2 y.$$

One can transfer from one model to the other simply by setting $x = y\mathrm{j}$.

$\mathbf{R}^3$: The universal Clifford algebra $\mathbf{R}_{0,3}$ is ${}^2\mathbf{H}$, while the universal Clifford algebra $\mathbf{R}_{3,0}$ is $\mathbf{C}(2)$, the even Clifford algebras $\mathbf{R}_{0,3}^0$ and $\mathbf{R}_{3,0}^0$ each being isomorphic to $\mathbf{H}$. Besides these, there are the non-universal algebras $\mathbf{R}_{0,3}(1,0)$ and $\mathbf{R}_{0,3}(0,1)$, also isomorphic to $\mathbf{H}$. Any of these may be used to represent the rotations of $\mathbf{R}^3$.

The simplest to use is $\mathbf{R}_{0,3}(1,0) \cong \mathbf{H}$, $\mathbf{R}^3$ being identified with the linear subspace of pure quaternions. An alternative is to use $\mathbf{R}_{0,3}^0 \cong \mathbf{H}$, in which case $\mathbf{R}^3$ may be identified, by Proposition 16.3, with the linear subspace $\mathbf{R}\{1, \mathrm{i}, \mathrm{k}\}$. In either case $Spin(3) = Sp(1) = S^3$.

In the first of these two cases the rotation of $\mathbf{R}^3$ represented by $g \in Spin(3)$ is the map

$$x \mapsto g x g^{-1} = g x \bar{g},$$

while in the second case the rotation is the map

$$y \mapsto g y \widehat{g}^{-1} = g y \widetilde{g}.$$

One can transfer from the one model to the other by setting $x = y\mathrm{j}$, compatibility being guaranteed by the equation

$$g y \widehat{g}^{-1} = g y \widetilde{g}\,\mathrm{j}.$$

$\mathbf{R}^4$: The universal Clifford algebras $\mathbf{R}_{0,4}$ and $\mathbf{R}_{4,0}$ are each isomorphic to $\mathbf{H}(2)$, the even Clifford algebra in either case being isomorphic to ${}^2\mathbf{H}$. There are various identifications of $\mathbf{R}^3$ with a linear subspace of ${}^2\mathbf{H}$ such that, for any $x \in \mathbf{R}^3$, $x^{(2)} = -x^2 = \bar{x}x$. Once one is chosen, $\mathbf{R}^4$ may be identified with $\mathbf{R} \oplus \mathbf{R}^3$, with $y^{(2)} = \bar{y}\,y$, for any $y \in \mathbf{R}^4$.

One method is to identify $\mathbf{R}^4$ with the linear subspace

$$\left\{\begin{pmatrix} y & 0 \\ 0 & \bar{y} \end{pmatrix} : y \in \mathbf{H}\right\}$$

of ${}^2\mathbf{H}$, $\mathbf{R}^3$ being identified with

$$\mathbf{R}\left\{\begin{pmatrix} \mathrm{i} & 0 \\ 0 & -\mathrm{i} \end{pmatrix}, \begin{pmatrix} \mathrm{j} & 0 \\ 0 & -\mathrm{j} \end{pmatrix}, \begin{pmatrix} \mathrm{k} & 0 \\ 0 & -\mathrm{k} \end{pmatrix}\right\}.$$

Then, for any $\begin{pmatrix} q & 0 \\ 0 & r \end{pmatrix} \in {}^2\mathbf{H}$,

$$\widehat{\begin{pmatrix} q & 0 \\ 0 & r \end{pmatrix}} = \begin{pmatrix} r & 0 \\ 0 & q \end{pmatrix},$$

while $Spin(4) = \{\begin{pmatrix} q & 0 \\ 0 & r \end{pmatrix} \in {}^2\mathbf{H} : |q| = |r| = 1\}$. The rotation of $\mathbf{R}^4$ represented by $\begin{pmatrix} q & 0 \\ 0 & r \end{pmatrix} \in Spin(4)$ is then, by Proposition 16.9, the map

$$\begin{pmatrix} y & 0 \\ 0 & \bar{y} \end{pmatrix} \mapsto \begin{pmatrix} q & 0 \\ 0 & r \end{pmatrix}\begin{pmatrix} y & 0 \\ 0 & \bar{y} \end{pmatrix}\widehat{\begin{pmatrix} q & 0 \\ 0 & r \end{pmatrix}}^{-1} = \begin{pmatrix} q\,y\,\bar{r} & 0 \\ 0 & r\,\bar{y}\,\bar{q} \end{pmatrix}.$$

This is essentially the map

$$y \mapsto q\,y\,\bar{r},$$

which is what we had before, in Chapter 8.

An alternative is to identify $\mathbf{R}^4$ with the linear subspace

$$\left\{\begin{pmatrix} y & 0 \\ 0 & \bar{y} \end{pmatrix} : y \in \mathbf{H}\right\}.$$

The rotation induced by $\begin{pmatrix} q & 0 \\ 0 & r \end{pmatrix} \in Spin(4)$ is then, by a similar argument, the map

$$\begin{pmatrix} y & 0 \\ 0 & \tilde{y} \end{pmatrix} \mapsto \begin{pmatrix} q\,y\,\tilde{r} & 0 \\ 0 & r\,\tilde{y}\,\tilde{q} \end{pmatrix}$$

and this reduces to the map $y \mapsto q\,y\,\tilde{r}$.

Proposition 17.3 $Spin(6) \cong SU(4)$.

Proof A proof of this may be based on Exercise 10.2. One proves first that if Y is the image of the injective real linear map $\gamma : \mathbf{C}^3 \to \mathbf{C}(4)$ constructed in that exercise then, for each $y \in Y$, $\bar{y}^\tau y \in \mathbf{R}$, and that if Y is assigned the quadratic form $Y \to \mathbf{R}$; $y \mapsto \bar{y}^\tau y$ then γ is an orthogonal map and T is the unit sphere in Y. The rest is then a straightforward checking of the things that have to be checked. Note that, for all $t \in Sp(2)$, $t^{\sim} = t^{-1}$. □

The algebras $\mathbf{R}_{p,q}$ *for small* p, $q \neq 0$

For any $g \in Spin(n)$, $N(g) = 1$. For $g \in Spin(p, q)$, on the other hand, with neither p nor q equal to zero, $N(g)$ can be equal either to 1 or to -1. The subgroup $\{g \in Spin(p, q) : N(g) = 1\}$ will be denoted by $Spin^+(p, q)$. The image of $Spin^+(p, q)$ in $SO(p, q)$ by ρ is a subgroup of $SO(p, q)$. This subgroup, called the (*proper*) *Lorentz group* of $\mathbf{R}^{p,q}$, will be denoted by

$SO^+(p,q)$. In Proposition 22.48 the Lorentz group of $\mathbf{R}^{p,q}$ will be shown to be the set of rotations of $\mathbf{R}^{p,q}$ that preserve the semi-orientations of $\mathbf{R}^{p,q}$.

Example 17.4 *Let $g \in Spin(1,1)$. Then the induced rotation ρ_g of $\mathbf{R}^{1,1}$ preserves the semi-orientations of $\mathbf{R}^{1,1}$ if and only if $N(g) = 1$, and reverses them if and only if $N(g) = -1$.*

The next proposition covers the cases of greatest interest in physics, the algebra $\mathbf{R}^0_{1,3} \cong \mathbf{C}(2)$ being known as the *Pauli algebra.*

Proposition 17.5

$$Spin^+(1,1) \cong \{\begin{pmatrix} a & 0 \\ 0 & d \end{pmatrix} \in {}^2\mathbf{R} : ad = 1\} \cong \mathbf{R}^\bullet \cong GL(1;\mathbf{R}),$$

$$Spin^+(1,2) \cong \{\begin{pmatrix} a & c \\ b & d \end{pmatrix} \in \mathbf{R}(2) : \det\begin{pmatrix} a & c \\ b & d \end{pmatrix} = 1\} = Sp(2;\mathbf{R})$$

and $$Spin^+(1,3) \cong \{\begin{pmatrix} a & c \\ b & d \end{pmatrix} \in \mathbf{C}(2) : \det\begin{pmatrix} a & c \\ b & d \end{pmatrix} = 1\} = Sp(2;\mathbf{C}).$$

Proof It is enough to give the proof for $Spin^+(1,3)$, which may be regarded as a subgroup of $\mathbf{R}_{1,2} \cong \mathbf{C}(2)$, since $\mathbf{R}^0_{1,3} = \mathbf{R}_{1,2}$. Now, by Proposition 16.15 and Proposition 15.34,

$$Spin^+(1,3) = \{g \in \mathbf{R}_{1,2} : g^- g = 1\},$$

so that the problem is reduced to determining the conjugation anti-involution of $\mathbf{R}_{1,2}$. To do so we have just to select a suitable copy of $\mathbf{R}^{1,2}$ in $\mathbf{R}_{1,2}$. Our choice is to represent e_0, e_1 and e_2 in $\mathbf{R}^{1,2}$ by $\begin{pmatrix} 1 & 0 \\ 0 & -1 \end{pmatrix}$, $\begin{pmatrix} 0 & -1 \\ 1 & 0 \end{pmatrix}$ and $\begin{pmatrix} 0 & i \\ i & 0 \end{pmatrix}$, respectively, in $\mathbf{C}(2)$, these matrices being mutually anti-commutative and satisfying the equations

$$\begin{pmatrix} 1 & 0 \\ 0 & -1 \end{pmatrix}^2 = 1, \begin{pmatrix} 0 & -1 \\ 1 & 0 \end{pmatrix}^2 = -1 \text{ and } \begin{pmatrix} 0 & i \\ i & 0 \end{pmatrix}^2 = -1,$$

as is necessary. Now the anti-involution $\begin{pmatrix} a & c \\ b & d \end{pmatrix} \mapsto \begin{pmatrix} d & -c \\ -b & a \end{pmatrix}$ sends each of these three matrices to its negative. This, by Proposition 2.8, is the conjugation anti-involution. Since,

$$\text{for any } \begin{pmatrix} a & c \\ b & d \end{pmatrix} \in \mathbf{C}(2), \begin{pmatrix} d & -c \\ -b & a \end{pmatrix}\begin{pmatrix} a & c \\ b & d \end{pmatrix} = \det\begin{pmatrix} a & c \\ b & d \end{pmatrix},$$

the proposition is proved. □

It is natural, therefore, to identify the spinor space $\mathbf{C}^2$ for $\mathbf{R}_{1,2}$ with the complex symplectic plane $\mathbf{C}^2_{\mathrm{sp}}$ and, similarly, to identify the spinor space $\mathbf{R}^2$ for $\mathbf{R}_{1,0}$ with ${}^2\mathbf{R}^\sigma$ and the spinor space $\mathbf{R}^2$ for $\mathbf{R}_{1,1}$ with $\mathbf{R}^2_{\mathrm{sp}}$. When this is done the induced adjoint anti-involution on the real algebra of endomorphisms of the spinor space coincides with the conjugation anti-involution on the Clifford algebra.

Note, incidentally the algebra injections

$$Spin(2) \to Spin^+(1,2) \text{ and } Spin(3) \to Spin^+(1,3)$$

induced by the standard (real orthogonal) injections

$$\mathbf{R}^{0,2} \to \mathbf{R}^{1,2} \text{ and } \mathbf{R}^{0,3} \to \mathbf{R}^{1,3},$$

the image of $Spin(2) = U(1)$ in $Spin^+(1,2)$ being SO(2) and the image of $Spin(3) = Sp(1)$ in $Spin^+(1,3)$ being $SU(2)$.

The isomorphisms $U(1) \cong SO(2)$ and $Sp(1) \cong SU(2)$ fit nicely, therefore, into the general scheme of things.

Proposition 17.6 is a step towards the determination and classification of the conjugation anti-involutions for the universal Clifford algebras $\mathbf{R}_{p,q}$ other than those already considered.

Proposition 17.6 *Let V be the spinor space for the orthogonal space $\mathbf{R}^{p,q}$ with $\mathbf{R}_{p,q} = \operatorname{End} V$. Then, if $p > 0$ and if $(p,q) \neq (1,0)$, the conjugation anti-involution on $\mathbf{R}_{p,q}$ coincides with the adjoint anti-involution on* End V *induced by a neutral semi-linear correlation on V.*

Proof By Theorem 13.8 there is a reflexive non-degenerate $\mathbf{A}^\psi$-linear correlation on the right $\mathbf{A}$-linear space V producing the conjugation anti-involution on $\mathbf{R}_{p,q}$ as its adjoint. What we prove is that the correlation must be neutral. This follows at once from the even-dimensionality of V over $\mathbf{A}$ unless $\mathbf{A}^\psi = \mathbf{R}, \mathbf{C}, \overline{\mathbf{H}}, {}^2\mathbf{R}$ or ${}^2\overline{\mathbf{H}}$. However, since $p > 0$, there exists in every case $t \in \operatorname{End} V$ such that $t^- t = -1$, namely $t = e_0$; for $e_0^- e_0 = -e_0^2 = e_0^{(2)} = -1$. The existence of such an element guarantees neutrality when $\mathbf{A}^\psi = \mathbf{R}$, by Theorem 5.28. The obvious analogue of Theorem 5.28 guarantees neutrality in each of the other cases. □

Analogous results hold for the algebras $\overline{\mathbf{C}}_{p,q}$, where, as has already been said in Chapter 15, $\overline{\mathbf{C}}_{p,q}$ is obtained by tensoring $\mathbf{R}_{p,q}$ with the superfield $\overline{\mathbf{C}}$, that is $\mathbf{C}$ assigned the conjugation involution, which fixes the field $\mathbf{C}$.

Proposition 17.7 *Conjugation on $\overline{\mathbf{C}}_{0,n}$ is the adjoint anti-involution induced by the standard positive-definite correlation on the spinor space. Conjugation on $\overline{\mathbf{C}}_{p,q}$, where $p > 0$ and $(p, q) \neq (1,0)$, is the adjoint anti-involution induced by a neutral semi-linear correlation on the spinor space.*

Completion of the classification of conjugation

Tables 3 to 8 of Theorem 17.8 complete the classification of the conjugation anti-involutions of the *five* Clifford algebras naturally associated to the orthogonal space $\mathbf{R}^{p,q}$, for each signature p, q. It is because there are five and not just the one of these that the notation $\mathbf{R}_{p,q}$ for the standard real Clifford algebra for $\mathbf{R}^{p,q}$ has been chosen in this book for the real Clifford algebras rather than $Cl(p, q)$, or some such.

Besides the algebras $\mathbf{R}_{p,q}$ themselves we have, first of all, the Clifford algebras $\mathbf{C}_n$, obtained by tensoring $\mathbf{R}_{p,q}$ with $\mathbf{C}$, with the identity as involution, where $p + q = n$, and $\overline{C}_{p,q}$, obtained by tensoring $\mathbf{R}_{p,q}$ with $\mathbf{C}$, with conjugation as involution, as in Proposition 17.7 above. Besides these we consider also the algebras ${}^2\mathbf{R}^{\sigma}_{p,q}$ obtained by tensoring $\mathbf{R}_{p,q}$ with the superfield ${}^2\mathbf{R}^{\sigma}$ of diagonal 2×2 real matrices, with *swapping* the slots as involution, which also fixes the field $\mathbf{R}$, and ${}^2\mathbf{C}^{\sigma}_n$, obtained by tensoring $\mathbf{R}_{p,q}$ with the superfield ${}^2\mathbf{C}^{\sigma}$, where swap fixes the field $\mathbf{C}$.

In each of these five cases, minus the identity on $\mathbf{R}^{p,q}$ extends both to an involution, the *main* or *grade* involution on the algebra, and also to an anti-involution of the algebra, namely *conjugation.* In this way each of the five should be regarded as a *superalgebra*, that is as a $\mathbf{Z}_2$-graded algebra, equipped also with the conjugation anti-involution as an integral part of its structure. Roughly speaking, $\mathbf{C}_n$ encodes the rank information in $\mathbf{R}_{p,q}$ while ${}^2\mathbf{R}^{\sigma}_{p,q}$ encodes the signature information. It is of course the case that as algebras $\overline{\mathbf{C}}_{p,q}$ and $\mathbf{C}_n$ are isomorphic, but with their assigned conjugations they are not at all the same.

By Table 15.27 each Clifford algebra of any of the five types described above is isomorphic to a full matrix algebra over $\mathbf{R}$, ${}^2\mathbf{R}$, $\mathbf{C}$, ${}^2\mathbf{C}$, $\mathbf{H}$, or ${}^2\mathbf{H}$, the module on which the matrix acts, identifiable with a minimal left ideal of the algebra, being the *spinor* space for the algebra.

Now by Chapter 13 any anti-involution of such an algebra (in particular conjugation) may be regarded as the anti-involution adjoint to some symmetric or skew sesqui-linear form on the spinor space. There are *ten* such classes and to appreciate the tables of Theorem 17.8 one must be familiar with all of them. It is appropriate to recall Theorem 13.7.

Theorem 13.7 *Let ξ be an irreducible correlation on a right* **A***-linear space of finite dimension > 1, and therefore equivalent to a symmetric or skew correlation. Then ξ is equivalent to one of the following ten types, these being mutually exclusive.*

0	*a symmetric* $\mathbf{R}$*-correlation;*
1	*a symmetric, or equivalently a skew,* ${}^2\mathbf{R}^\sigma$*-correlation;*
2	*a skew* $\mathbf{R}$*-correlation;*
3	*a skew* $\mathbf{C}$*-correlation;*
4	*a skew* $\widetilde{\mathbf{H}}$*-, or equivalently a symmetric,* $\overline{\mathbf{H}}$*-correlation;*
5	*a skew, or equivalently a symmetric,* ${}^2\overline{\mathbf{H}}^\sigma$*-correlation;*
6	*a symmetric* $\widetilde{\mathbf{H}}$*-, or equivalently a skew,* $\overline{\mathbf{H}}$*-correlation;*
7	*a symmetric* $\mathbf{C}$*-correlation;*
8	*a symmetric, or equivalently a skew,* $\overline{\mathbf{C}}$*-correlation;*
9	*a symmetric, or equivalently a skew,* ${}^2\overline{\mathbf{C}}^\sigma$*-correlation.*

The logic behind the numbering of these ten types derives from the order in which most of the cases appear in Table 3 below.

We begin by reminding the reader that universal Clifford algebras $\mathbf{R}_{p,q}$ for $0 \leq p, q < 7$ and $\mathbf{C}_n$ for $0 \leq n < 7$ are as in Tables 1 and 2, where $\mathbf{A}(m)$ denotes the algebra of $m \times m$ matrices over $\mathbf{A}$.

$p \downarrow$ \ $q \rightarrow$								
± 1	$\mathbf{R}$	$\mathbf{C}$	$\mathbf{H}$	${}^2\mathbf{H}$	$\mathbf{H}(2)$	$\mathbf{C}(4)$	$\mathbf{R}(8)$	${}^2\mathbf{R}(8)$
$\mathbf{R}$	${}^2\mathbf{R}$	$\mathbf{R}(2)$	$\mathbf{C}(2)$	$\mathbf{H}(2)$	${}^2\mathbf{H}(2)$	$\mathbf{H}(4)$	$\mathbf{C}(8)$	$\mathbf{R}(16)$
$\mathbf{C}$	$\mathbf{R}(2)$	${}^2\mathbf{R}(2)$	$\mathbf{R}(4)$	$\mathbf{C}(4)$	$\mathbf{H}(4)$	${}^2\mathbf{H}(4)$	$\mathbf{H}(8)$	$\mathbf{C}(16)$
$\mathbf{H}$	$\mathbf{C}(2)$	$\mathbf{R}(4)$	${}^2\mathbf{R}(4)$	$\mathbf{R}(8)$	$\mathbf{C}(8)$	$\mathbf{H}(8)$	${}^2\mathbf{H}(8)$	$\mathbf{H}(16)$
${}^2\mathbf{H}$	$\mathbf{H}(2)$	$\mathbf{C}(4)$	$\mathbf{R}(8)$	${}^2\mathbf{R}(8)$	$\mathbf{R}(16)$	$\mathbf{C}(16)$	$\mathbf{H}(16)$	${}^2\mathbf{H}(16)$
$\mathbf{H}(2)$	${}^2\mathbf{H}(2)$	$\mathbf{H}(4)$	$\mathbf{C}(8)$	$\mathbf{R}(16)$	${}^2\mathbf{R}(16)$	$\mathbf{R}(32)$	$\mathbf{C}(32)$	$\mathbf{H}(32)$
$\mathbf{C}(4)$	$\mathbf{H}(4)$	${}^2\mathbf{H}(4)$	$\mathbf{H}(8)$	$\mathbf{C}(16)$	$\mathbf{R}(32)$	${}^2\mathbf{R}(32)$	$\mathbf{R}(64)$	$\mathbf{C}(64)$
$\mathbf{R}(8)$	$\mathbf{C}(8)$	$\mathbf{H}(8)$	${}^2\mathbf{H}(8)$	$\mathbf{H}(16)$	$\mathbf{C}(32)$	$\mathbf{R}(64)$	${}^2\mathbf{R}(64)$	$\mathbf{R}(128)$

Table 1

$q \rightarrow$								
± 1	$\mathbf{C}$	${}^2\mathbf{C}$	$\mathbf{C}(2)$	${}^2\mathbf{C}(2)$	$\mathbf{C}(4)$	${}^2\mathbf{C}(4)$	$\mathbf{C}(8)$	${}^2\mathbf{C}(8)$

Table 2

Table 1 extends indefinitely either way with period 8, while Table 2 has period 2. In either case one obtains the even Clifford algebra in any location by moving one square to the left, if necessary into the additional column on the left.

Theorem 17.8 *Conjugation types for the algebras $\mathbf{R}_{p,q}$, and $\mathbf{C}_n$, to be overlaid on the relevant part of Tables 1 and 2, are shown as Tables 3 and 4.*

p mod 8 ↓ \ q mod 8 →									
	$^{2}0$	0	8	4	$^{2}4$	4	8	0	$^{2}0$
	0	1	2	3	4	5	6	7	0
	8	2	$^{2}2$	2	8	6	$^{2}6$	6	8
	4	3	2	1	0	7	6	5	4
	$^{2}4$	4	8	0	$^{2}0$	0	8	4	$^{2}4$
	4	5	6	7	0	1	2	3	4
	8	6	$^{2}6$	6	8	2	$^{2}2$	2	8
	0	7	6	5	4	3	2	1	0

Table 3

n mod 8 →								
$^{2}7$	7	9	3	$^{2}3$	3	9	7	$^{2}7$

Table 4

By contrast the table for $\overline{\mathbf{C}}_{p,q}$ is given in Table 5.

p mod 2 ↓ \ q mod 2 →			
	$^{2}8$	8	$^{2}8$
	8	9	8

Table 5

To complete the set we give as Tables 6 and 7 the tables for $^{2}\mathbf{R}^{\sigma}_{p,q}$ and $^{2}\mathbf{C}^{\sigma}{}_{n}$, the algebras for these, as algebras, being just the doubles of the algebras $\mathbf{R}_{p,q}$ and $\mathbf{C}_n$.

$-p+q$ mod 8 →								
$^{2}1$	1	9	5	$^{2}5$	5	9	1	$^{2}1$

Table 6

n mod 2 →		
$^{2}9$	9	$^{2}9$

Table 7

For the codes 0, 4 and 8 in Tables 3 and 5 there is a further classification by signature. The choice along the top row or down the extra column on the left is the positive-definite one. Elsewhere the choice is the neutral one.

Proof The first row of Table 3 is known to be correct by Proposition 17.1. Tensoring by **C**, assigned the identity (anti-)involution, then gives Table 4, by Theorem 13.11. The remainder of Table 3 then follows at once, also by Theorem 13.11. Tables 5, 6 and 7 follow in their turn by tensoring Table 3 by $\overline{\mathbf{C}}$ and using Theorem 13.11, by tensoring Table 3 by ${}^2\mathbf{R}^\sigma$ and using Theorem 13.12 and finally, by tensoring either Table 5 or Table 6 by **C** and using Theorem 13.11. The neutrality statement follows from Proposition 17.6. □

Tables 3 to 7 may be appreciated the more if the various code numbers are replaced by the classical groups that preserve the sesqui-linear forms on the spinor spaces. We give them here for $0 \le p, q < 8$ as Tables 8 to 12, contenting ourselves with the tables up to $n = p + q = 7$. To save space we abbreviate the notations slightly in obvious ways.

$p \downarrow$ $q \rightarrow$								
	O_1	U_1	Sp_1	${}^2Sp_1$	Sp_2	U_4	O_8	${}^2O_8$
	$GL_1(\mathbf{R})$	$Sp_2(\mathbf{R})$	$Sp_2(\mathbf{C})$	$Sp_{1,1}$	$GL_2(\mathbf{H})$	$O_4(\mathbf{H})$	$O_8(\mathbf{C})$	
	$Sp_2(\mathbf{R})$	${}^2Sp_2(\mathbf{R})$	$Sp_4(\mathbf{R})$	$U_{2,2}$	$O_4(\mathbf{H})$	${}^2O_4(\mathbf{H})$		
	$Sp_2(\mathbf{C})$	$Sp_4(\mathbf{R})$	$GL_4(\mathbf{R})$	$O_{4,4}$	$O_8(\mathbf{C})$			
	$Sp_{1,1}$	$U_{2,2}$	$O_{4,4}$	${}^2O_{4,4}$				
	$GL_2(\mathbf{H})$	$O_4(\mathbf{H})$	$O_8(\mathbf{C})$					
	$O_4(\mathbf{H})$	${}^2O_4(\mathbf{H})$						
	$O_8(\mathbf{C})$							

Table 8

$n \rightarrow$							
$O_1(\mathbf{C})$	$GL_1(\mathbf{C})$	$Sp_2(\mathbf{C})$	${}^2Sp_2(\mathbf{C})$	$Sp_4(\mathbf{C})$	$GL_4(\mathbf{C})$	$O_8(\mathbf{C})$	${}^2O_8(\mathbf{C})$

Table 9

$p \downarrow$ $q \rightarrow$								
	U_1	${}^2U_1$	U_2	${}^2U_2$	U_4	${}^2U_4$	U_8	${}^2U_8$
	$GL_1(\mathbf{C})$	U_2	$GL_2(\mathbf{C})$	U_4	$GL_4(\mathbf{C})$	U_8	$GL_8(\mathbf{C})$	
	U_2	${}^2U_2$	U_4	${}^2U_4$	U_8	${}^2U_8$		
	$GL_2(\mathbf{C})$	U_4	$GL_4(\mathbf{C})$	U_8	$GL_8(\mathbf{C})$			
	U_4	${}^2U_4$	U_8	${}^2U_8$				
	$GL_4(\mathbf{C})$	U_8	$GL_8(\mathbf{C})$					
	U_8	${}^2U_8$						
	$GL_8(\mathbf{C})$							

Table 10

	$q \rightarrow$							
p	$GL_1(\mathbf{R})$	$GL_1(\mathbf{C})$	$GL_1(\mathbf{H})$	${}^2GL_1(\mathbf{H})$	$GL_2(\mathbf{H})$	$GL_4(\mathbf{C})$	$GL_8(\mathbf{R})$	${}^2GL_8(\mathbf{R})$
$\downarrow$	${}^2GL_1(\mathbf{R})$	$GL_2(\mathbf{R})$	$GL_2(\mathbf{C})$	$GL_2(\mathbf{H})$	${}^2GL_2(\mathbf{H})$	$GL_4(\mathbf{H})$	$GL_8(\mathbf{C})$	$GL_{16}(\mathbf{R})$
	etc.							

Table 11

$n \rightarrow$							
$GL_1(\mathbf{C})$	${}^2GL_1(\mathbf{C})$	$GL_2(\mathbf{C})$	${}^2GL_2(\mathbf{C})$	$GL_4(\mathbf{C})$	${}^2GL_4(\mathbf{C})$	$GL_8(\mathbf{C})$	${}^2GL_8(\mathbf{C})$

Table 12

The dimensions of the groups in Table 8 are shown in Table 13.

0	1	3	6	10	16	28	56
1	3	6	10	16	28	56	
3	6	10	16	28	56		
6	10	16	28	56			
10	16	28	56				
16	28	56					
28	56						
56							

Table 13

These depend only on the rank $n = p + q$ and not on the index (p, q).

The dimensions of the groups in Table 9 are twice those of the groups in Table 8.

Tables of Spin groups

For each n the quadratic norm $g^- g$ of any element g of the group $Spin(n) = Spin(0, n) = Spin(n, 0)$ is equal to $+1$, the group being a subgroup of the *even* classical group associated in Table 8 to the index $(0, n)$ or $(n, 0)$. That group, being the part of the classical group that lies in the even Clifford algebra for the given index, lies in the table either in the position $p = 0$, $q = n - 1$ or in the position $p = n$, $q = -1$ (in an extra column on the left that matches the first row). For any p, q with neither p nor q equal to 0 the quadratic norm of an element of $Spin(p, q)$ may be equal either to $+1$ or to -1. It is then the subgroup $Spin^+(p, q)$, consisting of those elements of $Spin(p, q)$ with quadratic norm $+1$, that is a subgroup of the even classical group for the index, namely the classical group in the position $p, q-1$.

The group $Spin(n)$ or $Spin^+(p, q)$, with $n = p+q$, has dimension $\frac{1}{2}n(n-1)$. It is the whole group for $n = p+q \leq 5$, but is of dimension one less than this, namely of dimension 15, rather than 16, for $n = p+q = 6$. In that case it happens that each algebra has a real-valued determinant, and lowering the dimension by 1 corresponds to taking the determinant equal to 1. In the case of $GL_2(\mathbf{H})$ the determinant is defined by representing each quaternionic 2×2 matrix as a 4×4 complex matrix and then taking the determinant of that matrix, this necessarily being a positive real number.

Theorem 17.9 *The groups $Spin(n)$ for $n \leq 6$, as well as the groups $Spin^+(p,q)$ for $p+q \leq 6$, in the case that both p and q are non-zero, are shown in Table 14.*

$p\downarrow$ \ $q\rightarrow$						
± 1	O_1	U_1	Sp_1	${}^2Sp_1$	Sp_2	SU_4
O_1	$GL_1(\mathbf{R})$	$Sp_2(\mathbf{R})$	$Sp_2(\mathbf{C})$	$Sp_{1,1}$	$SL_2(\mathbf{H})$	
U_1	$Sp_2(\mathbf{R})$	${}^2Sp_2(\mathbf{R})$	$Sp_4(\mathbf{R})$	$SU_{2,2}$		
Sp_1	$Sp_2(\mathbf{C})$	$Sp_4(\mathbf{R})$	$SL_4(\mathbf{R})$			
${}^2Sp_1$	$Sp_{1,1}$	$SU_{2,2}$				
Sp_2	$SL_2(\mathbf{H})$					
SU_4						

Table 14

The groups $Spin(n;\mathbf{C})$, for $n \leq 6$, are shown in Table 15.

$n\rightarrow$						
± 1	$O_1(\mathbf{C})$	$GL_1(\mathbf{C})$	$Sp_2(\mathbf{C})$	${}^2Sp_2(\mathbf{C})$	$Sp_4(\mathbf{C})$	$SL_4(\mathbf{C})$

Table 15

(Most recent writers do not give the whole of these tables and some are in error. Note in particular that $Spin^+(3,3) \cong SL(4;\mathbf{R})$.)

Proof This theorem extends the results of Propositions 17.2 and 17.5. The only difficulty is with the cases where $p+q=6$. A proof of the isomorphism $Spin(6) \cong SU(4)$ has already been sketched as Proposition 17.3, and an alternative proof, based on Diagram 24.5, will be given later as a lead-up to the discussion of triality. The first of these proofs is the more explicit and may easily be adapted to provide proofs of the isomorphisms $Spin^+(3,3) \cong SL(4;\mathbf{R})$ and $Spin(6;\mathbf{C}) \cong SL(4;\mathbf{C})$, as we now show. These are of interest as exemplifying the point stressed at the outset that in the context of Clifford algebras the general linear groups are best thought of as generalised unitary groups, where the relevant conjugation involves the swapping of components of a module over a double field.

Consider the following diagram of 'left-coset exact' sequences of maps:

$$\begin{array}{ccccc} Sp(2;\mathbf{R}) = SL(2;\mathbf{R}) & \longrightarrow & SL(3;\mathbf{R}) & \longrightarrow & S(\mathbf{R}^3 \times \mathbf{R}^3) \\ \downarrow & & \downarrow & & \downarrow ? \\ Sp(4;\mathbf{R}) & \longrightarrow & SL(4;\mathbf{R}) & \xrightarrow{\pi} & Q \subset SL(4;\mathbf{R}) \\ \downarrow & & \downarrow & & \\ S(\mathbf{R}^4 \times \mathbf{R}^4) & \xrightarrow{1} & S(\mathbf{R}^4 \times \mathbf{R}^4), & & \end{array}$$

where, for any n, $S(\mathbf{R}^n \times \mathbf{R}^n) = \{(x,y) \in \mathbf{R}^n \times \mathbf{R}^n : x \cdot y = 1\}$.

What we prove is that the second row of this diagram may be identified with the left-coset exact sequence

$$Spin^+(3,2) \xrightarrow{\iota} Spin^+(3,3) \xrightarrow{\pi} S(\mathbf{R}^3 \times \mathbf{R}^3),$$

and hence that $Spin^+(3,3)$ may be identified with $SL(4;\mathbf{R})$, the identification of $Spin^+(3,2)$ with $Sp(4;\mathbf{R})$ having been previously established. Explicitly the second map π of this sequence is the obvious action of the group $Spin^+(3,3)$, through the group $SO^+(3,3)$, on the point

$$\left(\begin{pmatrix} 0 \\ 0 \\ 1 \end{pmatrix}, \begin{pmatrix} 0 \\ 0 \\ 1 \end{pmatrix} \right)$$

of the quadric $S(\mathbf{R}^3 \times \mathbf{R}^3)$ in $\mathbf{R}^3 \times \mathbf{R}^3$ with equation $x \cdot y = 1$, that is

$$-(x_1-y_1)^2-(x_2-y_2)^2-(x_3-y_3)^2+(x_1+y_1)^2+(x_2+y_2)^2+(x_3+y_3)^2 = 4.$$

The subgroup of the group $Spin^+(3,3)$ that leaves that point fixed is identifiable with $Spin^+(3,2)$, that identification being the map ι, and the fibres of the map π are all the left cosets of this subgroup in $Spin^+(3,3)$.

To return to the diagram, the map $SL(3;\mathbf{R}) \to S(\mathbf{R}^3 \times \mathbf{R}^3)$ is defined by

$$\begin{pmatrix} a_{00} & a_{01} & a_{02} \\ a_{10} & a_{11} & a_{12} \\ a_{20} & a_{21} & a_{22} \end{pmatrix} \mapsto \left(\begin{pmatrix} a_{02} \\ a_{12} \\ a_{22} \end{pmatrix}, \begin{pmatrix} A_{02} \\ A_{12} \\ A_{22} \end{pmatrix} \right),$$

the first component of the image being the last column of the matrix a and the second component the last column of the the matrix $(a^\tau)^{-1}$, with

$$a_{02}A_{02} + a_{12}A_{12} + a_{22}A_{22} = \det a = 1.$$

The fibres of this map are the left cosets in $SL(3;\mathbf{R})$ of the subgroup $SL(2;\mathbf{R})$, identified with the inverse image of $\left(\begin{pmatrix}0\\0\\1\end{pmatrix}, \begin{pmatrix}0\\0\\1\end{pmatrix}\right)$. The map $SL(4;\mathbf{R}) \to S(\mathbf{R}^4 \times \mathbf{R}^4)$ is analogously defined.

The map $\pi : SL(4;\mathbf{R}) \to Q \subset SL(4;\mathbf{R})$ is defined, as is the map π in Exercise 10.2, by

$$a \mapsto \begin{pmatrix} a_{00} & a_{01} & a_{02} & a_{03} \\ a_{10} & a_{11} & a_{12} & a_{13} \\ a_{20} & a_{21} & a_{22} & a_{23} \\ a_{30} & a_{31} & a_{32} & a_{33} \end{pmatrix} \begin{pmatrix} a_{11} & -a_{01} & a_{31} & -a_{21} \\ -a_{10} & a_{00} & -a_{30} & a_{20} \\ a_{13} & -a_{03} & a_{33} & -a_{23} \\ -a_{12} & a_{02} & -a_{32} & a_{22} \end{pmatrix},$$

with fibres the left cosets in $SL(4;\mathbf{R})$ of the group $Sp(4;\mathbf{R})$ identified with the inverse image by π of the unit matrix. Note that the inverse of any element of $Sp(4;\mathbf{R})$ is then explicitly given.

The map $Sp(4;\mathbf{R}) \to S(\mathbf{R}^4 \times \mathbf{R}^4)$ is defined by

$$a \mapsto \begin{pmatrix} a_{00} & a_{01} & a_{02} & a_{03} \\ a_{10} & a_{11} & a_{12} & a_{13} \\ a_{20} & a_{21} & a_{22} & a_{23} \\ a_{30} & a_{31} & a_{32} & a_{33} \end{pmatrix} \mapsto \left(\begin{pmatrix} -a_{12} \\ a_{02} \\ -a_{32} \\ a_{22} \end{pmatrix}, \begin{pmatrix} a_{03} \\ a_{13} \\ a_{23} \\ a_{33} \end{pmatrix} \right),$$

with
$$\begin{aligned} a_{02}a_{13} - a_{12}a_{03} + a_{22}a_{33} - a_{32}a_{03} \\ &= (-a_{12}, a_{02}, -a_{32}, a_{22}) \cdot (a_{03}, a_{13}, a_{23}, a_{33}) \\ &= 1. \end{aligned}$$

That the map between the two copies of $S(\mathbf{R}^4 \times \mathbf{R}^4)$ is the identity is then easily verified. By elementary diagram-chasing, as in Proposition 3.4, it follows that there is a linear map $\mathbf{R}^3 \times \mathbf{R}^3 \to \mathbf{R}(4)$, restricting to a bijection between $S(\mathbf{R}^3 \times \mathbf{R}^3)$ and the image Q of π, that makes the diagram commute. Explicitly this is the map

$$\left(\begin{pmatrix} x_0 \\ x_1 \\ x_2 \end{pmatrix}, \begin{pmatrix} y_0 \\ y_1 \\ y_2 \end{pmatrix} \right) \mapsto \begin{pmatrix} y_2 & 0 & x_0 & y_1 \\ 0 & y_2 & x_1 & -y_0 \\ -y_0 & -y_1 & x_2 & 0 \\ -x_1 & x_0 & 0 & x_2 \end{pmatrix}.$$

The image of this linear map may be taken to be the space of paravectors in the even Clifford algebra $\mathbf{R}^0_{3,3}$. Identification of $SL(4; \mathbf{R})$ with $Spin^+(3,3)$ follows directly.

The whole of the above argument goes through unchanged if the field $\mathbf{R}$ is replaced by the field $\mathbf{C}$. Accordingly $SL(4; \mathbf{C}))$ may be identified with $Spin(6; \mathbf{C})$.

As a matter of fact it is enough simply to prove this last result. Then all the real cases follow by restriction. □

Tables for anti-involutions other than conjugation

Anti-involutions of Clifford algebras other than conjugation may also be classified.

Theorem 17.10 *The reversion anti-involution of* $\mathbf{R}_{p,q}$, *induced by the identity endomorphism of the quadratic space* $\mathbf{R}^{p,q}$, *is classified by a table similar to Table 3 of Theorem 17.8 which classifies conjugation, namely Table 16.*

q mod 8 → p mod 8 ↓	0	7	6	5	4	3	2	1
	^{20}	0	8	4	^{24}	4	8	0
	0	1	2	3	4	5	6	7
	8	2	^{22}	2	8	6	^{26}	6
	4	3	2	1	0	7	6	5
	^{24}	4	8	0	^{20}	0	8	4
	4	5	6	7	0	1	2	3
	8	6	^{26}	6	8	2	^{22}	2

Table 16

In this case positive-definiteness of the spinor product and compactness of the associated classical group reign down the left-hand column, with neutrality everywhere else where there is a choice.

Note that Table 16, like Table 3 of period 8, is just Table 3 translated one square to the 'South-East'.

More generally any orthogonal involution of $\mathbf{R}^{p,q}$ induces an anti-involution of $\mathbf{R}_{p,q}$, and one would like to classify all such. The relevant table for the algebras $\mathbf{R}_{0,n}$ just looks like Table 3 turned upside down, as displayed in Table 17.

0	8	4	$^{2}4$	4	8	0	$^{2}0$
7	6	5	4	3	2	1	0
6	$^{2}6$	6	8	2	$^{2}2$	2	8
5	6	7	0	1	2	3	4
4	8	0	$^{2}0$	0	8	4	$^{2}4$
3	2	1	0	7	6	5	4
2	$^{2}2$	2	8	6	$^{2}6$	6	8
1	2	3	4	5	6	7	0

Table 17

Here the first row classifies conjugation and the first column classifies reversion. The transition from conjugation to reversion as fewer and fewer in turn of the standard basis vectors of $\mathbf{R}^{p,q}$ have their signs changed by the involution is made by traversing this table steadily from the first row to the first column in a South-Westerly direction. At all intermediate stages neutrality is the rule.

For the algebras $\mathbf{R}_{p,n-p}$ comprising the pth row of Table 3 there is a similar table. For example in the case that $p = 2$ we have, for the algebras $\mathbf{R}_{2,n-2}$, Table 18, which is just Table 17, bordered by *two* additional rows on the top and *two* additional rows on the left that respect the periodicity 8.

		2	$^{2}2$	2	8	6	$^{2}6$	6	8
	0	1	2	3	4	5	6	7	0
0	$^{2}0$	0	8	4	$^{2}4$	4	8	0	$^{2}0$
1	0	7	6	5	4	3	2	1	0
2	8	6	$^{2}6$	6	8	2	$^{2}2$	2	8
3	4	5	6	7	0	1	2	3	4
4	$^{2}4$	4	8	0	$^{2}0$	0	8	4	$^{2}4$
5	4	3	2	1	0	7	6	5	4
6	8	2	$^{2}2$	2	8	6	$^{2}6$	6	8
7	0	1	2	3	4	5	6	7	0

Table 18

Here also the transition from conjugation to reversion as fewer and fewer in turn of the standard basis vectors of $\mathbf{R}^{p,q}$ have their signs changed by the involution is made by traversing the table steadily from the first row to the first column in a South-Westerly direction. The spinor product is positive-definite if the involution changes the sign of all vectors of square -1 while all those of square $+1$ remain fixed. It is neutral otherwise.

For the complex Clifford algebras $\mathbf{C}_n$ the whole thing is so much easier!

Theorem 17.11 *The reversion classification of the algebras* $\mathbf{C}_n$ *coincides with the conjugation classification.*

Hurwitz pairs

Let A be a possibly non-universal Clifford algebra with conjugation for a finite-dimensional non-degenerate real quadratic space X, embedded in A in the standard way, and let the spinor space V be taken to be a minimal left ideal of A (for example, as those matrices of $A = \mathbf{K}(m)$, all of whose entries are zero, with the exception of the last column. Then, for any $x \in X$, $v, v' \in V$,

$$(xv)^{-}(xv') = v^{-}x^{-}x\,v' = (x^{-}x)(v^{-}v').$$

Such a pairing of correlated spaces X and V is known as a *Hurwitz pair*, the theory of such pairs having originated in a paper of A. Hurwitz in (1898) and subsequently developed in work that was published posthumously in (1923). It has recently become clear that *all* such pairings arise in such a Clifford algebra setting. See, for example, Ławrynowicz and Rembieliński (1986), Randriamihamison (1990) and Cnops (1994).

An important example, and one that involves the non-universal algebras in an essential way, is the construction of linear subspaces of the groups $GL(s; \mathbf{R})$, for finite s, a *linear subspace* of $GL(s; \mathbf{R})$ being, by definition, a linear subspace of $\mathbf{R}(s)$ all of whose elements, with the exception of the origin, are invertible.

For example, the standard copy of $\mathbf{C}$ in $\mathbf{R}(2)$ is a linear subspace of $GL(2; \mathbf{R})$ of dimension 2, while either of the standard copies of $\mathbf{H}$ in $\mathbf{R}(4)$ is a linear subspace of $GL(4; \mathbf{R})$ of dimension 4. On the other hand, when s is odd, there is no linear subspace of $GL(s; \mathbf{R})$ of dimension greater than 1. For if there were such a space of dimension greater than 1 then there would exist linearly independent elements a and b of $GL(s; \mathbf{R})$ such that, for all $\lambda \in \mathbf{R}$, $a + \lambda b \in GL(s; \mathbf{R})$ and therefore such that $c + \lambda\, 1 \in GL(s; \mathbf{R})$, where $c = b^{-1}a$. However, by the fundamental theorem of algebra, there is a real number λ such that $\det(c + \lambda\, 1) = 0$, the map $\mathbf{R} \to \mathbf{R};\ \lambda \mapsto \det(c + \lambda\, 1)$ being a polynomial map of odd degree. This provides a contradiction.

Proposition 17.12 provides a method of constructing linear subspaces of $GL(s; \mathbf{R})$.

Proposition 17.12 *Let* End $\mathbf{K}^m$ *be a possibly non-universal Clifford algebra for the positive-definite orthogonal space* $\mathbf{R}^n$, *for any positive integer* n. *Then* $\mathbf{R} \oplus \mathbf{R}^n$ *is a linear subspace of* Aut $\mathbf{K}^m = GL(m; \mathbf{K})$ *and therefore of* $GL(m; \mathbf{R})$, $GL(2m; \mathbf{R})$ *or* $GL(4m; \mathbf{R})$, *according as* $\mathbf{K} = \mathbf{R}$, $\mathbf{C}$ *or* $\mathbf{H}$. *Moreover the conjugate of any element of* $\mathbf{R} \oplus \mathbf{R}^n$ *is the conjugate transpose of the representative in* $GL(m; \mathbf{K})$ *or, equivalently, the transpose of its representative in* $GL(m; \mathbf{R})$, $GL(2m; \mathbf{R})$ *or* $GL(4m; \mathbf{R})$.

Proof Let $y = \lambda + x \in \mathbf{R} \oplus \mathbf{R}^n$, where $\lambda \in \mathbf{R}$ and $x \in \mathbf{R}^n$. Then $y^- y = (\lambda - x)(\lambda + x) = \lambda^2 + x^{(2)}$ is real, and is zero if and only if $y = 0$. Therefore y is invertible if and only if $y \neq 0$.

The last statement of the proposition follows at once from Proposition 17.1. □

The following is an immediate corollary of Proposition 16.1 coupled with the explicit information concerning the Clifford algebras $\mathbf{R}_{0,n}$ contained in Table 15.27 and its extension by Corollary 15.26.

Theorem 17.13 *Let* $(\chi(k))$ *be the sequence of positive integers defined by* $\chi(8p+q) = 4p+j$, *where* $j = 0$ *for* $q = 0$, 1 *for* $q = 1$, 2 *for* $q = 2$ *or* 3 *and* 3 *for* $q = 4, 5, 6$ *or* 7. *Then if* $2^{\chi(k)}$ *divides* s *there exists a* k*-dimensional linear subspace* X *of* $GL(s; \mathbf{R})$ *such that*

(i) *for each* $x \in X$, $x^\tau = -x$, $x^\tau x = -x^2$ *being a non-negative real multiple of* ${}^s1$, *zero only if* $x = 0$,
(ii) $\mathbf{R} \oplus X$ *is a* $(k+1)$*-dimensional linear subspace of* $GL(s; \mathbf{R})$.

The sequence χ is called the *Radon-Hurwitz sequence* (Radon (1923) and Hurwitz (1923)). It can be proved that there is no linear subspace of $GL(s; \mathbf{R}))$ of dimension greater than that asserted by Theorem 17.13(ii).

As a particular case of Proposition 17.12 there is an eight-dimensional linear subspace of $GL(8; \mathbf{R})$, since $\mathbf{R}(8)$ is a (non-universal) Clifford algebra for $\mathbf{R}^7$. This remark provides a route in to the study of the algebra of Cayley numbers which we undertake in Chapter 19.

Terminology

Because the theory of spinors has developed piecemeal, driven by applications in theoretical physics, an extensive range of different types of spinor have appeared in the literature – Dirac spinors, Majorana spinors, Weyl spinors and the like – in the description of which both the complexifications of the 'natural' real, complex or quaternionic spinor spaces play important roles. See for example Benn and Tucker (1987) or Trautman (1993). My personal view (as a pure mathematician) is that given the overview of all possible cases given by the tables of this chapter, most of this often very confusing terminology can with advantage be dropped!

18
2×2 Clifford matrices

Let A be a universal real Clifford algebra for a finite-dimensional non-degenerate real quadratic space X, with Clifford group Γ. Then by Proposition 15.17 the real algebra $A(2)$ of 2×2 matrices with entries in A is a universal real Clifford algebra for the real quadratic space $X \oplus \mathbf{R}^{1,1}$, where elements of $X \oplus \mathbf{R}^{1,1}$ are represented in $A(2)$ by matrices of the form $\begin{pmatrix} x & v \\ \mu & -x \end{pmatrix}$, where $x \in X$ and $\mu, v \in \mathbf{R}$, such matrices being referred to below as *vectors* in $A(2)$. Let $\Gamma(2)$ then denote the Clifford group of $A(2)$. For many applications one would like to characterise the elements of $\Gamma(2)$ in terms of X and Γ. In the case that X is positive-definite such a characterisation was given by Vahlen (1902), and his work was re-presented in a series of papers by Ahlfors in the early 1980's. for example (1985), (1986). The indefinite case is somewhat trickier to handle. The characterisation we give here is that of Jan Cnops (1994), developed from earlier work of Maks (1989) and Fillmore and Springer (1990). For a parallel account, involving paravectors, see Elstrodt, Grunewald and Mennicke (1987). See also Waterman (1993).

The characterisation of Cnops

We begin by characterising conjugation and reversion on $A(2)$.

Proposition 18.1 *Let A be a universal Clifford algebra for a finite-dimensional non-degenerate real quadratic space X and let $A(2)$ be the universal Clifford algebra for $X \oplus \mathbf{R}^{1,1}$ as in Proposition 15.17. Then*

$$\begin{pmatrix} a & c \\ b & d \end{pmatrix}^{-} = \begin{pmatrix} d^{\sim} & -c^{\sim} \\ -b^{\sim} & a^{\sim} \end{pmatrix} \quad \text{and} \quad \begin{pmatrix} a & c \\ b & d \end{pmatrix}^{\sim} = \begin{pmatrix} d^{-} & c^{-} \\ b^{-} & a^{-} \end{pmatrix}.$$

Proof These hold for vectors in $A(2)$ and so for the whole of $A(2)$. □

The *pseudo-determinant* Δ of the matrix $\begin{pmatrix} a & c \\ b & d \end{pmatrix}$ is defined to be $a d^{\sim\!-} - c b^{\sim\!-}$. For any element of the Clifford group $\Gamma(2)$ of $A(2)$ its pseudo-determinant is equal to the product

$$\begin{pmatrix} a & c \\ b & d \end{pmatrix}\begin{pmatrix} a & c \\ b & d \end{pmatrix}^{-} = \begin{pmatrix} a & c \\ b & d \end{pmatrix}\begin{pmatrix} d^{\sim\!-} & -c^{\sim} \\ -b^{\sim} & a^{\sim} \end{pmatrix}.$$

Proposition 18.2 *Let* $\begin{pmatrix} a & c \\ b & d \end{pmatrix}$ *be an element of the Clifford group* $\Gamma(2)$ *of the Clifford algebra* $A(2)$, *with pseudo-determinant* Δ. *Then*

$$\Delta = a d^{\sim\!-} - c b^{\sim\!-} = d^{\sim\!-} a - c^{\sim\!-} b.$$

Proof For such a matrix

$$\begin{pmatrix} a & c \\ b & d \end{pmatrix}\begin{pmatrix} d^{\sim\!-} & -c^{\sim} \\ -b^{\sim} & a^{\sim} \end{pmatrix} = \begin{pmatrix} d^{\sim\!-} & -c^{\sim} \\ -b^{\sim} & a^{\sim} \end{pmatrix}\begin{pmatrix} a & c \\ b & d \end{pmatrix},$$

either side of the equation being equal to Δ. □

By Corollary 16.8 any element g of the Clifford group Γ of a finite-dimensional non-degenerate real quadratic space X is representable as the product of a finite number of invertible elements of X. Following Cnops (1994) we introduce the monoid Θ of all finite products of elements of X, whether invertible or not, a *monoid* being a set furnished with an associative product. In the case that X is positive-definite $\Theta = \Gamma \cup \{0\}$. In the next proposition we list several elementary properties of Θ.

Proposition 18.3 *For each* $a \in \Theta$ *and each* $x, y, z \in X$,

(i) $a a^{\sim\!-} = a^{\sim\!-} a \in \mathbf{R}$, *being non-zero if and only if a is invertible,*

(ii) $a x a^{\sim\!-} \in X$,

(iii) *either* $a u a^{\sim\!-} = 0$ *for all* $u \in X$ *or* $a = v g$ *for some* $v \in X$ *and* $g \in \Gamma$,

(iv) $1 + x y \in \Theta$,

(v) $xy + yz \in \Theta$,

(vi) *if there exists* $u \in X$ *such that* $a u a^{\sim\!-} \neq 0$ *then* $x a + a y \in \Theta$,

(vii) $x + yxz \in \Theta$.

Proof

(i) This is clear.

(ii) It is sufficient to prove this when $a \in X$. But then

$$a\,x\,a^{\sim} = a\,x\,a = (a\,x + x\,a)a - x(a^2) \in X.$$

(iii) This is obvious when a is invertible, when the second alternative is true. Otherwise a can be written as a product $v_1 v_2 ... v_k$ of elements of X, where we can arrange for all the non-invertible factors to come first. We prove the proposition explicitly for a product of two non-invertible vectors, the induction being easy. So let $a = v_1 v_2$, where v_1 and v_2 are not invertible. If these are not mutually orthogonal then $v_1 + v_2$ is invertible and $v_1 v_2 = v_1(v_1 + v_2)$, giving the second alternative. Otherwise $v_1 v_2 + v_2 v_1 = 0$, and then, for any $u \in X$, $v_2\,u\,v_2 = (v_2\,u + u\,v_2)v_2$, while $v_1 v_2 v_1 = (v_1 v_2 + v_2 v_1)v_1 = 0$, from which it follows that $a\,u\,a^{\sim} = 0$, the first alternative.

(iv) Suppose first that at least one of x and y, say x, is invertible. Then $1 + x\,y = x(x^{-1} + y)$. Secondly, suppose that neither x nor y is invertible, but that $x \cdot y \neq 0$. Then

$$(x\,y + y\,x)(1 + x\,y) = x\,y + y\,x + x\,y\,x\,y = (x + y + x\,y\,x)(x + y).$$

Finally, suppose that neither x nor y is invertible, but that $x \cdot y = 0$, though $x \neq y$. Now X is non-degenerate, so by Proposition 5.4 there exists a non-invertible element z of X such that $z \cdot y = 0$ but $z \cdot x \neq 0$, and we may choose z so that $z \cdot x = 1$. Then $x^2 = y^2 = z^2 = 0$, $x\,z + z\,x = -2$ and

$$\begin{aligned}&(x + z)(x + z - y)(x - z - y)(x - z)\\ &\quad = (-2 - (x + z)y)(2 - y(x - z)) = -4(1 + x\,y).\end{aligned}$$

(v) Since $x\,y + y\,z = (x\,y + y\,x) + y(z - x)$ it follows at once from (iv) that $x\,y + y\,z \in \Theta$.

(vi) By (iii) $a = v\,g$, where v is a vector and g is invertible. Then

$$x\,a + a\,y = (x\,v + v\,g\,y\,g^{-1})g = (x\,v + v\,y')g,$$

where $y' = g\,y\,g^{-1} \in X$, by (v).

(vii) Suppose first that x is invertible. It then follows that $x + y\,x\,z = x(1 + x^{-1}\,y\,x\,z)$, which is in Γ by (iv).

Next, suppose that x is not invertible and that $x \cdot z = 0$. Then $x + y\,x\,z = (1 - y\,z)x$, which is in Γ by (iv).

Finally, suppose that x is not invertible and that $x \cdot z \neq 0$. Choose $k \in \mathbf{R}$ so that $kx + z$ is invertible. Then

$$(x + y\,x\,z)(kx + z) = xz + y(k(x\,z + z\,x) + z^2)x,$$

which is in Γ by (v). Since $kx + z$ is invertible it follows that $x + y\,x\,z$ also is in Γ.

□

We shall require Lemma 18.4 during the proof of Theorem 18.5.

Lemma 18.4 *Let A be a universal real Clifford algebra for a finite-dimensional real quadratic space X, not necessarily non-degenerate, and let u, v, w, $y \in X$. Then*

$$u\,v\,y\,w\,v - v\,y\,w\,v\,u \in X.$$

Proof Choose an orthonormal basis for X such that v, w and y are in the subspace spanned by e_0, e_1 and e_2. Then the even element $v\,w\,y\,v$ must be of the form $a + \sum_{i,j \in 3} b_{ij} e_i e_j$, where a and $b_{ij} \in \mathbf{R}$. Now e_k commutes with $e_i e_j$ if i, j, k are distinct, and otherwise $e_k e_i e_j = \pm e_i e_j e_k \in X$. The assertion follows. □

Theorem 18.5 *Let A be a universal Clifford algebra for a finite-dimensional non-degenerate real quadratic space X, with Clifford group Γ and Clifford monoid Θ, let $A(2)$ be the universal Clifford algebra for $X \oplus \mathbf{R}^{1,1}$ constructed as in Proposition 15.17, and let G be the set of all matrices $\begin{pmatrix} a & c \\ b & d \end{pmatrix}$ of $A(2)$ such that*

$$(a)\ a, b, c, d \in \Theta, \quad (b)\ a\,b^{\sim}, c\,d^{\sim}, a^{\sim}c, b^{\sim}d \in X, \quad (c)\ \Delta = a\,d^{\sim} - c\,b^{\sim} \in \mathbf{R}^{\bullet}.$$

Then G is the Clifford group $\Gamma(2)$ of $A(2)$.

Proof We prove first that each entry of a matrix $\begin{pmatrix} a & c \\ b & d \end{pmatrix}$ in G can be written in the form $v\,g$, where $v \in X$ and $g \in \Gamma$, and that there exist p, q, r, $s \in X$ such that in each case one of the following alternatives holds:

$$\begin{array}{llllll} a = b\,p & \text{or} & b = a\,p; & a = q\,c & \text{or} & c = q\,a; \\ c = d\,r & \text{or} & c = d\,r; & b = s\,d & \text{or} & d = s\,b. \end{array}$$

Suppose first that either a or b, say a, is invertible. Then $b = v\,a$ with $v = a\,b^{\sim}/(a\,a^{\sim}) \in X$, and $b = a\,p$, where $p = a^{-1}v\,a$. Otherwise neither a nor b is invertible, but neither is zero, with

$$(a\,d^{\sim} - c\,b^{\sim})b = a(d^{\sim}b) \quad \text{and} \quad (d\,a^{\sim} - b\,c^{\sim})a = b(-c^{\sim}a),$$

neither of these being zero. Now $a(d^{\sim}b) = a(b^{\sim}d) = (a\,b^{\sim})d$. So $a\,b^{\sim} \neq 0$. It follows that

$$a(b^{\sim}d)a^{\sim} = a\,b^{\sim}(d\,a^{\sim} - b\,c^{\sim}) \neq 0,$$

from which it follows from (iii) of Proposition 18.3 that $a = w\,g$ for $w \in X$ and $g \in \Gamma$, a similar result clearly holding for b.

The case of c and d is similar. To prove the statement for a with c or for b with d one uses the alternative expression for the pseudo-determinant Δ given in Proposition 18.2.

Next we prove that, for any $x \in X$, $a\,x\,d^{\sim} + c\,x\,b^{\sim} \in X$. Suppose firstly that all four entries in the matrix are non-invertible and, necessarily, non-zero. Then in particular there exist $v \in X$ and $g \in \Gamma$ such that $a = v\,g$ and explicit computation shows that

$$\begin{pmatrix} a & c \\ b & d \end{pmatrix} = \begin{pmatrix} v\,g & u\,v \\ v\,w & -u\,v\,w \end{pmatrix} \begin{pmatrix} g & 0 \\ 0 & g \end{pmatrix},$$

where $w = \Delta^{-1}g(d^{\sim}b)g^{-1}$ and $u = c\,d^{\sim}$ both are in X. Then

$$a\,x\,d^{\sim} + c\,x\,b^{\sim} = -(v\,y\,w\,v)u + u(v\,y\,w\,v),$$

where $y = g\,x\,g^{\sim}$, and this is in X, by Lemma 18.4.

Alternatively, at least one of the entries of the matrix, say a, is invertible. Then $b = a\,p = a\,p\,a^{-1}\,a$, $c = ra$ and $d = (\lambda + a\,p\,a^{-1}\,r)a$, where $p' = a\,p\,a^{-1} \in X$, $r \in X$ and $\Delta = \lambda a$. Then

$$a\,x\,d^{\sim} + c\,x\,b^{\sim} = y(\lambda + r\,p') + r\,y\,p' = \lambda + (y\,r + r\,y)p',$$

where $y = a\,x\,a^{-1}$, and this is in X.

Similarly it may be verified that, for any $x \in X$, $a\,x\,c^{\sim} + c\,x\,a^{\sim} \in \mathbf{R}$ and $b\,x\,d^{\sim} + d\,x\,b^{\sim} \in \mathbf{R}$.

Now consider the product

$$\begin{pmatrix} a & c \\ b & d \end{pmatrix} \begin{pmatrix} x & v \\ \mu & -x \end{pmatrix} \begin{pmatrix} d^{-} & c^{-} \\ b^{-} & a^{-} \end{pmatrix},$$

where $x \in X$ and μ, $v \in \mathbf{R}$. Now, by property (b) of G either a and d are in Θ^0 and b and c are in Θ^1, or vice versa. So up to sign this product is equal to

$$\begin{pmatrix} a & c \\ b & d \end{pmatrix}\begin{pmatrix} x & \nu \\ \mu & -x \end{pmatrix}\begin{pmatrix} d^{\sim} & -c^{\sim} \\ -b^{\sim} & a^{\sim} \end{pmatrix},$$

which is equal to

$$\begin{pmatrix} axd^{\sim}+\mu cd^{\sim}-\nu ab^{\sim}+cxb^{\sim} & -axc^{\sim}-\mu cc^{\sim}+\nu aa^{\sim}-cxa^{\sim} \\ bxd^{\sim}+\mu dd^{\sim}-\nu bb^{\sim}+dxb^{\sim} & -bxc^{\sim}-\mu dc^{\sim}+\nu ba^{\sim}-dxa^{\sim} \end{pmatrix},$$

and this is the matrix of a vector in $A(2)$. From this it follows at once that $G \subset \Gamma(2)$.

Finally, we have to prove that $\Gamma(2) \subset G$. For this it is enough to prove that the product

$$\begin{pmatrix} a & c \\ b & d \end{pmatrix}\begin{pmatrix} x & \nu \\ \mu & -x \end{pmatrix}$$

is in G, where the first matrix is in G and the second is an invertible vector in $A(2)$, the condition for this being that $-x^2 - \nu\mu \neq 0$. None of the verifications causes any trouble, though at one point one requires again to use the fact that $a\,x\,d^{\sim} + c\,x\,b^{\sim}$ is in X, for any $x \in X$. □

An example of an element of $\Gamma(2)$ none of whose entries is invertible has been given by Maks (1989). Consider generators e_0 and e_1 of $\mathbf{R}^{1,1}$ with $e_0^2 = -1$ and $e_1^2 = 1$, and take the standard model of $\mathbf{R}^{2,2}$ in $\mathbf{R}_{1,1}(2)$. Then the rotation of $\mathbf{R}^{2,2}$ that sends e_0 to $\begin{pmatrix} 0 & -1 \\ 1 & 0 \end{pmatrix}$, and that vector to $-e_0$, and similarly sends e_1 to $\begin{pmatrix} 0 & 1 \\ 1 & 0 \end{pmatrix}$, and that vector to $-e_1$, is induced by the matrix

$$\begin{pmatrix} 1 & e_0 \\ e_0 & 1 \end{pmatrix}\begin{pmatrix} 1 & -e_1 \\ e_1 & 1 \end{pmatrix} = \begin{pmatrix} 1+e_0e_1 & e_0-e_1 \\ e_0+e_1 & 1-e_0e_1 \end{pmatrix}$$

of $\Gamma(2)$, none of whose entries is invertible.

Cnops (1994) has given examples to show that none of the requirements (a), (b), (c) in the characterisation of $\Gamma(2)$ can be derived from the other two. His examples are

$$\begin{pmatrix} 1+e_0 & 0 \\ 0 & 1-e_0 \end{pmatrix}, \quad \begin{pmatrix} 1 & e_0e_1 \\ 0 & 1 \end{pmatrix} \quad \text{and} \quad \begin{pmatrix} 1 & 0 \\ 0 & e_0e_1e_2 \end{pmatrix},$$

which violate (a), (b) and (c), respectively, but not the other two. None is in $\Gamma(2)$, since each is the sum of an even part and an odd part, both different from zero.

Theorem 18.6 *Let $A = \mathbf{R}_{p,q}$, with $X = \mathbf{R}^{p,q}$, Clifford monoid Θ, and Clifford group Γ. Then the Clifford group $\Gamma(p, q+1)$ is representable as the subset of $A(2)$ consisting of matrices of the form $\begin{pmatrix} a & -\widehat{b} \\ b & \widehat{a} \end{pmatrix}$, where*

(a) *a and b are in Θ, with either a or b in Γ,*
(b) *$a b^{\sim} \in X$ and (c) $a a^{-} + \widehat{b} b^{\sim} = a a^{-} + b b^{-} \in \mathbf{R}^{\bullet}$.*

For matrices in $\Gamma^0(p, q+1)$, $\widehat{a} = a \in \Theta^0$ and $\widehat{b} = -b \in \Theta^1$.

Moreover, in the case that $a \in \Gamma^0$, there exists $p \in X$ such that $b = a\,p$ and, for any sufficiently small $\lambda \in \mathbf{R}$, $\begin{pmatrix} a & -\lambda b \\ \lambda b & a \end{pmatrix} \in \Gamma(2)^0$.

Proof Everything follows directly from Theorem 18.5. □

Groups of motions

Of obvious practical importance is the group of *rigid motions* of $\mathbf{R}^3$, that is the group of rotations of $\mathbf{R}^3$ extended by the group of translations of $\mathbf{R}^3$. More generally we may consider the group of rigid motions of any finite-dimensional quadratic space X, this being the group of rotations of X extended by the group of translations of X. Such rigid motions are readily representable with the aid of the Clifford group $\Gamma = \Gamma(X)$.

Theorem 18.7 *Let $A = \mathbf{R}_{p,q}$, with $X = \mathbf{R}^{p,q}$ and Clifford group Γ, and let $\begin{pmatrix} a & b \\ b & a \end{pmatrix}$ in $A(2)$ represent an element of the Clifford group $\Gamma^0(p, q+1)$, with $a \in \Gamma^0$. Then the map $X \to X;\ x \mapsto axa^{-1} + ba^{-1}$ is a rigid motion of X and any rigid motion of X may be so represented, the representation being unique up to non-zero real multiples of a and b.*

Strictly speaking what is involved here is the subgroup of $A(2)$ consisting of all matrices of the form $\begin{pmatrix} a & b \\ 0 & a \end{pmatrix}$, with $a \in \Gamma^0$ and $b = a\,p$, where $p \in X$. In the particular case that $X = \mathbf{R}^3$ $Spin(4)$ is most frequently identified with the group $S^3 \times S^3 \subset {}^2\mathbf{H}$. An alternative to ${}^2\mathbf{H}$ consists of the matrices of $\mathbf{H}(2)$ of the form $\begin{pmatrix} a & b \\ b & a \end{pmatrix}$ with the injective algebra map ${}^2\mathbf{H} \to \mathbf{H}(2)$ sending $\begin{pmatrix} q & 0 \\ 0 & r \end{pmatrix}$ to $\begin{pmatrix} \frac{1}{2}(q+r) & \frac{1}{2}(q-r) \\ \frac{1}{2}(q-r) & \frac{1}{2}(q+r) \end{pmatrix}$, it then being the case that if $|q| = |r|$ and $r \neq -q$ then $(q-r)(q+r)^{-1}$ is a pure quaternion, that is an element of $\mathbf{R}^3$. It is to be noted for future reference (cf. Exercise 8.2) that α and β are quaternions such that $\beta\alpha^{-1}$ is pure if and only if α and β as elements of $\mathbf{R}^4$ are mutually orthogonal.

The subalgebra of $\mathbf{H}(2)$ consisting of all matrices of the form $\begin{pmatrix} a & b \\ 0 & a \end{pmatrix}$ is known as Clifford's algebra (1873) of *biquaternions* . Elements of it are all of the form $a + be$, where a and b are quaternions and $e^2 = 0$, e being represented in $\mathbf{H}(2)$ by the matrix $\begin{pmatrix} 0 & 1 \\ 0 & 0 \end{pmatrix}$. We discuss this important special case further at the end of Chapter 24.

Pfaffian charts

By Proposition 14.15 for any $s \in \mathrm{End}_-(\mathbf{R}^n)$ the endomorphism $1 - s$ is invertible, with $(1 + s)(1 - s)^{-1} \in SO(n)$. Moreover the map

$$\mathrm{End}_-(\mathbf{R}^n) \to SO(n); s \mapsto (1 + s)(1 - s)^{-1}$$

is injective, this being the *Cayley chart* on $SO(n)$ at ${}^n1$. The question naturally arises, what is the corresponding chart on $Spin(n)$?

A start to the answer is provided by Proposition 18.8.

Proposition 18.8 *Let $g \in \Gamma^0(n)$ be such that $\rho_g = (1 + s)(1 - s)^{-1}$. Then the real part of g is non-zero, and if this is taken to be* 1 *then $g = 1 + \sum_{i<j} s_{ij} e_i e_j +$ higher order terms.*

Proof Let $g = a + \sum_{i<j} r_{ij} e_i e_j +$ higher order terms, where a, $t_{ij} \in \mathbf{R}$, and let $x = \sum_{i \in n} x_i e_i \in \mathbf{R}^n$, with $x' = g\,x\,g^{-1}$ also in $\mathbf{R}^n$. Then the coefficients of e_i on either side of the equation $x'\,g = g\,x$ are equal; that is

$$a\,x_i' - \sum_{i<j} x_i' = a x_i + \sum_{i<j} x_i;$$

that is $(a - r)X_i' = (a + r)x_i$.

So $(a - r)(1 + s)(1 - s)^{-1} = a + r$; that is $(a - r)(1 + s) = (a + r)(1 - s)$, implying that $a\,s = r$.

So $a \neq 0$ and if $a = 1$ then $r = s$. □

There is more than one way of presenting the complete answer. The one we have chosen to present involves the *complete Pfaffian* of the skew-symmetric matrix s.

So let $s \in \mathrm{End}_{-}(\mathbf{R}^n)$, that is $s \in \mathbf{R}(n)$ and $s^\tau = -s$. The *Pfaffian* of s, $\mathrm{pf}\, s$, is defined to be 0 if n is odd and to be the real number

$$\sum_{\pi \in P} \operatorname{sgn} \pi \prod_{i \in \mathbf{m}} s_{\pi(2k),\pi(2k+1)}$$

if $n = 2m$ is even, P being the set of all permutations of $2m$ for which

(i) for any $h, k \in \mathbf{m}$, $h < k \Rightarrow \pi(2h) < \pi(2k)$,
(ii) for any $k \in \mathbf{m}$, $\pi(2k) < \pi(2k+1)$.

For example, if $n = 4$, $\mathrm{pf}\, s = s_{01}s_{23} - s_{02}s_{13} + s_{03}s_{12}$. By convention, $\mathrm{pf}\, s = 1$ if $n = 0$, in which case $s = {}^0 1 = 0$.

For any $I \subset \mathbf{n}$, let s_I denote the matrix $(s_ij \;:\; i, j \in I)$. Then $s_I \in \mathrm{End}_{-}(\mathbf{R}^n)$, where $k = \#I$. The *complete Pfaffian* of s, $\mathrm{Pf}\, s$, is, by definition, the element

$$\sum_{I \subset \mathbf{n}} \mathrm{pf}\, s_I e_I$$

of the Clifford algebra $\mathbf{R}_{0,n}$. Since $\mathrm{pf}\, s_I = 0$ for $\#I$ odd, $\mathrm{Pf}\, s \in \mathbf{R}_{0,n}^0$.

In fact $\mathrm{Pf}\, s \in \Gamma^0(n)$, as we now state formally.

Theorem 18.9 *Let* $s \in \mathrm{End}_{-}(\mathbf{R}^n)$. *Then* $\mathrm{Pf}(s) \in \gamma^0(n)$ *and is the unique element of* $\Gamma^0(n)$, *with real part* 1, *that induces the rotation* $(1+s)(1-s)^{-1}$.

Proof A start to the proof has already been provided by Proposition 18.8. It then follows by application of part of Theorem 18.6 that the coefficient of e_I is a polynomial in the terms of the matrix s_I, obtained from s by deleting all the rows and columns of s except for those with $i, j \in I$. One proceeds by induction, and by verifying at each stage that the coefficient of highest degree in either $e_I\overline{e_I}$ or $e_I e_0 \overline{e_I}$ is zero. Moreover, by the final part of Theorem 18.6, this polynomial contains exactly one term from each row and each column of s_I, so that the terms of the polynomial are, up to real multiples, the terms of $\mathrm{pf}\, s_I$.

Finally, consider any one such term,

$$\lambda s_{01}s_{23}s_{45}e_0e_1e_2e_3e_4e_5, \text{ for example.}$$

This term will be equal to the corresponding term in $\mathrm{pf}\, s'$ where $s' \in \mathrm{End}_{-}(\mathbf{R}^n)$ is defined by

$$s'_{01} = s_{01} = -s'_{10},\ s'_{23} = s_{23} = -s'_{32} \text{ and } s'_{45} = s_{45} = -s'_{54},$$

all the other terms being zero. However,

$$\begin{aligned}\text{Pf}\, s' &= (1 + s_{01}e_0e_1)(1 + s_{23}e_2e_3)(1 + s_{45}e_4e_5) \\ &= 1 + s_{01}e_0e_1 + s_{23}e_2e_3 + s_{45}e_4e_5 + \ldots + s_{01}s_{23}s_{45}e_0e_1e_2e_3e_4e_5,\end{aligned}$$

since each of the factors is in $\Gamma^0(6)$, the real part is 1 and the coefficients of the terms e_ie_j are correct. So, in this case, $\lambda = 1$ in accordance with the statement of the theorem. The other terms are handled analogously. □

The map

$$\text{End}_-(\mathbf{R}^n) \to Spin(n); \; s \mapsto \text{Pf}\, s/|\,\text{Pf}\, s)|$$

will be called the *Pfaffian*, or *Lipschitz, chart* at 1 on $Spin(n)$. Lipschitz (1880), (1884) shares with Clifford the discovery of the Clifford algebras. See, for example, 'Correspondence from an ultramundane correspondent' (1959). For further details of the early history of Clifford algebras see Van der Waerden (1985). and a forthcoming book by Pertti Lounesto.

For the alternative presentation of the Lipschitz chart that involves the exterior exponential of the *bivector* $\sum_{i<j} s_{ij}e_ie_j$, see, for example, Lounesto (1987) and Ahlfors and Lounesto (1989).

The following property of the Pfaffian is sometimes used to characterise it. See for example Artin (1957).

Theorem 18.10 *For any* $s \in \text{End}_-(\mathbf{R}^n)$, $(\text{pf}\, s)^2 = \det s$.

Proof Let $s \in \text{End}_-(\mathbf{R}^n)$, Then, for any $t \in \mathbf{R}(n)$, $t^\tau st \in \text{End}_-(\mathbf{R}^n)$. Now, for any such s and t,

$$\text{pf}(t^\tau st) = \det t\, \text{pf}\, s.$$

To show this it is enough, by Proposition 1.8, to verify that pf s is invariant under an elementary column operation coupled with a matching elementary row operation.

The matrix s induces a skew correlation on $\mathbf{R}^n$ with product

$$\mathbf{R}^n \times \mathbf{R}^n \to \mathbf{R}; \; (x, x') \mapsto x^\tau s x'.$$

Let $2m$ be the rank of this correlation. Then, by a slight generalisation of Theorem 6.7 to include the degenerate case, there exists $u \in GL(n; \mathbf{R})$ such that

$$(u^\tau su)_{2k,2k+1} = 1 = -(u^\tau su)_{2k,2k+1}$$

for all $k \in \mathbf{m}$, and $(u^\tau su)_{i,j} = 0$ otherwise. It follows from this that

$$\operatorname{Pf}(u^\tau su) = \prod_{k\in\mathbf{m}} (1 + e_{2k}e_{2k+1}).$$

There are two cases. If $2m < n$, $\operatorname{pf}(u^\tau su) = 0$, implying that $\operatorname{pf} s = 0$, since $\det u \neq 0$, while $\det(u^\tau su) = 0$, implying that $\det s = 0$. If $2m = n$, $\operatorname{pf}(u^\tau su) = 1$ and $\det(u^\tau su) = 0$, implying that

$$(\det u)^2(\operatorname{pf} s)^2 = 1 = (\det u)^2 \det s.$$

In either case, $(\operatorname{pf} s)^2 = \det s$. □

Exercises

18.1 Let ρ_g be the rotation of $\mathbf{R}^4$ induced by an element g of $\Gamma^0(\mathbf{R}^4)$ with real part equal to 1. Prove that ρ_g is expressible as the composite of *two* hyperplane reflections (cf. Theorem 5.15) if and only if g is of the form

$$1 + s_{01}e_0e_1 + s_{02}e_0e_2 + s_{03}e_0e_3 + s_{12}e_1e_2 + s_{13}e_1e_3 + s_{23}e_2e_3$$

where (e_0, e_1, e_2, e_3) is the standard basis for $\mathbf{R}^4$. Deduce that

$$1 + s_{01}e_0e_1 + s_{02}e_0e_2 + s_{03}e_0e_3 + s_{12}e_1e_2 + s_{13}e_1e_3 + s_{23}e_2e_3$$

is the product in the Clifford algebra $\mathbf{R}_{0,4}$ of two elements of $\mathbf{R}^4$ if and only if

$$\operatorname{pf} s = s_{01}s_{23} - s_{02}s_{13} + s_{03}s_{12} = 0.$$

18.2 Prove that an invertible element of the Clifford algebra $\mathbf{C}_4$

$$1 + s_{01}e_0e_1 + s_{02}e_0e_2 + s_{03}e_0e_3 + s_{12}e_1e_2 + s_{13}e_1e_3 + s_{23}e_2e_3$$

is the product of two elements of $\mathbf{C}^4$ if and only if $\operatorname{pf} s = 0$.

18.3 Prove that an element

$$s_{01}e_0e_1 + s_{02}e_0e_2 + s_{03}e_0e_3 + s_{12}e_1e_2 + s_{13}e_1e_3 + s_{23}e_2e_3$$

of $\bigwedge^2(\mathbf{K}^4)$, where $\mathbf{K} = \mathbf{R}$ or $\mathbf{C}$, is the product of two elements of $\mathbf{K}^4$ if and only if $\operatorname{pf} s = 0$. Deduce that the image constructed in Exercise 15.7 of the Grassmannian $\mathscr{G}_2(\mathbf{R}^4)$ in the projective space $\mathscr{G}_1(\bigwedge^2(\mathbf{R}^4))$ is the projective quadric with equation

$$s_{01}s_{23} - s_{02}s_{13} + s_{03}s_{12} = 0.$$

18.4 Define an analogue of the Pfaffian chart for $Spin(p,q), pq \neq 0$.

19
The Cayley algebra

In this chapter we take a brief look at a non-associative algebra over **R** that nevertheless shares many of the most useful properties of **R**, **C** and **H**. Though it is rather esoteric, it often makes its presence felt in classification theorems and can ultimately be held 'responsible' for a rich variety of exceptional cases. Most of these lie beyond our scope, but the existence of the algebra and its main properties are readily deducible from our work on Clifford algebras in previous chapters.

Real division algebras

A *division algebra* over **R** or *real division algebra* is, by definition, a finite-dimensional real linear space X with a bilinear product $X^2 \to X$; $(a, b) \mapsto ab$ such that, for all $a, b \in X$, the product $ab = 0$ if and only if $a = 0$ or $b = 0$, or, equivalently, if and only if the linear maps

$$X \to X;\ x \mapsto xb \text{ and } x \mapsto ax$$

are injective when a and b are non-zero, and therefore bijective.

We are already familiar with three associative real division algebras, namely **R** itself, **C**, the field of complex numbers, representable as a two-dimensional subalgebra of **R**(2), and **H**, the non-commutative field of quaternions, representable as a four-dimensional subalgebra of **R**(4). Each has a unit element and for each there is an anti-involution, namely conjugation, which may be made to correspond to transposition in the matrix algebra representation, such that the map of the algebra to **R**,

$$N;\ a \mapsto N(a) = \bar{a}\,a,$$

is a real-valued positive-definite quadratic form that respects the algebra product, that is, is such that, for all a, b in the algebra,

$$N(a\,b) = N(a)\,N(b).$$

A division algebra X furnished with a positive-definite quadratic form $N : X \to \mathbf{R}$ such that, for all $a, b \in X$, $N(a\,b) = N(a)\,N(b)$ is said to be a *normed division algebra.*

Alternative division algebras

An algebra X such that, for all $a, b \in X$, $a(a\,b) = a^2 b$ and $(a\,b)b = a\,b^2$ is said to be an *alternative* algebra. For example, any associative algebra is an alternative algebra.

Proposition 19.1 *Let X be an alternative algebra. Then, for all $a, b \in X$, $(a\,b)a = a(b\,a)$.*

Proof For all $a, b \in X$,

$$\begin{aligned}(a+b)^2\,a &= (a+b)((a+b)\,a)\\ &\Rightarrow (a^2 + a\,b + b\,a + b^2)a = (a+b)(a^2 + b\,a)\\ &\Rightarrow a^2 a + (a\,b)a + (b\,a)a + b^2 a = a\,a^2 + a(b\,a) + b\,a^2 + b(b\,a)\\ &\Rightarrow (a\,b)a = a(b\,a).\end{aligned}$$

□

Proposition 19.2 *Let X be an alternative division algebra. Then X has a unit element and each non-zero $a \in X$ has an inverse.*

Proof If X has a single element there is nothing to be proved. So suppose that it has more than one element. Then there is an element $a \in X$, with $a \neq 0$. Let e be the unique element such that $\mathrm{e}\,a = a$. This exists, since the map $x \mapsto x\,a$ is bijective. Then $\mathrm{e}^2 a = \mathrm{e}(\mathrm{e}\,a) = \mathrm{e}\,a$. So $\mathrm{e}^2 = \mathrm{e}$. Therefore, for all $x \in X$, $\mathrm{e}(\mathrm{e}\,x) = \mathrm{e}^2 x = \mathrm{e}\,x$ and $(x\,\mathrm{e})\mathrm{e} = x\,\mathrm{e}^2 = x\,\mathrm{e}$. So $\mathrm{e}\,x = x$ and $x\,\mathrm{e} = x$. That is, e is a unit element, necessarily unique.

Again let $a \neq 0$ and let b be such that $a\,b = \mathrm{e}$. Then $a(b\,a) = (a\,b)a = \mathrm{e}\,a = a\,\mathrm{e}$. So $b\,a = \mathrm{e}$. That is, b is inverse to a. □

The Cayley algebra

There are many non-associative division algebras over $\mathbf{R}$. Such an algebra may fail even to be *power-associative*, that is, it may contain an element

a such that, for example, $(a^2)a \neq a(a^2)$. A less exotic example is given in Exercise 19.1. However, only one of the non-associative division algebras is of serious interest. This is the alternative eight-dimensional *Cayley algebra* or *algebra of Cayley numbers* (1845), also discovered independently in 1843 by John Graves (1848) and known by him as the algebra of *octaves* or *octonions.* Despite the lack of associativity and commutativity there is a unit element, the subalgebra generated by any two of its elements is isomorphic to **R**, **C** or **H** and so is associative, and there is a conjugation anti-involution sharing the same properties as conjugation for **R**, **C** or **H**.

The existence of the Cayley algebra depends on the fact that the matrix algebra $\mathbf{R}(8)$ may be regarded as a (non-universal) Clifford algebra for the positive-definite orthogonal space $\mathbf{R}^7$ in such a way that conjugation of the Clifford algebra corresponds to transposition in $\mathbf{R}(8)$. For then, as was noted following Theorem 17.13, the images of **R** and $\mathbf{R}^7$ in $\mathbf{R}(8)$ together span an eight-dimensional linear subspace, passing through ^{81}, such that each of its elements, other than zero, is invertible. This eight-dimensional subspace of $\mathbf{R}(8)$ will be denoted by **Y**.

Proposition 19.3 *Let* $\mu : \mathbf{R}^8 \to \mathbf{Y}$ *be a linear isomorphism. Then the map* $\mathbf{R}^8 \times \mathbf{R}^8 \to \mathbf{R}^8$; $(a, b) \mapsto ab = (\mu(a))(b)$ *is a bilinear product on* $\mathbf{R}^8$ *such that, for all* $a, b \in \mathbf{R}^8$, $ab = 0$ *if and only if* $a = 0$ *or* $b = 0$. *Moreover, any non-zero element* $\mathrm{e} \in \mathbf{R}^8$ *can be made the unit element for such a product by choosing* μ *to be the inverse of the isomorphism*

$$\mathbf{Y} \to \mathbf{R}^8;\ y \mapsto y\mathrm{e}.$$

The division algebra with unit element introduced in Proposition 19.3 is called the *Cayley algebra* on $\mathbf{R}^8$ with unit element e. It is rather easy to see that any two such algebras are isomorphic. We shall therefore speak simply of *the* Cayley algebra, denoting it by **O** (for octonions). Though the choice of e is essentially unimportant, it is advantageous to select an element of length 1 in $\mathbf{R}^8$. For definiteness we select e_0, the zeroth element of the standard basis for $\mathbf{R}^8$. We then denote by υ (upsilon): $\mathbf{R}^r \to \mathbf{Y}$ the inverse of the linear isomorphism $\mathbf{Y} \to \mathbf{R}^8$; $y \mapsto y\,e_0$, which associates to each $y \in \mathbf{Y}$ its zeroth column.

Here we have implicitly assigned to $\mathbf{R}^8$ its standard positive-definite structure, with quadratic form

$$N : \mathbf{R}^8 \to \mathbf{R};\ a \mapsto N(a) = a \cdot a = a^\tau a.$$

The space **Y** also has an orthogonal structure, induced by conjugation, namely transposition, on the Clifford algebra $\mathbf{R}(8)$, with quadratic form

$$\mathbf{Y} \to \mathbf{R}\{\mathrm{e}\};\ y \mapsto y^\tau y.$$

The Cayley algebra $\mathbf{O}$ inherits both, the one directly and the other via the isomorphism υ. As the next proposition shows, the choice of e as an element of length 1 guarantees that these two structures coincide.

Proposition 19.4 *For all* $a \in \mathbf{R}^8$, $(\upsilon(a))^\tau \upsilon(a) = N(a)({}^8 1)$.

Proof For all $a \in \mathbf{R}^8$, $a = \upsilon(a)\mathrm{e}$. So $N(a) = a^\tau a = \mathrm{e}^\tau(\upsilon(a))^\tau \upsilon(a)\mathrm{e}$. Since $y^\tau y \in \mathbf{R}({}^8 1)$ for all $y \in \mathbf{Y}$ and since $\mathrm{e}^\tau \mathrm{e} = 1$, it follows that

$$\upsilon(a)^\tau \upsilon(a) = N(a)({}^8 1).$$

□

Conjugation on $\mathbf{R}(8)$ induces a *linear* involution

$$\mathbf{O} \to \mathbf{O};\ a \mapsto \bar{a} = (\upsilon(a))^\tau \mathrm{e}$$

which we shall call *conjugation* on $\mathbf{O}$. This involution induces a direct sum decomposition $\mathbf{O} = \mathbf{R}\{\mathrm{e}\} \oplus \mathbf{O}'$ in which $\mathbf{O}' = \{b \in \mathbf{O} : \bar{b} = -b\}$.

The following proposition lists some important properties both of the quadratic form and of conjugation on $\mathbf{O}$. The product on $\mathbf{R}(8)$ and the product on $\mathbf{O}$ will both be denoted by juxtaposition, as will be the action of $\mathbf{R}(8)$ on $\mathbf{O}$. It is important to remember throughout the discussion that, though the product on $\mathbf{R}(8)$ is associative, the product on $\mathbf{O}$ need not be.

Proposition 19.5 *For all* $a, b \in \mathbf{O}$,
$N(a\,b) = N(a)N(b)$, implying that $\mathbf{O}$ is a normed division algebra,
$(a \cdot b)\mathrm{e} = \frac{1}{2}(\bar{a}\,b + \bar{b}\,a)$, implying that $\mathbf{O}' = (\mathbf{R}\{\mathrm{e}\})^\perp$,
$(N(a))\mathrm{e} = \bar{a}\,a = a\,\bar{a}$,
and $\overline{a\,b} = \bar{b}\,\bar{a}$, implying that, conjugation is an algebra anti-involution. Moreover, for all $a, b, c \in \mathbf{O}$, $\bar{a} \cdot (b\,c) = \bar{b} \cdot (c\,a) = \bar{c} \cdot (a\,b)$.

Proof For all $a, b \in \mathbf{O}$,

$$N(a\,b) = N(\upsilon(a)b) = b^\tau \upsilon(a)^\tau \upsilon(a)b = b^\tau(N(a)({}^8 1))b = N(a)N(b).$$

Also $\bar{a}\,b + \bar{b}\,a = \bar{a}(b\,\mathrm{e}) + \bar{b}(a\,\mathrm{e}) = \upsilon(a)^\tau \upsilon(b)\mathrm{e} + \upsilon(b)^\tau \upsilon(a)\mathrm{e} = 2(a \cdot b)\mathrm{e}$, implying that, if $a \in \mathbf{R}\{\mathrm{e}\}$ and if $b \in \mathbf{O}'$, then $2(a \cdot b)\mathrm{e} = a\,b - b\,a = 0$, since e, and therefore any real multiple of e, commutes with any element of

O. It implies, secondly, since $N(a) = a \cdot a$, that $N(a) = \bar{a}a$ and, since $v(a)v(a)^{\tau} = v(a)^{\tau}v(a)$, that $a\bar{a} = N(a)$.

Next we prove that, for all $a, b, c \in \mathbf{O}$, $\bar{a} \cdot (bc) = \bar{b} \cdot (ca)$ and this we do by proving that each is equal to $(\bar{b}\,\bar{a}) \cdot c$. Firstly

$$\bar{a} \cdot (bc) = \bar{a}^{\tau}v(b)c = \bar{a}^{\tau}v(\bar{b})^{\tau}c = (v(\bar{b})\bar{a})^{\tau}c = (\bar{b}\,\bar{a}) \cdot c.$$

Secondly, $(\bar{b}\,\bar{a}) \cdot c = \bar{b} \cdot (ca)$ when $a \in \mathbf{R}\{e\}$. On the other hand, when $a \in \mathbf{O}'$, $a^{\tau}e = a \cdot e = 0$ and

$$
\begin{aligned}
b \cdot (ca) - (\bar{b}\,\bar{a}) \cdot c &= (ca) \cdot \bar{b} + (\bar{b}\,a) \cdot c \\
&= a \cdot (\bar{c}\,\bar{b}) + a \cdot (bc), \text{ by the argument used above,} \\
&= a \cdot (\bar{c}\,\bar{b} + bc) \\
&= a \cdot 2(\bar{c} \cdot b)e \\
&= 0.
\end{aligned}
$$

So, for all $a \in \mathbf{O}$, $\bar{a} \cdot (bc) = \bar{b} \cdot (ca)$. Permuting a, b and c cyclically we also obtain $\bar{b} \cdot (ca) = \bar{c} \cdot (ab)$. Finally we set $c = e$ in the equation $\bar{c} \cdot (ab) = \bar{a} \cdot (bc)$. Then

$$e(ab) + \overline{ab}\,\bar{e} = a(be) + \overline{be}\,\bar{a}.$$

That is, $ab + \overline{ab} = ab + \bar{b}\,\bar{a}$, so that $\overline{ab} = \bar{b}\,\bar{a}$. ☐

The real number $\bar{a} \cdot (bc)$ is said to be the *scalar triple product* of the Cayley numbers a, b and c, in that order. This generalises the scalar product on $\mathbf{H}$, defined after Proposition 8.17.

The algebra $\mathbf{O}$ is clearly not commutative, since $\dim \mathbf{O}' > 1$. Nor is it associative, as we shall see. Nevertheless we have the following.

Proposition 19.6 *The Cayley algebra $\mathbf{O}$ is alternative.*

Proof For any $a, b \in \mathbf{O}$, $\bar{a}(ab) = v(\bar{a})v(a)b = v(a)^{\tau}v(a)b = (\bar{a}\,a)b$. So

$$
\begin{aligned}
a(ab) &= (a + \bar{a})ab - \bar{a}(ab) \\
&= ((a + \bar{a})a)b - (\bar{a}\,a)b, \text{ since } a + \bar{a} \in \mathbf{R}\{e\}, \\
&= a^2 b.
\end{aligned}
$$

By proving that their conjugates are equal it follows likewise that

$$(ab)b = ab^2.$$

☐

Throughout the remainder of this chapter we shall identify **R** with **R**{e}. In particular, we shall write 1 in place of e for the unit element in **O**, being careful to distinguish the numeral 1 from the letter l.

Hamilton triangles

It has been remarked that two elements $a, b \in \mathbf{O}'$ are orthogonal as elements of $\mathbf{R}^8$ if and only if they anti-commute. An orthonormal ordered subset (i, j, k) of $\mathbf{O}'$, with $\mathrm{i} = \mathrm{j}\,\mathrm{k}$, $\mathrm{j} = \mathrm{k}\,\mathrm{i}$ and $\mathrm{k} = \mathrm{i}\,\mathrm{j}$, therefore spans, with 1, a subalgebra of **O** isomorphic with the quaternion algebra **H**. Such a subset will be said to be a *Hamilton triangle* in **O** and will be denoted by the diagram

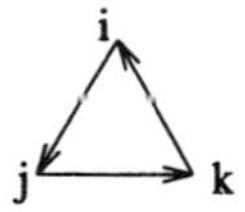

in which each vertex is the product of the other two in the order indicated by the arrows.

Proposition 19.7 *Let a and b be mutually orthogonal elements of $\mathbf{O}'$ and let $c = a\,b$. Then $c \in \mathbf{O}'$ and is orthogonal both to a and to b.*

Proof First

$$\begin{aligned} a \cdot b = 0 &\Rightarrow a\,b + b\,a = 0 \\ &\Rightarrow \bar{c} = \overline{ab} = \bar{b}\,\bar{a} = (-b)(-a) = -c \\ &\Rightarrow c \in \mathbf{O}'. \end{aligned}$$

Also
$$\begin{aligned} a \cdot c &= \tfrac{1}{2}(\bar{a}(a\,b) + (\overline{ab})a) \\ &= \tfrac{1}{2}(N(a)b + \bar{b}\,N(a)), \text{ by Proposition 19.6,} \\ &= 0, \text{ since } b + \bar{b} = 0. \end{aligned}$$

Similarly, $b \cdot c = 0.$

□

Corollary 19.8 *Let* (i, j) *be an orthonormal ordered pair of elements of* $\mathbf{O}'$ *and let* $\mathrm{k} = \mathrm{i}\,\mathrm{j}$. *Then* (i, j, k) *is a Hamilton triangle in* $\mathbf{O}'$.

From this follows the assertion made earlier that the subalgebra generated by any two element of **O** is isomorphic to **R**, **C** or **H** and so is, in particular, associative.

Cayley triangles

Finally, any Hamilton triangle in $\mathbf{O}'$ may be extended to a useful orthonormal basis for $\mathbf{O}'$. We begin by defining a *Cayley triangle* in $\mathbf{O}'$ to be an orthonormal ordered triple (a, b, c) in $\mathbf{O}'$ such that c is also orthogonal to $a\,b$.

Proposition 19.9 *Let (a, b, c) be a Cayley triangle in $\mathbf{O}'$. Then*

(i) $a(b\,c) + (a\,b)c = 0$, *exhibiting the non-associativity of* $\mathbf{O}$,

(ii) $a\cdot(b\,c) = 0$, *implying that the elements a, b, c form a Cayley triangle in whatever order they are listed,*

(iii) $a\,b \cdot b\,c = 0$, *implying that $(a, b, b\,c)$ is a Cayley triangle,*

(iv) $(a\,b)(b\,c) = a\,c$, *implying that $(a\,b, b\,c, a\,c)$ is a Hamilton triangle.*

Proof

(i) Since (a, b, c) is a Cayley triangle,

$$a\,b + b\,a = a\,c + c\,a = b\,c + c\,b = (a\,b)c + c(a\,b) = 0.$$

So

$$\begin{aligned} a(b\,c) + (a\,b)c &= -a(c\,b) - c(a\,b) \\ &= (a^2 + c^2)b - (a + c)(a\,b + c\,b) \\ &= (a + c)^2 b - (a + c)((a + c)b) \\ &= 0. \end{aligned}$$

(ii) From (i) it follows by conjugation that $(\bar{c}\,\bar{b})\bar{a} + \bar{c}(\bar{b}\,\bar{a}) = 0$ and therefore that $(b\,c)a + c(a\,b) = 0$. Since $(a\,b)c = +c(a\,b) = 0$, it follows that $a(b\,c) + (b\,c)a = 0$, implying that $a \cdot (b\,c) = 0$.

(iii)

$$\begin{aligned} 2\,a\,b \cdot b\,c &= (b\,a)(b\,c) + (b\,c)(b\,a) \\ &= (b\,a)^2 + (b\,c)^2 = (b(a - c))^2 \\ &= -b^2 a^2 - b^2 c^2 + b^2(a - c)^2 \\ &= -b^2(a\,c + c\,a) \\ &= 2\,b^2 a \cdot c \\ &= 0. \end{aligned}$$

(iv) Apply (i) to the Cayley triangle $(a, b, b\,c)$.
Then $(a\,b)(b\,c) = -a(b(b\,c)) = a\,c$, since $b^2 = -1$.

□

We can reformulate this as follows ('l' being the letter 'l').

Proposition 19.10 *Let* (i, j, l) *be a Cayley triangle in* $\mathbf{O}'$ *and let* k = i j. *Then* {i, j, k, l, i l, jl, kl} *is an orthonormal basis for* $\mathbf{O}'$, *and if these seven elements are arranged in a regular heptagon as shown then each of the*

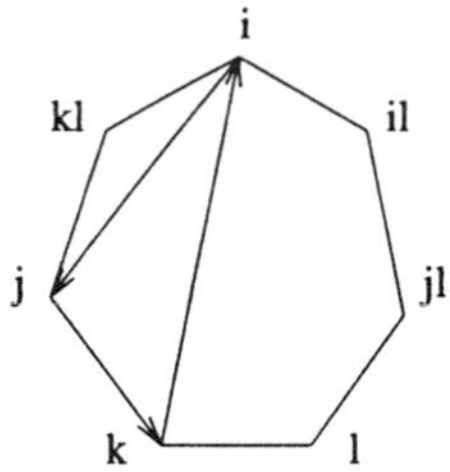

seven triangles obtained by rotating the triangle (i, j, k) *through an integral multiple of* $2\pi/7$ *is a Hamilton triangle, that is, each vertex is the product of the other two vertices in the appropriate order.*

This heptagon is essentially the multiplication table for the Cayley algebra **O**.

From this it is easy to deduce that there cannot be any division algebra over **R** of dimension greater than 8 such that the subalgebra generated by any three elements is isomorphic to **R**, **C**, **H** or **O**. Such an algebra A, if it exists, has a conjugation anti-involution, inducing a direct sum decomposition $\mathbf{R} \oplus A'$ of A in which A' consists of all the elements of A which equal the negative of their conjugate. Further details are in Exercise 19.3. The following proposition then settles the matter.

Proposition 19.11 *Let* (i, j, l) *be any Cayley triangle in* A', *let* k = i j *and let* m *be an element orthogonal to each of the seven elements* i, j, k, l, i l, j l *and* k l *of the Cayley heptagon. Then* m = 0.

Proof We remark first that parts (i) and (ii) of Proposition 19.9 hold for any a, b, $c \in A'$ such that $a \cdot b = a \cdot c = b \cdot c = a\,b \cdot c = 0$. Using this several times, we find, on making a circuit of the 're-bracketing pentagon', that

$$(\mathrm{i\,j})(\mathrm{l\,m}) = -((\mathrm{i\,j})\mathrm{l})\mathrm{m} = (\mathrm{i}(\mathrm{j\,l}))\mathrm{m} = -\mathrm{i}((\mathrm{j\,l})\mathrm{m}) = \mathrm{i}(\mathrm{j}(\mathrm{l\,m})) = -(\mathrm{i\,j})(\mathrm{l\,m}).$$

So $(\mathrm{i\,j})(\mathrm{lm}) = 0$. But $\mathrm{i\,j} \neq 0$; so $\mathrm{l\,m} = 0$, and therefore, since $\mathrm{l} \neq 0$, $\mathrm{m} = 0$. □

Further results

There are various stronger results, for example

(i) Frobenius' theorem (1878) that any associative division algebra over **R** is isomorphic to **R**, **C** or **H**,
(ii) Hurwitz' theorem (1898) that any normed division algebra over **R**, with unit element, is isomorphic to **R**, **C**, **H** or **O**,
(iii) the theorem of Skornyakov (1950) and Bruck, Kleinfeld (1951) that any alternative division algebra over **R** is isomorphic to **R**, **C**, **H** or **O**; and
(iv) the theorem of Kervaire (1958), Milnor, Bott (1958), and Adams (1958), that any division algebra over **R** has dimension 1, 2, 4 or 8.

The first two of these are little more difficult to prove than what we have proved here and can be left as exercises. The starting point in the proof of (i) is the remark that any element of an associative n-dimensional division algebra must be a root of a polynomial over **R** of degree at most n and, therefore, by the fundamental theorem of algebra, must be the solution of a quadratic equation. From this it is not difficult to define the conjugation map and to prove its linearity. Result (iii) is harder to prove. The discussion culminates in the following.

Theorem 19.12 *Any real non-associative alternative division algebra is a Cayley algebra.*

Proof Let A be a real non-associative alternative division algebra, and, for any $x, y, z \in A$, let

$$[x, y] = x\,y - y\,x \quad \text{and} \quad [x, y, z] = (x\,y)z - x(y\,z).$$

It can be shown that if x and y are such that $u = [x\,y] \neq 0$, then there exists z such that $v = [x, y, z] \neq 0$. It can then be shown that $u\,v + v\,u = 0$ and therefore, by the previous remark, that there exists t such that $w = [u\,v\,t] \neq 0$. One can now verify that u^2, v^2 and w^2 are negative real numbers and that

$$\mathrm{i} = u/\sqrt{-u^2}, \ \mathrm{j} = v/\sqrt{-v^2} \ \text{ and } \ \mathrm{l} = w/\sqrt{-w^2}$$

form a Cayley triangle. Then A contains a Cayley algebra as a sub-algebra. It follows, essentially by Proposition 19.11, that A coincides with the Cayley algebra.

The details are devious and technical, and the reader is referred to Kleinfeld (1963) for a full account. □

Finally, (iv) is very hard indeed. Its proof uses the full apparatus of algebraic topology. See Adams (1958), (1960), Kervaire (1958) and Milnor and Bott (1958).

The Cayley projective line and plane

Most of the standard results of linear algebra do not generalise over the Cayley algebra **O**, for the very definition of a linear space involves the associativity of the field of scalars. Nevertheless we can regard the map

$$\mathbf{O}^n \times \mathbf{O} \to \mathbf{O}^n;\ ((y_i : i \in n), y) \mapsto (y_i y : i \in n))$$

as a quasi-linear structure for the additive group $\mathbf{O}^n$.

It is also possible to define a 'projective line' and a 'projective plane' over **O**.

The *Cayley projective line* $\mathbf{O}P^1$ is constructed by fitting together two copies of **O** in the manner of Example 14.2 for the projective line $\mathbf{K}P^1$, for $\mathbf{K} = \mathbf{R}$, **C** or **H**. Any point is represented either by $[1, y]$ or by $[x, 1]$, with $[1, y] = [x, 1]$ if and only if $y = x^{-1}$, the square brackets here having their projective-geometry connotation. There is even a 'Hopf map' $h : \mathbf{O}^2\backslash(0, 0) \to \mathbf{O}P^1$ defined by $h(y_0, y_1) = [y_0 y_1^{-1}, 1]$, whenever $y_1 \neq 0$, and by $h(y_0, y_1) = [1, y_1, y_0^{-1}]$, whenever $y_0 \neq 0$. Since any two elements of **O** (for example, y_0 and y_1) generate an associative subalgebra, it is true that $y_0 y_1^{-1} = (y_1 y_0^{-1})^{-1}$, and so the two definitions agree, whenever y_0 and y_1 are both non-zero.

The *Cayley projective plane* $\mathbf{O}P^2$ is similarly constructed by fitting together three copies of $\mathbf{O}^2$. Any point is represented in at least one of the forms $[1, y_0, z_0]$, $[x_1, 1, z_1]$ or $[x_2, y_2, 1]$. The obvious identifications are compatible, though this requires careful checking because of the general lack of associativity. What we require is that the equations

$$x_1 = y_0^{-1},\ z_1 = z_0 y_0^{-1} \quad \text{and} \quad x_2 = x_1 z_1^{-1},\ y_2 = z_1^{-1}$$

be compatible with the equations

$$x_2 = z_0^{-1},\ y_2 = y_0 z_0^{-1}.$$

But all is well, since

$$x_1 z_1^{-1} = y_0^{-1}(z_0 y_0^{-1})^{-1} = z_0^{-1} \quad \text{and} \quad z_1^{-1} = (z_0 y_0^{-1})^{-1} = y_0 z_0^{-1},$$

once again because the subalgebra generated by any two elements is associative.

Useful analogues over **O** of projective spaces of dimension greater than 2 do not exist. The reader is referred to Bruck (1955) for a discussion.

Exercises

19.1 Let X be a four-dimensional real linear space with basis elements denoted by 1, i, j and k, and let a product be defined on X by prescribing that

$$\mathrm{i}^2 = \mathrm{j}^2 = \mathrm{k}^2 = -1, \ \mathrm{jk} + \mathrm{kj} = \mathrm{ki} + \mathrm{ik} = \mathrm{ij} + \mathrm{ji} = 0$$

and

$$\mathrm{jk} = \alpha\mathrm{i}, \ \mathrm{ki} = \beta\mathrm{j}, \ \mathrm{ij} = \gamma\mathrm{k},$$

where α, β, γ are non-zero real numbers, all of the same sign. Prove that X, with this product, is a real division algebra and that X is associative if and only if $\alpha = \beta = \gamma = 1$ or $\alpha = \beta = \gamma = -1$.

19.2 Prove that if $a, b \in \mathbf{O}'$ then $ab - ba \in \mathbf{O}'$.

19.3 Let X be a division algebra over $\mathbf{R}$ such that for each $x \in X$ there exist $\alpha, \beta \in \mathbf{R}$ such that $x^2 - 2\alpha x + \beta = 0$ and let X' consist of all $x \in X$ for which there exists $\beta \in \mathbf{R}$ such that $x^2 + \beta = 0$, with $\beta \geq 0$.

Prove that X' is a linear subspace of X, that $X = \mathbf{R} \oplus X'$ and that the map

$$\mathbf{R} \oplus X' \to \mathbf{R} \oplus X'; \ \lambda + x' \mapsto \lambda - x',$$

where $\lambda \in \mathbf{R}$, $x' \in X'$, is an anti-involution of X.

19.4 Let A be a real alternative division algebra, and, for any x, y, $z \in A$, let

$$[x, y] = xy - yx \quad \text{and} \quad [x, y, z] = (xy)z - x(yz).$$

Prove that interchanging any two letters in $[x, y, z]$ changes the sign, and that

$$[xy, z] - x[y, z] - [x, z]y = 3[x, y, z].$$

Hence show that, if A is commutative, then A also is associative. (For all $x, y, z \in A$, $[x + y, x + y, z] = 0 = [x, y + z, y + z]$.)

19.5 Let A be a real alternative division algebra, and, for any $w, x, y, z \in A$, let

$$[w, x, y, z] = [wx, y, z] - x[w, y, z] - [x, y, z]w.$$

Prove that the interchange of any two letters in $[w, x, y, z]$ changes the sign.

(For all $w, x, y, z \in A$,
$w[x, y, z] - [wx, y, z] + [w, xy, z] - [w, x, yz] + [w, x, y]z = 0$.)

19.6 Let A be a real alternative division algebra, let $x, y \in A$ and let $u = [x, y]$, $v = [x, y, z]$. Prove that $[v, x, y] = vu = -uv$.

19.7 Prove that the real linear involution $\mathbf{O} \to \mathbf{O}$; $a \mapsto \tilde{a}$, sending j, l, jl to −j, −l, −jl, respectively, and leaving 1, i, k, il and kl fixed, is an algebra anti-involution of **O**.

19.8 Verify that the map $\beta : \mathbf{H}^2 \to \mathbf{O}$; $x \mapsto x_0 + \mathrm{l}\, x_1$ is a right **H**-linear isomorphism and compute $\beta^{-1}(\widetilde{\beta(x)}\, \beta(y))$, for any $x, y \in \mathbf{H}^2$. (Cf. Exercise 19.7.)

Let $Q = \{(x, y) \in (\mathbf{H}^2)^2 : \tilde{x}_0 y_0 + \tilde{x}_1 y_1 = 1\}$. Prove that for any $(a, b) \in \mathbf{O}^\bullet \times \mathbf{H}$, $(\beta^{-1}(\tilde{a}), \beta^{-1}(a^{-1}(1 + \mathrm{l}\, b))) \in Q$ and that the map

$$\mathbf{O}^\bullet \times \mathbf{H} \to Q;\ (a, b) \mapsto (\beta^{-1}(\tilde{a}), \beta^{-1}(a^{-1}(1 + \mathrm{l}\, b)))$$

is bijective. (Cf. Exercise 9.4.)

19.9 Verify that the map $\gamma : \mathbf{C}^4 \to \mathbf{O}$; $x \mapsto x_0 + \mathrm{j}\, x_1 + \mathrm{l}\, x_2 + \mathrm{j\,k}\, x_3$ is a right **C**-linear isomorphism and compute $\gamma^{-1}(\widetilde{\gamma(x)}\, \gamma(y))$, for any $x, y \in \mathbf{C}^4$.

Let $Q = \{(x, y) \in (\mathbf{C}^4)^2 : \sum_{i \in \mathbf{4}} x_i y_i = 1\}$. Prove that, for any $(a, (b, c, d)) \in \mathbf{O}^\bullet \times \mathbf{C}^3$,

$$(\gamma^{-1}(\tilde{a}), \gamma^{-1}(a^{-1}(1 + \mathrm{j}\, b + \mathrm{l}\, c + \mathrm{j\,k}\, d))) \in Q$$

and that the map

$$\mathbf{O}^\bullet \times \mathbf{C}^3 \to Q;$$

$$(a, (b, c, d)) \mapsto (\gamma^{-1}(\tilde{a}), \gamma^{-1}(a^{-1}(1 + \mathrm{j}\, b + \mathrm{l}\, c + \mathrm{j\,k}\, d)))$$

is bijective.

19.10 Show that the fibres of the restriction of the Hopf map

$$\mathbf{O}^2 \to \mathbf{O}P^1;\ (y_0, y_1) \mapsto [y_0, y_1]$$

to the sphere $S^{15} = \{(y_0, y_1) \in \mathbf{O}^2 : \tilde{y}_0 y_0 + \tilde{y}_1 y_1 = 1\}$ are 7-spheres, any two of which link. (Cf. Exercises 14.1 and 14.2.)

19.11 Let a, b, $c \in \mathbf{O}$. Prove that

$$a(b(a\,c)) = ((a\,b)a)c, \quad ((a\,b)c)b = a(b(c\,b)), \quad a(b\,c)a = (a\,b)(c\,a).$$

These are known as the *Moufang identities* (1935) for an alternative product. They are most easily proved, for **O**, as exercises on the rebracketing pentagon.

20
Topological spaces

In this chapter we shall simply summarise those topological concepts that will be required in the sequel. For most of the proofs the reader is referred to this book's parent, Porteous (1981).

Topological spaces

Cohesion may be given to any set X by singling out a subset $\mathscr{T}$ of the set Sub X of subsets of X such that

(i) $\emptyset \in \mathscr{T}$ and $X \in \mathscr{T}$;
(ii) for all $A, B \in \mathscr{T}$, $A \cap B \in \mathscr{T}$;
(iii) for all $\mathscr{S} \subset \mathscr{T}$, the union of all the elements of $\mathscr{S}$ is in $\mathscr{T}$.

The set $\mathscr{T}$ is said to be a *topology* for X, and the elements of $\mathscr{T}$ are called the *open sets* of the topology. A *topological space* consists of a set X and a topology $\mathscr{T}$ for X.

A set X may have many topologies. For example, for any set X, both Sub X and $\{\emptyset, X\}$ are topologies for X. However, most of the sets that we shall be concerned with will be subsets of a finite-dimensional real linear space, and for any such set there is a natural choice for its topology.

We start by remarking that any *norm* on a finite-dimensional real linear space determines a topology for the space. In Chapter 4 we defined the *norm* $|x|$ of an element x of a positive-definite real quadratic space X to be $\sqrt{|x^{(2)}|}$, and in Proposition 5.31 we listed some of the properties of the map

$$X \to \mathbf{R};\ x \mapsto |x|.$$

These included the following:

(i) for all $x \in X$, $|x| \geq 0$, with $|x| = 0$ if and only if $x = 0$;

(ii) for all $x \in X$ and all $\lambda \in \mathbf{R}$, $|\lambda x| = |\lambda|\,|x|$;

(iii) for all $x, x' \in X$, $|x + x'| \leq |x| + |x'|$.

This last condition, known as the *triangle inequality*, is equivalent, by (ii), with $\lambda = -1$, to

(iv) for all $x, x' \in X$, $|\,|x| - |x'|\,| \leq |x - x'|$.

In practice one does not want to be restricted to norms induced by a scalar product, any map $X \to \mathbf{R}$; $x \mapsto |x|$ satisfying all these properties being said to be a *norm* on the real linear space X.

On occasion a norm will be denoted by $\|\ \ \|$ rather than by $|\ \ |$.

A *normed linear space* consists of a real linear space X and a norm $|\ \ |$ on X.

The *distance* between any two points a and b of a normed linear space X with norm $|\ \ |$ is defined to be the non-negative real number $|a - b|$. It follows from (i) that $a = b$ if and only if $|a - b| = 0$.

Let a be a point of a normed linear space X. Then a *neighbourhood* of a is a subset A of X such that, for some positive real number δ, all points of X within a distance δ of a belong to A. A subset A of X is said to be *open* in X if it is a neighbourhood of each of its points. A point b of X is said to be a *boundary point* of A if, for any positive δ, there are within the distance δ of b at least one point of A and one point of the complement of A. A subset A of X is open if and only if none of its boundary points is in A.

A subset B of X is *closed* in X if its complement in X is open in X. A subset B of X is closed if and only if all of its boundary points are in B.

Both the null set $\emptyset$ and the whole normed linear space X are both open and closed in X.

Proposition 20.1 *For any normed linear space X the set of subsets of X open with respect to the norm is a topology for X.*

Normed linear spaces are examples of Hausdorff spaces, a topological space X being said to be *Hausdorff* if any two distinct points a and b of X have disjoint neighbourhoods A and B.

We shall see presently that any two norms on a finite-dimensional real linear space determine the *same* topology. This is the topology assumed to be chosen if nothing is said to the contrary.

Continuity

Let X and Y be topological spaces. A map $f : X \to Y$ is said to be *continuous* if the inverse image in X of any open set of Y is open in X. For a map $f : X \to Y$ between normed linear spaces X and Y this is equivalent to the statement that, for any point of X and for any positive real number ε (in general depending on a), there is a positive real number δ such that

$$|x - a| < \delta \Rightarrow |f(x) - f(a)| < \varepsilon.$$

Proposition 20.2 *Let X and Y be topological spaces and let $f : X \to Y$ be a constant map. Then f is continuous.*

Proposition 20.3 *Let X be a topological space. Then the identity map 1_X is continuous.*

Proposition 20.4 *Let W, X and Y be topological spaces and let $g : W \to X$ and $f : X \to Y$ be continuous maps. Then the composite map $fg : W \to X$ is continuous.*

The inverse of a bijective continuous map need not be continuous. For example, let X be any set with more than one element. Then the map

$$1_X : X, \operatorname{Sub} X \to X, \{\emptyset, X\}$$

is continuous, but its inverse is not continuous.

A bijective continuous map whose inverse is also continuous is said to be a *homeomorphism*.

Two topological spaces X and Y are said to be *homeomorphic*, $X \cong Y$, if there exists a homeomorphism $f : X \to Y$. The relation $\cong$ is an equivalence on any set of topological spaces.

Subspaces

Let W be a subset of a topological space X. Then the *induced topology*, or *subspace topology*, on W is the smallest topology on W for which the inclusion map is continuous, a subset C of W being *open* in W if and only if there is an open set A in X such that $C = A \cap W$. A subset C that is open in W need not be open in X.

Proposition 20.5 *Let $f : X \to Y$ be a continuous map and let W be a subspace of X. Then the map $f|W : W \to Y$ is continuous.*

In thc scquel it will bc often be convenient to denote by $f : X \rightarrowtail Y$ a map between topological spaces X, the *source*, and Y, the *target*, whose domain, $\operatorname{dom} f$, is a subset of X, not necessarily the whole of X. In most cases $\operatorname{dom} f$ will be an open subset of X, but whether this is the case or not $\operatorname{dom} f$ will be supposed to have assigned to it the topology induced from X. One then has the following extension of Proposition 20.4.

Proposition 20.6 *Let W, X and Y all be topological spaces and let $g : W \rightarrowtail X$ and $f : X \rightarrowtail Y$ be continuous maps. Then the composite map $f g : W \rightarrowtail X$, with domain $g^{-1}(\operatorname{dom} f)$, is continuous.*

More on normed linear spaces

Proposition 20.7 *Let X and Y be normed linear spaces. Then the map*

$$X \times Y \to \mathbf{R};\ (x, y) \mapsto \max\{|x|, |y|\}$$

is a norm on $X \times Y$.

This norm is called the *product norm* on $X \times Y$.

A linear map $t : X \to Y$ between normed linear spaces X and Y is said to be *bounded* if there is a real number K such that, for all $x \in X$, $|t(x)| \leq K|x|$. When such a number K exists the set $\{|t(x)| : |x| \leq 1\}$ is bounded above by K. This set is non-null, since it contains 0, so it has a supremum. The supremum is denoted by $|t|$ and is called the *gradient norm* of t.

Proposition 20.8 *Let $t : X \to Y$ be a bounded linear map between normed linear spaces X and Y. Then, for all $x \in X$, $|t(x)| \leq |t|\,|x|$, $|t|$ being the smallest real number K such that, for all $x \in X$, $|t(x)| \leq K\,|x|$.*

For any normed linear spaces X and Y the set of bounded linear maps is denoted by $L(X, Y)$.

Proposition 20.9 *Let X and Y be normed linear spaces. Then the gradient norm is a norm on $L(X, Y)$.*

Norms $|\ \ |$ and $||\ \ ||$ on a real linear space X are said to be *equivalent* if there exist positive real numbers H and K such that, for all $x \in X$,

$$||x|| \leq H\,|x| \quad \text{and} \quad |x| \leq K\,||x||.$$

Proposition 20.10 *Equivalent norms on a real linear space X define the same topology on X.*

Inversion

Let X and Y be finite-dimensional linear spaces. Then the set of linear maps of maximal rank from X to Y will be denoted by $GL(X, Y)$. When $\dim X = \dim Y$ this is just the set of invertible maps from X to Y. Proposition 20.12 is concerned with the continuity of the inversion map $L(X, Y) \rightarrowtail L(Y, X);\ t \mapsto t^{-1}$ with domain $GL(X, Y)$. First we have a preparatory lemma.

Lemma 20.11 *Let X be a finite-dimensional linear space with norm $|\ \ |$, let $U \in L(X, X)$ and let $|u| < 1$. Then $1_X - u \in GL(X, X)$ and $|(1_X - u)^{-1}| \le (1 - |u|)^{-1}$. Moreover the map*

$$L(X, X) \rightarrowtail L(X, X);\ t \mapsto t^{-1}$$

is defined on a neighbourhood of $1\ (= 1_X)$ and is continuous at 1.

Proposition 20.12 *Let X and Y be linear spaces of the same finite dimension. Then the map $L(X, Y) \rightarrowtail L(Y, X);\ t \mapsto t^{-1}$ is continuous, its domain, $GL(X, Y)$, being open in $L(X, Y)$.*

Proof This follows at once from Lemma 20.11 and the decomposition

$$\begin{array}{ccccccc} L(X, Y) & \longrightarrow & L(X, X) & \rightarrowtail & L(X, X) & \longrightarrow & L(Y, X) \\ t & \mapsto & u^{-1}t & \mapsto & t^{-1}u & \mapsto & t^{-1}, \\ u & \mapsto & 1 & \mapsto & 1 & \mapsto & u^{-1}. \end{array}$$

□

Even if $\dim X \neq \dim Y$ we still have the following.

Proposition 20.13 *Let X and Y be any two finite-dimensional linear spaces. Then $GL(X, Y)$ is an open subset of $L(X, Y)$.*

Quotient spaces and product spaces

A map $f : X \rightarrow Y$ is said to be a *partition* of X, and Y to be the *quotient* of X by f, if f is surjective, if each element of Y is a subset of X, and if the fibre of f over any $y \in Y$ is the set y itself. Any map $f : X \rightarrow Y$ with domain a given set X induces a partition of X, f_{par} defined by

$f_{par}(a) = \{x \in X : f(x) = f(a)\}$. Any topology $\mathcal{T}$ on X then induces a topology on Y called the *quotient topology* for Y, this being the largest topology on Y for which the partition is continuous, a subset B of Y being *open* in Y if and only if its inverse image by f is open in X.

The subspace and quotient topologies have particular relevance to the canonical decomposition

$$X \xrightarrow{f_{par}} \operatorname{coim} f \xrightarrow{f_{bij}} \operatorname{im} f \xrightarrow{f_{inc}} Y$$

of a continuous map $f : X \to Y$, the subspace $\operatorname{im} f$ of Y being assigned the subspace topology and $\operatorname{coim} f$ the quotient topology.

The map f_{bij} need not be a homeomorphism. A continuous injection $f : X \to Y$ such that f_{bij} is a homeomorphism is said to be a *(topological) embedding* of X in Y. A continuous surjection $f : X \to Y$ such that f_{bij} is a homeomorphism is said to be a *(topological) projection* of X on to Y.

Proposition 20.14 *Let W, X and Y all be topological spaces and let $g : W \to X$ and $f : X \to Y$ be maps, whose composite $fg : W \to Y$ is continuous. Then, if f is an embedding, g is continuous and, if g is a projection, f is continuous.*

For any continuous map $f : X \to Y$ the inverse image of any open set is open and the inverse image of any closed set is closed. Such a map is said to be *open* if the forward image of any open set of X is open in Y and to be *closed* if the forward image of any closed set in X is closed in Y.

The following proposition generalises the construction of the subspace topology.

Proposition 20.15 *Let W be a set, X and Y topological spaces and $p : W \to X$ and $q : W \to Y$ maps. Define a subset C of W to be* open *in W if and only if C is the union of a set of subsets of W each of the form $p^{-1}A \cap q^{-1}B$, where A is open in X and B is open in Y. Then the set of open subsets of W is a topology, being the smallest topology for W such that both p and q are continuous.*

The topology so defined is said to be the topology for W *induced* by the maps p and q from the topologies for X and Y.

When $W = X \times Y$ and $(p, q) = 1_W$ the topology induced on W by p and q is called the *product topology* for W.

Proposition 20.16 *Let X and Y be normed linear spaces. Then the product norm on $X \times Y$ induces the product topology on $X \times Y$.*

Proposition 20.17 *A map $(f, g) : W \to X \times Y$ is continuous if and only if each of its components $f : W \to X$ and $g : W \to Y$ is continuous, W, X and Y being topological spaces.*

Proposition 20.18 *Let X and Y be topological spaces and let $y \in Y$. Then the injection $X \to X \times Y$; $x \mapsto (x, y)$ is an embedding and the product projection $X \times Y$; $(x, y) \mapsto x$ is a topological projection.*

Compact sets

An *(open) cover* for a topological space $X, \mathcal{T}$ is, by definition, a subset $\mathcal{S}$ of $\mathcal{T}$ such that $\bigcup \mathcal{S} = X$.

Proposition 20.19 *Let B be a subset of a topological space X and let $\mathcal{S}$ be a cover for X. Then B is open in X if and only if, for each $A \in \mathcal{S}$, $B \cap A$ is open in A.*

Corollary 20.20 *Two topologies on a set X are the same if and only if the induced topologies on each of the elements of some cover for X are the same.*

It follows that in studying a topological space X nothing is lost by choosing as cover for X and studying separately each element of the cover.

Let W be a subspace of a topological space X. A set $\mathcal{S}$ of open sets of X such that $W \subset \bigcup \mathcal{S}$ will be called an *X-cover* for W. The set $\{A \cap W : A \in \mathcal{S}\}$ is then a cover for W, called the *induced* cover.

For example, the set $\{]-1, 1[,]0, 2[\}$ is an $\mathbf{R}$-cover for the closed interval $[0, 1]$. The induced cover is the set $\{[0, 1[,]0, 1]\}$.

It follows from the definition of the induced topology that every cover for W is indexed by some X-cover for W (generally not unique).

Theorem 20.21 (The Heine-Borel Theorem.) *Let $\mathcal{S}$ be an $\mathbf{R}$-cover of a bounded closed interval $[a, b] \subset \mathbf{R}$. Then a finite subset $\mathcal{S}'$ of $\mathcal{S}$ covers $[a, b]$.*

Corollary 20.22 *Let $\mathcal{P}$ be any cover for $[a, b]$. Then there exists a finite subset $\mathcal{P}'$ of $\mathcal{P}$ covering $[a, b]$.*

A topological space X is said to be *compact* if for *each* cover $\mathcal{S}$ for X a finite subset $\mathcal{S}'$ of $\mathcal{S}$ covers X. Theorem 20.21 states that every bounded closed interval of $\mathbf{R}$ is compact. By contrast, the interval $]0, 1[$ is not compact, since no finite subset of the cover $\{](n+1)^{-1}, 1];\ n \geq 0\}$ covers $]0, 1]$. More generally one has the following.

Theorem 20.23 *A subset of $\mathbf{R}^n$ is compact if and only if it is closed and bounded.*

There is no short proof! One route is provided by the following sequence of propositions. The first two are easy.

Proposition 20.24 *A compact subspace A of a normed linear space X is bounded.*

Proof Consider the set $\mathcal{S}$ of all balls in X of radius 1 with centre some point of A. □

Proposition 20.25 *A closed subset A of a compact space X is compact.*

Proof Let $\mathcal{S}$ be a cover of A by open subsets of X. Then consider the cover $\mathcal{S} \cup \{X \backslash A\}$ of X. □

The next relates compactness to continuity.

Proposition 20.26 *Let $f : X \to Y$ be a continuous surjection and let X be compact. Then Y is compact.*

Corollary 20.27 *Let X be a compact space and let $f : X \to Y$ be a partition of X. Then the quotient Y is compact.*

Proposition 20.28 *Let W be a compact subspace of a Hausdorff space X. Then W is closed in X.*

Corollary 20.29 *Any compact subset of a normed linear space is closed.*

Putting together Corollary 20.22, Proposition 20.24, Proposition 20.25 and Corollary 20.29 we obtain the following characterisation of compact subsets of $\mathbf{R}$.

Proposition 20.30 *A subset of $\mathbf{R}$ is compact if and only if it is closed and bounded.*

Also from Proposition 20.24 and Corollary 20.29 we have part of Theorem 20.23.

Corollary 20.31 *Let* $\mathbf{R}^n$ *be furnished with the product norm. Then any compact subset of* $\mathbf{R}^n$ *is closed and bounded.*

The inverse image by a continuous map $f : X \to Y$ of a compact subset of Y need not be compact in X. When this is so we say that the map f is *compact*.

Proposition 20.32 *A closed continuous map* $f : X \to Y$ *is compact if and only if each fibre is compact.*

Proposition 20.33 *Let* X *and* Y *be topological spaces,* Y *being compact. Then the projection* $X \times Y \to X;\ (x, y) \mapsto x$ *is* closed, *that is the image of any closed set is closed.*

Theorem 20.34 *Let* X *and* Y *be non-null compact topological spaces. Then* $X \times Y$ *is compact.*

Proof By Proposition 20.33 the projection $X \times Y;\ (x,y) \mapsto x$ is closed, and therefore compact, by Proposition 20.32. But X is compact. So $X \times Y$ is compact. □

It follows at once that the product of any finite number of closed bounded intervals in **R** is compact.

The final stage in the proof of Theorem 20.23 is then provided by the following proposition.

Proposition 20.35 *Any bounded closed subset of* $\mathbf{R}^n$ *is compact.*

Connectedness

The simplest intuitive example of a disconnected set is the set $\mathbf{2} = \{0, 1\}$, the standard set with two elements. Of the four topologies for this set only the discrete topology, namely that in which each subset is both open and closed, is Hausdorff. Let it be assigned this topology.

A non-null topological space X is said to be *disconnected* if there is a continuous surjection $\pi : X \to \mathbf{2}$, and to be *connected* if every continuous map $f : X \to \mathbf{2}$ is constant.

Any non-null topological space is easily seen to be either connected or disconnected, but not both. The null space is considered to be neither connected nor disconnected.

Proposition 20.36 *A topological space X is disconnected if and only if it is the union of two disjoint non-null open sets of X.*

Proposition 20.37 *Any bounded closed interval $[a, b]$ of $\mathbf{R}$ is connected.*

Theorem 20.38 *A non-null subset C of $\mathbf{R}$ is connected if and only if it is an interval.*

In particular $\mathbf{R}$ is connected.

Proposition 20.39 *Let $f : X \to Y$ be a continuous surjection, and suppose that X is connected. Then Y is connected.*

Corollary 20.40 *Let $f : X \to Y$ be a continuous map and let A be a connected subset of X. Then* $\operatorname{graph}(f|A)$ *is a connected subset of $X \times Y$ and $f(A)$ is a connected subset of Y.*

Theorem 20.41 (The intermediate value theorem.) *Let $f : X \to \mathbf{R}$ be a continuous map, let X be connected and let $c, d \in f(X)$. Then the interval $[c, d]$ is a subset of $f(X)$.*

Proposition 20.42 *Let X be a topological space such that for any $a, b \in X$ there exists a continuous map $f : [0, 1] \to X$ such that $f(0) = a$ and $f(1) = b$. Then X is connected.*

Proposition 20.43 *For any positive integer n the unit sphere S^n is a connected subset of $\mathbf{R}^{n+1}$.*

Proposition 20.44 *Let X and Y be non-null topological spaces. Then $X \times Y$ is connected if and only if X is connected and Y is connected.*

Proposition 20.44 may also be regarded as a particular case of the following proposition, whose proof recalls the proof of Proposition 1.5.

Proposition 20.45 *Let $f : X \to Y$ be a topological projection of a topological space X on to a connected topological space Y, each of the fibres of f being connected. Then X is connected.*

Proof Let $h : X \to \mathbf{2}$ be a continuous map. Since the fibres of f are connected, the restriction of h to any fibre is constant. So there exists a map $g : Y \to \mathbf{2}$ defined, for all $y \in Y$, by the formula $g(y) = h(x)$, for any $x \in f^{-1}\{y\}$, such that $h = gf$. Since h is continuous and since f is a projection, g is continuous, by Proposition 20.14, and therefore constant, since Y is connected. So h is constant.

Therefore X is connected. □

Exercises

20.1 Prove that the set $\mathbf{2} = \{0, 1\}$ has four different topologies that may be assigned to it and that the set $\mathbf{3} = \{0, 1, 2\}$ has twenty-nine. Which of these are Hausdorff and in which is each subset of the space either open or closed?

20.2 Prove that if $x \in]0, \infty[$ then $\dfrac{x-1}{x+1} \in]-1, 1[$ and that the map $]0, \infty[\to]-1, 1[;\ x \mapsto \dfrac{x-1}{x+1}$ is a homeomorphism.

20.3 Prove that the intervals $]-1, 1[$ and $[-1, 1]$ are not homeomorphic.

(There are various proofs. One uses compactness. Another, which considers the complements of points of the space, uses connectedness.)

20.4 Let $X = \{-1, 1\} \times]-1, 1[\subset \mathbf{R}^2$, and consider the partition $\pi : X \to Y$ of X which identifies $(-1, x)$ with $(1, x)$, for all $x \in]-1, 0[$, with Y assigned the quotient topology. Prove that Y is not a Hausdorff space.

21
Manifolds

Manifolds are sets that locally are like linear spaces. In particular they have dimension, non-singular curves being one-dimensional manifolds and non-singular surfaces being two-dimensional manifolds. There are several levels of sophistication in their definition. Our interest here will be in smooth manifolds, so we start by reviewing differentiability. This done, we introduce smooth submanifolds of linear spaces, and then finally smooth manifolds more generally, and their tangent spaces.

Tangency

Let f and $t : X \rightarrowtail Y$ be maps between finite-dimensional real linear spaces X and Y, and let $a \in X$. We say that f is *tangent to t at a*, or that f and t are *mutually tangent at a*, if (i) $\operatorname{dom} f$ and $\operatorname{dom} t$ are neighbourhoods of a in X, (ii) $f(a) = t(a)$ and (iii) $\lim_{x \to a} \dfrac{|f(x) - t(x)|}{|x - a|} = 0$, where the modulus signs in (iii) denote the assigned norms.

Note that the particular positions of the origins in X and Y are not relevant to the definition. Accordingly in assessing the tangency of a pair of maps $f : X \rightarrowtail Y$ and $t : X \rightarrowtail Y$ at a particular point $a \in X$ there is normally no loss of generality in assuming that $a = 0$ in X and that $f(a) = t(a) = 0$ in T.

The following results are all basic to the theory. Most of the proofs will be omitted. The interested reader is referred to Porteous (1981) for these. It will be assumed tacitly here that W, X, Y and Z are all *finite-dimensional*. Most of the story does generalise to infinite-dimensional complete normed linear spaces (Banach spaces), but we shall not require this generalisation here. For this also the reader is referred to the 1981 edition of my former book.

Proposition 21.1 *Let f, g and h be maps from X to Y, and let f be tangent to g and g tangent to h at $a \in X$. Then f is tangent to h at a.*

Proposition 21.2 *Let f and t be maps from X to Y, tangent at $a \in X$. Then f is continuous at a if and only if t is continuous at a.*

Proposition 21.3 *Maps (f, g) and $(t, u) : W \rightarrowtail X \times Y$ are tangent at $c \in W$ if and only if f and t are tangent at c and g and u are tangent at c.*

The generalisation to products with any number of factors is obvious.

In the next proposition it is convenient to introduce the notations $(-, b)$ and $(a, -)$ for the affine maps

$$X \to X \times Y;\ x \mapsto (x, b) \text{ and } Y \to X \times Y;\ y \mapsto (a, y),$$

a being any point of X and b any point of Y.

Proposition 21.4 *Let $f : X \times Y \rightarrowtail Z$ be tangent to $t : X \times Y \rightarrowtail Z$ at (a, b). Then $f(-, b)$ is tangent to $t(-, b)$ at a and $f(a, -)$ is tangent to $t(a, -)$ at b.*

Theorem 21.5 (The chain rule.) *Let $f : X \rightarrowtail Y$ be tangent to an affine map $t : X \to Y$ at $a \in X$ and let $g : Y \rightarrowtail Z$ be tangent to an affine map $u : Y \to Z$ at $b = f(a)$. Then $g : X \rightarrowtail Z$ is tangent to $ut : X \rightarrowtail Z$ at a.*

The proof is not straightforward, and is best broken into two stages, in the first of which (the easier half) one proves that uf is tangent to ut at a and in the second of which (the harder half) one proves that gf is tangent to uf at a.

Until now we have assumed that the sources and targets of the various maps involved have assigned norms. The next proposition shows that the concept of tangency does not depend on the actual norms employed.

Proposition 21.6 *Let X' and X'' be finite-dimensional real normed linear spaces with the same underlying linear space X, let Y' and Y'' be finite-dimensional real normed linear spaces with the same underlying linear space Y, and let f and t be maps from X to Y. Then the maps f and $t : X' \rightarrowtail Y'$ are tangent at a point $a \in X$ if and only if the maps f and $t : X'' \rightarrowtail Y''$ are tangent at a.*

Proof The map $f : X'' \rightarrowtail Y''$ admits the decomposition

$$X'' \overset{1_X}{\rightarrowtail} X' \overset{f}{\rightarrowtail} Y' \overset{1_Y}{\rightarrowtail} Y''$$

and the map $t : X'' \rightarrowtail Y''$ the decomposition

$$X'' \overset{1_X}{\rightarrowtail} X' \overset{t}{\rightarrowtail} Y' \overset{1_Y}{\rightarrowtail} Y''.$$

Since the identity linear maps are certainly tangent to each other it follows, by Theorem 21.5, that if f and $t : X' \rightarrowtail Y'$ are tangent at $a \in X$ then f and $t : X'' \rightarrowtail Y''$ are tangent at a. □

In the infinite-dimensional case things are different – linear maps, even the identity map, need not be continuous, if the norms on either side are chosen appropriately.

The next theorem is a preliminary theorem on inverse maps. The *inverse function theorem* comes later!

Theorem 21.7 *Let $f : X \rightarrowtail Y$ be an injective map, tangent at a point $a \in X$ to a bijective affine map $t : X \rightarrowtail Y$ (so $\dim Y = \dim X$), and let $f^{-1} : Y \rightarrowtail X$ be defined in a neighbourhood of $b = f(a) = t(a)$ and be continuous at b. Then f^{-1} is tangent to t^{-1} at b.*

Theorems 21.5 and 21.7 indicate the special role played by affine maps in the theory of tangency. This role is further clarified by the following intuitively obvious proposition.

Proposition 21.8 *Let t and $u : X \to Y$ be affine maps, mutually tangent at a point a of X. Then $t = u$.*

Corollary 21.9 *A map $f : X \rightarrowtail Y$ is tangent at a point a to at most one affine map $t : X \to Y$.*

It may seem from this that Theorem 21.7 is nothing more than a corollary to Theorem 21.5. For if $f : X \rightarrowtail Y$ is an injective map, tangent at $a \in X$ to the affine map $t : X \to Y$, and if $f^{-1} : Y \rightarrowtail X$ is tangent at $b = f(a)$ to the affine map $u : Y \to X$ it follows, by Theorem 21.5, that $f^{-1}f$ is tangent to ut at a and ff^{-1} is tangent to tu at b. Now $f^{-1}f$ is also tangent to 1_X at a, and ff^{-1} is tangent to 1_Y at b, and therefore, by the above corollary, $ut = 1_X$ and $tu = 1_Y$. That is $u = t^{-1}$. However, Theorem 21.5 does not prove the *existence* of an affine map u tangent to f^{-1} but only determines it if it does exist.

By Corollary 21.9 a map $f : X \rightarrowtail Y$ is tangent at any point $a \in X$ to *at most one* affine map $t : X \to Y$, this map being uniquely determined by its linear part, since necessarily $t(a) = f(a)$. This linear part is called the *differential*, or more strictly the *value of the differential* of f at a, and will be denoted by dfa, the map f then being said to be *differentiable at* a. The *differential*, df of f is the map

$$df : X \rightarrowtail L(X, Y);\ x \mapsto dfx,$$

the map f being said to be *differentiable* if $\text{dom}(df) = \text{dom}(f)$, that is, if f is differentiable at every point of its domain.

Note that it follows from condition (i) in the definition of tangency that according to this definition the domain of a differentiable map $f : X \rightarrowtail Y$ is an *open* subset of the linear space X.

The differential df of a differentiable map $f : X \rightarrowtail Y$ need not be continuous. If it is the map f is said to be *continuously differentiable* or C^1.

Frequently in applications $X = \mathbf{R}^n$ and $Y = \mathbf{R}^p$. In that case the linear map dfx at a point x of $\text{dom} f$ may be represented by its matrix, a $p \times n$ matrix over $\mathbf{R}$ known as the *Jacobian matrix* of f at x.

In computational work the notations of Leibniz are frequently in use, the equation $y' = dfx(x')$, where $x \in X$, $x' \in X$ and $y' \in Y$, being often written as $dy = (dy/dx)dx$, the (i, j)th entry in the Jacobian matrix of dy/dx being denoted by $\partial y_i/\partial x_j$.

Numerous properties of differentials follow from the propositions and theorems already stated. In stating these the letters W, X, Y and Z will continue to denote finite-dimensional real linear spaces.

Proposition 21.10 *An affine map* $t : X \to Y$ *is continuously differentiable, its differential* $dt : X \to L(X, Y)$ *being constant, with constant value the linear part of* t.

The next proposition is just a restatement of Proposition 21.3, in the case where t and u are affine maps, the extension to continuously differentiable maps following at once from Proposition 20.17.

Proposition 21.11 *A map* $(f, g) : W \rightarrowtail X \times Y$ *is continuously differentiable at a point* $w \in W$ *if and only if each of the maps* $f : W \rightarrowtail X$ *and* $g : W \rightarrowtail Y$ *is continuously differentiable at* w, *with*

$$d(f, g)w = (dfw, dgw).$$

Next, a restatement of Proposition 21.4.

Proposition 21.12 *Let* $f : X \times Y \rightarrowtail Z$ *be continuously differentiable at a point* $(a, b) \in X \times Y$. *Then the map* $f(-, b) : X \rightarrowtail Z$ *is continuously differentiable at* a *and the map* $f(a, -) : Y \rightarrowtail Z$ *is continuously differentiable at* b, *with, for all* $x \in X$, $y \in Y$,

$$df(a, b)(x, y) = u(x) + v(y),$$

where $u = d(f(-, b))a$ *and* $v = d(f(a, -))b$.

Both the last two propositions have obvious generalisations to the case where the product of two linear spaces is replaced by a product of n spaces, for any positive number n.

The converse to Proposition 21.12 is true, but not quite immediate, since it is does not necessarily hold if the condition that the differentials are continuous is dropped. An example is provided by the map $f : \mathbf{R}^2 \rightarrow \mathbf{R}$ defined by the formula

$$f(0, 0) = 0 \quad \text{and} \quad f(x, y) = 2xy/(x^2 + y^2), \text{ for } (x, y) \neq (0, 0),$$

for the partial differentials of f exist at (0, 0) although f is not differentiable there.

The proof of the converse requires the increment formula, given below as Theorem 21.27.

The next proposition will be of frequent application.

Proposition 21.13 *Let* $\beta : X \times Y \rightarrow Z$ *be a bilinear map. Then, for any* $(a, b) \in X \times Y$, β *is tangent at* (a, b) *to the affine map*

$$X \times Y \rightarrow Z; \; (x, y) \mapsto \beta(x, b) + \beta(a, y) - \beta(a, b),$$

that is, β *is differentiable, and, for all* $(a, b), (x, y) \in X \times Y$,

$$d\beta(a, b)(x, y) = \beta(x, b) + \beta(a, y),$$

$d\beta$ *being linear, and* β *continuously differentiable.*

In the above do not confuse the linearity of $d\beta(a, b)$ with the linearity of $d\beta$.

Proposition 21.13 has, by Theorem 21.5, the following corollary.

Corollary 21.14 *Let* $\beta : X \times X \rightarrow Z$ *be a bilinear map. Then, for any* $a \in X$, *the induced quadratic map* $\eta : X \rightarrow Z; \; x \mapsto \beta(x, x)$ *is tangent at* a *to the affine map*

$$X \rightarrow Z; \; x \mapsto \beta(x, a) + \beta(a, x) - \beta(a, a),$$

that is η is differentiable, with, for all $a, x \in X$,

$$d\eta a(x) = \beta(x, a) + \beta(a, x),$$

$d\eta$ being linear and η continuously differentiable.

This generalises the familiar formula $\frac{d}{dx}x^2 = 2x$ of elementary real calculus, to which it reduces if $X = \mathbf{R}$ and $\cdot$ is ordinary multiplication.

There is a similar formula for the differential of a multilinear map.

Proposition 21.15 *Let X be a finite-dimensional* **K***-linear space, where* $\mathbf{K} = \mathbf{R}$ *or* $\mathbf{C}$. *Then the map*

$$\det : L(X, X) \to \mathbf{K};\ t \mapsto \det t$$

is continously differentiable, $d(\det)t$ being surjective if and only if $\operatorname{rk} t = \dim X$ *or* $\dim X - 1$.

The chain rule, Theorem 21.5, may be restated in terms of differentials and extended as follows.

Theorem 21.16 *Let $f : X \rightarrowtail Y$ be continuously differentiable at $a \in X$ and let $g : Y \rightarrowtail Z$ be continuously differentiable at $f(a) \in Y$. Then $gf : X \rightarrowtail Z$ is continuously differentiable at a.*

Proof The part of the theorem that concerns differentiability of gf is just a restatement of Theorem 21.5. The continuity of $d(gf)$ follows from Propositions 20.5 and 20.17, since the restriction of $d(gf)$ to $(\operatorname{dom} df) \cap f^{-1}(\operatorname{dom}(dg))$ decomposes as follows:

$$X \overset{(df,(dg)f)}{\rightarrowtail} L(X, Y) \times L(Y, Z) \overset{\text{composition}}{\longrightarrow} L(X, Z),$$

composition being bilinear and so continuous. □

The formula in Theorem 21.16 may be abbreviated to

$$d(gf) = ((dg)f) \circ df,$$

$\circ$ denoting composition of values.

In the Leibniz notation this says that if f and g are differentiable maps and if $y = f(x)$ and $z = g(y)$ then

$$\frac{dz}{dx} = \frac{dz}{dy}\frac{dy}{dx}.$$

The next proposition complements Proposition 20.12.

Proposition 21.17 *Let X and Y be* **K***-linear spaces of the same finite dimension, where* $\mathbf{K} = \mathbf{R}$, $\mathbf{C}$ *or* $\mathbf{H}$. *Then the map*

$$\psi : L(X, Y) \rightarrowtail L(Y, X);\ t \mapsto t^{-1}$$

is continuously differentiable and, for all $u \in GL(X, Y)$, and all $t \in L(X, Y)$,

$$d\psi u(t) = du^{-1}(t) = -u^{-1}\, t\, u^{-1}.$$

This generalises the formula $\frac{d}{dx}x^{-1} = -x^{-2}$ of elementary real calculus, to which it reduces when $X = Y = \mathbf{R}$.

Theorem 21.7 may also be restated and extended.

Theorem 21.18 *Let $f : X \rightarrowtail Y$ be an injective map, differentiable at $a \in X$, $dfa : X \to Y$ being bijective (so $\dim Y = \dim X$), and let $f^{-1} : Y \rightarrowtail X$ be defined in a neighbourhood of $f(a)$ in Y and continuous at $f(a)$. Then f^{-1} is differentiable at $f(a)$ and*

$$d(f^{-1})(f(a)) = (dfa)^{-1}.$$

Moreover, if df is continuous and if f^{-1} is continuous with open domain then $d(f^{-1})$ is continuous.

Proof The first part is Theorem 21.7. The second part follows, by Propositions 20.12 and 20.5, from the following decomposition of $d(f^{-1})$.

$$Y \overset{f^{-1}}{\rightarrowtail} X \overset{df}{\rightarrowtail} L(X, Y) \overset{\text{inversion}}{\rightarrowtail} L(Y, X) \quad .$$

□

The differential of a more complicated map can often be computed by decomposing the map in some manner and then applying several of the above propositions and theorems. The next proposition is a simple example of this.

Proposition 21.19 *Let $|\ \ |$ be the quadratic norm on a finite-dimensional real quadratic space X; that is, for any $x \in X$, $|x| = \sqrt{x \cdot x}$. Then the map $X \rightarrowtail \mathbf{R};\ x \mapsto |x|$, with domain $X \setminus \{0\}$, is continuously differentiable, with $dfx(x') = \dfrac{x \cdot x'}{|x|}$.*

Higher differentials

Until now we have only been concerned with the first differential of a map $f : X \rightarrowtail Y$ between finite-dimensional real linear spaces. Higher-order differentials are defined recursively, by the formula

$$d^{n+1}f = d(d^n f), \text{ for all } n,$$

where, by convention $d^0 f = f$. In general the targets of these differentials become progressively more complicated. For example, the first three differentials of the map $f : X \rightarrowtail Y$ are of the form

$$\begin{aligned} df : X &\rightarrowtail L(X, Y), \\ d^2 f : X &\rightarrowtail L(X, L(X, Y)), \\ d^3 f : X &\rightarrowtail L(X, L(X, L(X, Y))), \end{aligned}$$

respectively, though in the particular case that $X = \mathbf{R}$ each of the targets has a natural identification with Y. The map f is said to be *k-smooth*, or C^k, at a point $a \in X$, if $d^k f$ is defined on a neighbourhood of a and is continuous at a and to be *(infinitely) smooth*, or C^∞, at a if, for each k, $d^k f$ is defined in a neighbourhood of a.

When f is C^∞ at a there is, for each $x \in X$, a sequence on Y:

$$n \mapsto \sum_{m \in n} \frac{1}{m!}(d^m f a)(x)^k,$$

where $(x)^k$ denotes $(x)(x)...(x)$, with (x) occurring k times. This sequence is known as the *Taylor series* of f at a with *increment* x. The map f is said to be *analytic* at a if, for some $\delta > 0$, this sequence is convergent whenever $|x| < \delta$, with limit $f(a + x)$.

In real analysis a smooth map is not necessarily analytic. We shall only require at most smoothness in what follows, and will not be concerned with analyticity. The linear spaces involved will all be assumed to be real finite-dimensional linear spaces.

Proposition 21.20 *Any linear or bilinear map is C^∞.*

Proposition 21.21 *Let $(f, g) : W \rightarrowtail X \times Y$ be any map, W, X and Y being linear spaces. Then (f, g) is C^k at a point $a \in W$ if and only if f and g are each C^k, where k is a positive integer or ∞.*

Proposition 21.22 *Let $f : X \rightarrowtail Y$ be C^k at $a \in X$ and let $g : Y \rightarrowtail Z$ be C^k at $b = f(a)$, where k is a positive integer or ∞, X, Y and Z being linear spaces. Then $gf : X \rightarrowtail Z$ is C^k at a.*

Though we are omitting proofs here it is perhaps worth remarking that the proof of Proposition 21.21 uses a special case of Proposition 21.22 and conversely. Both inductions should therefore be carried out simultaneously.

Proposition 21.23 *For any linear spaces X and Y of the same finite dimension the inversion map*

$$L(X, Y) \rightarrowtail L(Y, X);\ t \mapsto t^{-1}$$

is C^∞.

Proposition 21.24 *Let $f : X \rightarrowtail Y$ be a map satisfying at a point $a \in X$ the same conditions as in Theorem 21.18 and suppose that f is C^k at a, where k is a positive integer or ∞. Then f^{-1} is C^k at $b = f(a)$.*

Finally the second differential of a map is *symmetric* in the following sense.

Proposition 21.25 *Let $f : X \rightarrowtail Y$ be a twice-differentiable map, X and Y being linear spaces. Then, for any $a \in \dim f$, and any $x, x' \in X$,*

$$(d^2 f a(x'))(x) = (d^2 f a(x))(x').$$

The proof of this is technical and is another example of one that requires the increment formula of Theorem 21.27 below.

What this means is that the *slots* in the twice-linear map $d^2 fa \in L(X, L(X, Y))$ can be filled in either order by vectors of X without affecting the result. The higher derivatives likewise are symmetric.

Corollary 21.26 *Let $f : X_0 \times X_1 \rightarrowtail Y$ be a twice-differentiable map, X_0, X_1 and Y being finite-dimensional linear spaces. Then, for any $(a_0, a_1) \in \operatorname{dom} f$ and any $x_0 \in X$, $x_1 \in X_1$,*

$$d_1 d_0 f(a_0, a_1)(x_1)(x_0) = d_0 d_1 f(a_0, a_1)(x_0)(x_1).$$

In all that follows *smooth* will mean C^k for any appropriate $k \geq 1$, including $k = \infty$. On a first reading, however, one should take $k = 1$, as this is always the first case to establish, the extension to greater values of k being routine.

The inverse function theorem

Let X and Y be real linear spaces of the same finite dimension and let A be an open subset of X and B an open subset of Y. Then a map $f : A \to B$ is said to be a *diffeomorphism* if it is a homeomorphism and if each of the maps $X \rightarrowtail Y$; $x \mapsto f(x)$ and $Y \rightarrowtail X$; $y \mapsto f^{-1}(y)$ is smooth. A map $f : X \rightarrowtail Y$ is said to be *locally a diffeomorphism at a* if there are open neighbourhoods A of a and B of $b = f(a)$ such that $f(A) = B$ and the map $A \to B$; $x \mapsto f(x)$ is a diffeomorphism.

The main theorem of this section, the *inverse function theorem*, is a criterion for a map $f : X \rightarrowtail Y$ to be locally a diffeomorphism at a point a of its domain. Important corollaries are preliminary to the study of smooth manifolds. As in the previous section many proofs are omitted.

One of the main tools used in the proof is the *increment formula.* This inequality, which we have twice referred to already, replaces the *mean value theorem* which occurs at this stage in the calculus of real-valued functions of one real variable.

Theorem 21.27 (The increment formula.) *Let a and b be points of the domain of a differentiable map $f : X \rightarrowtail Y$ such that the line-segment $[a, b]$ is a subset of* $\operatorname{dom} f$, *X and Y being finite-dimensional linear spaces with assigned norms, and suppose that M is a real number such that the gradient norm $|dfx|$ of the differential is $\leq M$ for all $x \in [a, b]$. Then*

$$|f(b) - f(a)| \leq M|b - a|.$$

The other tool in the proof is the *contraction lemma.*

Theorem 21.28 The contraction lemma.) *Let A be a non-null closed subset of a finite-dimensional normed linear space X, and suppose that $f : A \to A$ is a map such that, for some non-negative real number $M < 1$ and for all $a, b \in A$,*

$$|f(b) - f(a)| \leq M|b - a|.$$

Then there is a unique point $x \in A$ such that $f(x) = x$.

Theorem 21.29 (The inverse function theorem.) *Let $f : X \rightarrowtail Y$ be a smooth map and suppose that, at some point $a \in X$, f is tangent to an affine bijection $t : X \to Y$, the dimensions of the real linear spaces X and Y being equal. Then f is locally a diffeomorphism at a.*

Proof It is enough to consider the case that $a = 0 \in X$ and $f(a) = 0 \in Y = X$, with $t = 1_X$. The increment formula is applied to the map $h = f - 1_X$ near the origin, using the continuity of the differential there, and the contraction lemma is then applied to the map $x \mapsto y - h(x)$, where y is a point close to the origin in the target. This provides an inverse map which is readily proved to be continuous. Finally one applies Theorem 21.18 to prove that this inverse map is smooth. □

Submanifolds of $\mathbf{R}^n$

The inverse function theorem has direct application to the description of *submanifolds* of a finite-dimensional real linear space X.

An *affine subspace* of a finite-dimensional linear space X is a parallel in X to a linear subspace, its *dimension* being the dimension of the linear subspace to which it is parallel. A subset M of X is said to be *smooth* at a point $a \in M$ if there are an affine subspace T of X passing through a, an open neighbourhood A of a in X and a diffeomorphism $h : X, a \rightarrowtail X, a$ with domain A and $dha = 1_X$, the identity map on X, such that $h(A \cap T) = h(A) \cap M$. The affine space T, uniquely determined if it exists, is called the *tangent space* to M at a. It is generally given the structure of a linear space by taking a as origin. The subset N is said to be a *smooth submanifold* of X if it is smooth at each of its points. For a connected submanifold the dimension of the tangent space is constant. This dimension is said to be the *dimension* of the submanifold.

In practice a subset of X is often presented either *explicitly* parametrically as the image of a map or *implicitly* as a *fibre* or *level set* (most frequently the set of zeros) of a map.

Example 21.30 *Consider a map* $f : \mathbf{R} \rightarrowtail \mathbf{R};\ x \mapsto f(x)$. *Then* graph f *is both the image of the map* $\mathbf{R} \rightarrowtail \mathbf{R}^2 : x \mapsto (x, f(x))$ *and the fibre over* 0 *of the map* $\mathbf{R}^2 \rightarrowtail \mathbf{R};\ (x, y) \mapsto y - f(x)$.

Example 21.31 *The image of the map* $\mathbf{R} \to \mathbf{R}^2;\ t \mapsto (t^2 - 1, t(t^2 - 1))$ *is also the fibre over* 0 *of the map* $\mathbf{R}^2 \to \mathbf{R};\ (x, y) \mapsto y^2 - (1 + x)x^2$.

Two corollaries of the inverse function theorem relevant to the determination of smooth submanifolds are as follows.

Theorem 21.32 (The injective criterion.) *Let* $f : W \rightarrowtail X$ *be a smooth map, with* dfc *injective for some* $c \in \operatorname{dom} f$, W *and* X *being finite-*

dimensional linear spaces. Then there exists an open neighbourhood C of c in W such that the image of $f|C$ is smooth at $a = f(c)$, with tangent space the parallel through a of the image of dfc.

Proof Consider the case that c is the origin in W and a the origin in X and $df0 : W \to X$ is the inclusion in X of a linear subspace W. Let Y be a complementary linear subspace, and identify X with the product space $W \times Y$. Consider the map $h : W \times Y \rightarrowtail W \times Y$; $(w, y) \mapsto f(w) + (0, y)$, with derivative at $(0,0)$ the identity on $X = W \times Y$. Accordingly, by the inverse function theorem, h is a local diffeomorphism at $(0, 0)$. It follows at once that the image of the restriction of f to some open neighbourhood of the origin in W is C^1 at the origin in X with $W \times \{0\}$ as tangent space.

This completes the proof in this special case. The general case reduces at once to this one if the origin in X is set at a and new bases for the linear spaces W and X are chosen appropriately. □

By the above 'injective' corollary of the inverse function theorem the image of a parametric curve $\mathbf{r} : \mathbf{R} \rightarrowtail \mathbf{R}^2$, *regular* in the sense that $d\mathbf{r}$ is continuous and $d\mathbf{r}t$ is injective for all $t \in \mathbf{R}$, is locally everywhere a one-dimensional smooth submanifold of $\mathbf{R}^2$. It is necessary to insist on the word 'locally' here, for a regular curve need not be injective. See Exercise 21.6 for an example. More subtly, as Exercise 21.7 shows, even the injectivity of the parametrisation is not enough.

The following jargon is in common use. A smooth map $F : W \rightarrowtail X$ is said to be *immersive* at a point a of its domain if dfa is injective, to be an *immersion* if it is everywhere immersive, and to be an *embedding* if also the map $\operatorname{dom} f \to \operatorname{im} f$; $w \mapsto f(w)$ is a homeomorphism, where the topology on $\operatorname{dom} f$ is induced from the topology on W and the topology on $\operatorname{im} f$ is induced from the topology on X.

Theorem 21.33 (The surjective criterion.) *Let $f : X \rightarrowtail Y$ be a smooth map, with dfa surjective for some $a \in \operatorname{dom} f$, X and Y being finite-dimensional linear spaces. Then there exists an open neighbourhood A of a in X such that $A \cap f^{-1}(f(a))$ is smooth at a, with tangent space the parallel through a of the kernel of dfa.*

Proof Consider first the case that Y is a subspace of X with a and $f(a)$ both the origin in X, $df0$ being the projection of X on to Y with kernel a linear subspace W. Then identify X with the product space $W \times Y$.

Consider the map $h : W \times Y \rightarrowtail W \times Y$; $(W, y) \mapsto (w, f(w, y))$. This has as derivative at $(0, 0)$ the identity on $X = W \times Y$. Accordingly, by the inverse function theorem, h is a local diffeomorphism at $(0, 0)$. It follows at once that $f^{-1}\{0\}$ is smooth at the origin in X with $W \times \{0\}$ as tangent space.

This completes the proof in this special case. The general case reduces at once to this one if the origin in X is set at a and new bases for X and Y are chosen appropriately. □

A smooth map $f : X \rightarrowtail Y$ is said to be *submersive* at a point a of its domain if dfa is surjective, and to be a *submersion* if it is everywhere submersive.

We shall often have occasion to use the surjective criterion in the sequel. The following examples are typical of many.

Example 21.34 *The unit sphere S^n is a smooth submanifold of $\mathbf{R}^{n+1}$.*

Proof Consider the smooth map $F : \mathbf{R}^{n+1} \to \mathbf{R}$; $x \mapsto x \cdot x$, where n is any non-negative integer. Then, for any $x, x' \in \mathbf{R}^{n+1}$, $dFx(x') = 2x \cdot x'$, of rank 1 and therefore surjective unless $x = 0$. In particular it is surjective for any x on $S^n = F^{-1}(\{1\})$. Therefore S^n is a smooth submanifold of $\mathbf{R}^{n+1}$. It has dimension n since the kernel rank of dFx is $(n+1) - 1 = n$, for any $x \in S^n$. □

Example 21.35 *The group $O(n)$ is a smooth submanifold of $\mathbf{R}(n)$, of dimension $\frac{1}{2}n(n-1)$.*

Proof Let $\mathbf{R}_+(n)$ denote the linear subspace of $\mathbf{R}(n)$ consisting of all the symmetric $n \times n$ matrices c, that is matrices c such that $c = c^\tau$. Clearly this is a linear subspace of $\mathbf{R}(n)$ of dimension $\frac{1}{2}n(n+1)$. Consider the smooth map $\mathbf{R}(n) \to \mathbf{R}_+(n)$; $a \mapsto a^\tau a$. Now, since the map $\mathbf{R}(n) \to \mathbf{R}(n)$; $a \mapsto a^\tau$ is linear, it follows, by Proposition 21.13, that the differential of the former map at a is the linear map $\mathbf{R}(n) \to \mathbf{R}_+(n)$; $b \mapsto b^\tau a + a^\tau b$. This is surjective, when a belongs to $O(n)$, for let c be any element of $\mathbf{R}_+(n)$ and consider the equation $a^\tau b = \frac{1}{2}ac$. But then also $b^\tau a = \frac{1}{2}c^\tau = \frac{1}{2}c$ so that $b^\tau a + a^\tau b = c$. Now apply the surjective criterion, implying that $O(n)$ is smooth at a, the tangent space being the kernel of the linear map, of dimension $n - \frac{1}{2}n(n+1) = \frac{1}{2}n(n-1)$. □

Note in particular that the tangent space to $O(n)$ at the identity matrix 1 consists of all matrices b such that $b^\tau 1 + 1^\tau b = b^\tau + b = 0$, the linear subspace of *skew-symmetric* $n \times n$ matrices. The submanifold $O(n)$ is an example of a *Lie group*, the tangent space at the origin being its *Lie algebra*. The further study of such groups and their algebras is the subject of Chapter 22.

Manifolds

A topological space X is said to be *locally euclidean* if there is a cover $\mathcal{S}$ for X such that $A \in \mathcal{S}$ is homeomorphic to an open subset of a finite-dimensional real linear space.

The definition may be reformulated as follows. A pair (E, i), where E is a finite-dimensional real linear space, and $i : E \rightarrowtail X$ is an open embedding with open domain, will be called a *chart* on X, and a set $\mathcal{S}$ of charts whose images form a cover for X will be called an *atlas* for X. Clearly the topological space X is locally euclidean if and only if there is an atlas for X.

A chart *at* a point $x \in X$ is a chart (E, i) on X such that $X \in \operatorname{im} i$.

A locally euclidean space need not be Hausdorff. See Exercise 20.4 for an example. A Hausdorff locally euclidean space is said to be a *topological manifold*.

In the sequel we shall only be interested in *smooth manifolds*, where, as always, smooth means C^k for k a positive integer or ∞.

Consider again the definition above of a smooth submanifold M of a finite-dimensional real linear space. The subset M of X is said to be *smooth* at $a \in M$ if there are a linear subspace W of X passing through a, open neighbourhoods A and B of $a \in X$, and a diffeomorphism $h : A \to B$, tangent to 1_X at a, such that $h(A \cap W) = B \cap M$.

Let i in such a case denote the map $W \rightarrowtail M;\ w \mapsto h(w)$. Its domain is $A \cap W$, which is open in W, and it is an open embedding, since h is a homeomorphism and $B \cap M$ is open in M. So (W, i) is a chart on M. Such charts will be called the *standard charts* on M.

The following proposition follows at once from these remarks, and the fact that any topological subspace of a real linear subspace is Hausdorff, any real linear subspace being automatically Hausdorff.

Proposition 21.36 *Let M be a smooth submanifold of a finite-dimensional real linear space X. Then M is a topological manifold.*

Now consider two standard charts on a smooth submanifold.

Proposition 21.37 *Let $i : V \rightarrowtail M$ and $j : W \rightarrowtail M$ be standard charts on a smooth submanifold M of a finite-dimensional real linear space X. Then the map $j^{-1}i : \operatorname{dom} i \to \operatorname{dom} j$ is a diffeomorphism.*

These propositions provide the motivation for the definitions of the following section and their subsequent development.

Smooth manifolds

Let X be a topological manifold. Then a *smooth atlas* for X, where k is a positive integer or ∞, consists of an atlas $\mathscr{S}$ for X such that for each $(E, i), (F, j) \in \mathscr{S}$ the map

$$j^{-1}i : E \rightarrowtail F;\ a \mapsto j^{-1}i(a)$$

is smooth.

Example 21.38 *Let X be a finite-dimensional real linear space X. Then $\{(X, 1_A)\}$ is a C^∞ atlas for A.*

Example 21.39 *Let M be a smooth submanifold of a finite-dimensional real linear space X. Then the set of standard charts on M is a smooth atlas for M.*

Example 21.39 provides a C^∞ atlas for the sphere S^n, for any non-negative integer n, for, by Example 21.34, S^n is a C^∞ submanifold of $\mathbf{R}^{n+1}$. The next example also provides an atlas for the sphere $\mathscr{S}^n$.

Example 21.40 *Let i and j : $\mathbf{R}^n \to S^n$ be the inverses of stereographic projection on to the equatorial plane of the sphere S^n from its North and South poles, respectively* (cf. Proposition 5.34 and Exercise 5.3). *Then $\{(\mathbf{R}^n, i), (\mathbf{R}^n, j)\}$ is a C^∞ atlas for S^n.*

Proof The maps

$$i^{-1}j = j^{-1}i : \mathbf{R}^n \rightarrowtail \mathbf{R}^n;\ x \mapsto x/x^{(2)}$$

are C^∞. □

The Grassmannians, and in particular the projective spaces, also have C^∞ atlases.

Two smooth atlases on a topological manifold X are said to be *equivalent*, or to define the same *smooth structure* on X if their union is a smooth atlas for X.

A topological manifold with a smooth atlas is said to be a *smooth manifold*, smooth manifolds with the same underlying topological space and with equivalent atlases being said to be *equivalent*.

A chart (E, i) on a smooth manifold X, with atlas $\mathcal{S}$, is said to be *admissible* if $\mathcal{S} \cup \{(E, i)\}$ is a smooth atlas for X.

For most purposes the distinction between different but equivalent manifolds is unimportant and will be ignored. One place where it is logically important to have a particular atlas in mind is in the construction of the tangent bundle of a smooth manifold, but even this will turn out in the end to matter little, since by Corollary 21.60 the tangent bundles of equivalent manifolds are naturally isomorphic.

Submanifolds and products of manifolds are defined in the obvious ways.

Proposition 21.41 is of importance both in defining the dimension of a smooth manifold and in defining its tangent spaces.

Proposition 21.41 *Let (E, i) and (F, j) be admissible charts on a smooth manifold X such that $\operatorname{im} i \cap \operatorname{im} j \neq \emptyset$. Then, for any $x \in \operatorname{im} i \cap \operatorname{im} j$, the map*

$$d((j^{-1}i)(i^{-1}(x)) : E \to F$$

is a linear isomorphism. In particular, $\dim E = \dim F$.

A smooth manifold X is said to have *dimension n*, $\dim X = n$, if the dimension of the source of every admissible chart on X is n. Any connected smooth manifold has a well-defined dimension.

The dimension of any open subset of a finite-dimensional real linear space X is equal to $\dim X$.

Proposition 21.42 *Let W be a connected smooth submanifold of a connected smooth manifold X. Then $\dim W \leq \dim X$.*

Proposition 21.43 *Let X and Y be connected smooth manifolds. Then $X \times Y$ is a smooth manifold, with $\dim(X \times Y) = \dim X + \dim Y$.*

Maps between smooth manifolds

Proposition 21.44 *Let $f : X \to Y$ be a map between smooth manifolds X and Y, let (E, i) and (E', i') be admissible charts on X, let (F, j) and (F', j') be admissible charts on Y, and suppose that x is a point of $\operatorname{im} i \cap \operatorname{im} i'$ such* *that* $f(x) \in \operatorname{im} j \cap \operatorname{im} j'$. *Then the map* $j'^{-1} f\, i' : E' \rightarrowtail F'$ *is smooth at* $i'^{-1}(x)$ *if and only if the map* $j^{-1} f\, i : E \rightarrowtail F$ *is smooth at* $i^{-1}(x)$.

Proof Apply the chain rule (Theorem 21.16) to the equation

$$j'^{-1} f\, i'(a) = (j'^{-1} j)(j^{-1} f\, i)(i^{-1} i')(a)$$

for all $a \in E$ sufficiently close to $i^{-1}(a)$. □

A map $f : X \rightarrow Y$ between smooth manifolds X and Y is said to be *smooth at* a point x if the map $j^{-1} f\, i : E \rightarrowtail F$ is smooth at $i^{-1}(x)$, for some, and therefore, by Proposition 21.44, for any, admissible chart $(E < i)$ at x and any admissible chart (F, j) at $f(x)$. The map is said to be *smooth* if it is smooth at each of its points.

Clearly the definition of the smoothness of a map $f : X \rightarrow Y$ depends only on the smooth structures for X and Y and not on any particular choice of an atlas of admissible charts.

A bijective smooth map $f : X \rightarrow Y$ whose inverse $f^{-1} : Y \rightarrow X$ also is smooth is said to be a *diffeomorphism.*

Many of the earlier theorems on differentiable maps have analogues for smooth maps between manifolds.

Proposition 21.45 *Let* W, X *and* Y *be smooth manifolds. Then a map* $(f, g) : W \rightarrow X \times Y$ *is smooth if and only if its components* f *and* g *are smooth.*

Proposition 21.46 *Let* X, Y *and* Z *be smooth manifolds, and suppose that* $f : X \times Y \rightarrow Z$ *is a smooth map. Then, for any* $(a, b) \in X \times Y$, *the maps* $f(-, b) : X \rightarrow Z$ *and* $f(a, -) : Y \rightarrow Z$ *are smooth.*

Proposition 21.47 *Let* X, Y *and* Z *all be smooth manifolds, and let* $f : X \rightarrow Y$ *and* $g : Y \rightarrow Z$ *be smooth maps. Then the composite map* $g f : X \rightarrow Z$ *is smooth.*

Proposition 21.48 *Let* V *be a smooth submanifold of a smooth manifold* X, *let* W *be a smooth submanifold* Y, *and let* $f : X \rightarrow Y$ *be a smooth map, with* $f(V) \subset W$. *Then the restriction of* f *with domain* V *and target* W *is smooth.*

Tangent bundles and maps

The concept of the differential of a smooth map $f : X \succ\!\!\rightarrow Y$, where X and Y are finite-dimensional real linear spaces, does not generalise directly to the case where X and Y are smooth manifolds. What does generalise is the concept of the *tangent map* of the map f, as defined below.

The *tangent bundle* of a finite-dimensional real linear space X is, by definition, the topological space $TX = X \times X$, together with the projection $\pi_{TX} : TX \to X;\ (x, a) \mapsto a$, the fibre $\pi_{TX}^{-1}\{a\} = TX_a = X \times \{a\}$, for any $a \in X$, being assigned the linear structure with (a, a) as origin. That linear space is the *tangent space* to W *at a*. The *tangent bundle space* TX may therefore be thought of as the union of all the tangent spaces to X, each labelled by the point at which it is tangent, with the obvious topology, the product topology.

The *tangent bundle* of an open subset A of a finite dimensional real linear space X is, by definition, the open subset $TA = A \times A$ of TX, together with the projection $\pi_{TA} = \pi_{TX} | TA$, the fibres of π_{TA} being regarded as linear spaces, as above.

Now suppose that a map $f : X \succ\!\!\rightarrow Y$ between finite-dimensional linear spaces X and Y is tangent at a point a of X to an affine map $t : X \to Y$. Then, instead of representing the map t by its linear part dfa we may equally well represent it by the linear map

$$Tf_a : TX_a \to TY_{f(a)};\ (x, a) \mapsto (t(x), f(a)),$$

the *tangent map* of f at a. Its domain is the tangent space to X at a and its target the tangent space to Y at $f(a)$. If the map f is differentiable everywhere there is then a map

$$Tf : TX_a \to TY_{f(a)};\ (x, a) \mapsto Tf_a(x, a),$$

with domain $T\operatorname{dom} f$, called, simply, the *tangent map* of f.

Notice that, for any $(x, a) \in TX$, $Tf_a(x, a)$ may be abbreviated to $Tf(x, a)$. Notice also that the maps df and Tf are quite distinct. The maps dfa and Tf_a may be identified, for any $a \in X$, but not the maps df and Tf.

Proposition 21.49 *Let $f : X \succ\!\!\rightarrow Y$ be a smooth map, X and Y being finite-dimensional real linear spaces. Then the map Tf is continuous, with open domain.*

Proposition 21.50 *Let X be any finite-dimensional real linear space. Then $T1_X = 1_{TX}$.*

The following two propositions are corollaries of the chain rule.

Proposition 21.51 *Let $f : X \rightarrowtail Y$ and $g : W \rightarrowtail X$ be smooth maps, W, X and Y being finite-dimensional real linear spaces. Then, for any $a \in \operatorname{dom} f\,g$ and each $w \in W$,*

$$T(f\,g)(w, a) = Tf\,Tg(w, a).$$

Proposition 21.52 *Let $f : X \rightarrowtail Y$ and $g : Y \rightarrowtail X$ be smooth maps, with $g = f^{-1}$, X and Y being finite-dimensional real linear spaces. Then, for each $a \in \operatorname{dom} f$ and any $x \in X$,*

$$Tg\,Tf(x, a) = (x, a)$$

and, for any $b \in \operatorname{dom} g$ and any $y \in Y$,

$$Tf\,Tg(y, b) = (y, b).$$

Proposition 21.53 *Let X be smooth manifold with atlas $\mathscr{S}$ and let $\mathscr{S}' = \bigcup\{T \operatorname{dom} i \times \{i\} : (E, i) \in \mathscr{S}\}$ (to be thought of as the disjoint union of the $T \operatorname{dom} i$). Then the relation $\sim$ on $\mathscr{S}'$, given by the formula $((a', a), i) \sim ((b', b), j)$ if and only if $j(b) = i(a)$ and $T(j^{-1}i)(a', a) = (b', b)$, is an equivalence.*

The *tangent bundle* of X, with atlas $\mathscr{S}$, is defined to be the quotient TX of the set $\mathscr{S}'$ defined in Proposition 21.53 by the equivalence $\sim$, together with the surjection $\pi_{TX} : TX \to X;\ [((a', a), i)]_\sim \mapsto i(a)$.

Proposition 21.54 *The set of maps $\{Ti : (E, i) \in \mathscr{S}'\}$, where $T\,i$ is the map $T \operatorname{dom} i \to TX;\ (a', a) \mapsto [((a'\,a)\,i)]_\sim$, is an atlas for the set TX.*

The set TX is assigned the topology induced by this atlas and called the *tangent bundle space* of X.

Proposition 21.55 *For any smooth manifold X the map π_{TX} is locally trivial.*

The map π_{TX} will be referred to as the *tangent projection* on X. The next proposition examines the structure of the fibres of the tangent projection.

Proposition 21.56 *Let X be a smooth manifold with atlas $\mathcal{S}$. Then, for any $x \in X$, the fibre $\pi_{TX}^{-1}\{x\}$ is the quotient of the set*

$$\mathcal{S}'_x = \bigcup\{TE_a \times \{i\} : (E, i) \in \mathcal{S} \text{ and } i(a) = x\}$$

by the restriction to this set of the equivalence $\sim$ discussed above. Moreover there is a unique linear structure for the fibre such that each of the maps

$$TE_a \to \pi_{TX}^{-1}\{x\};\ (a', a) \mapsto [((a', a), i)]_\sim$$

is a linear isomorphism.

The fibre $\pi_{TX}^{-1}\{x\}$ is assigned the linear structure defined in Proposition 21.56 and is called the *tangent space* to X at x. It will be denoted also by TX_x. Its elements are the *tangent vectors* to X at a.

Notice that the definitions of tangent bundle, tangent projection and tangent space for a smooth manifold agree with the corresponding definitions given earlier for a finite-dimensional real linear space X, or an open subset A of X, provided that X, or A, is assigned the single chart atlas of Example 21.38.

A smooth map $f : X \to Y$ induces in a natural way a continuous map $Tf : TX \to TY$, the *tangent (bundle) map* of f. Special cases include the tangent map of a smooth map with source and target finite-dimensional real linear spaces, and also the map Ti induced by an admissible chart(E, i) on a smooth manifold X, as previously defined.

Proposition 21.57 *Let $f : X \to Y$ be a smooth map. Then, for any $x \in X$, $(Tf)(TX_x) \subset TY_{f(x)}$. Moreover, for any $x \in X$, the map $Tf_x : TX_x \to TY_{f(x)};\ v \mapsto Tf(v)$ is linear.*

The map Tf is easily computed in the following case.

Proposition 21.58 *Let X be a smooth submanifold of a linear space V, let Y be a smooth submanifold of a linear space W and let $g : V \rightarrowtail W$ be a smooth map, with $X \subset \operatorname{dom} g$, such that $g(X) \subset Y$. Then the restriction $f : X \to Y;\ x \mapsto g(x)$ of g is smooth and, for any $a \in X$, $Tf_a : TX_a \to TY_{f(a)}$ is the restriction of Tg_a with domain TX_a and target $TY_{f(a)}$.*

Finally, Propositions 21.50, 21.51 and 21.52 extend to smooth maps between smooth manifolds.

Proposition 21.59 *Let W, X and Y be smooth manifolds. Then*

$$T1_X = 1_{TX},$$
$$T(f\,g) = Tf\;Tg, \text{ for any smooth maps } g : W \to X,\ f : X \to Y,$$
and $$Tf^{-1} = (Tf)^{-1}, \text{ for any diffeomorphism } f : X \to Y.$$

Corollary 21.60 *Let X' and X'' be equivalent smooth manifolds with underlying topological manifold X. Then $T1_X : TX' \to TX''$ is a* tangent bundle isomorphism.

Corollary 21.61 *Let W be a smooth submanifold of a smooth manifold X, let W be assigned any admissible atlas, and let $i : W \to X$ be the inclusion. Then the tangent map $Ti : TW \to TX$ is a topological embedding whose image is independent of the atlas chosen for W.*

The tangent bundle of W, in such a case, is normally identified with its image by Ti in TX.

For example, for any non-negative integer n the sphere S^n may be thought of as a smooth submanifold of $\mathbf{R}^{n+1}$, TS^n being identified with the subspace

$$\{(x, a) \in \mathbf{R}^{n+1} \times S^n : x \cdot a = 0\}$$

of $T\mathbf{R}^{n+1} = \mathbf{R}^{n+1} \times \mathbf{R}^{n+1}$. A smooth map $f : X \to Y$ between smooth manifolds X and Y is said to be a *smooth embedding* if $Tf : TX \to TY$ is a topological embedding, and to be *smooth projection* if tf is a topological projection.

A smooth map $f : X \to Y$ is said to be an *immersion* if, for each $x \in X$, Tf_x is injective. An immersion need not be injective, nor need an injective immersion be a topological embedding.

A smooth map $f : X \to Y$ is said to be a *submersion* if, for each $x \in X, Tf_x$ is surjective.

Proposition 21.62 *A submersion $f : X \to Y$ is an open map. Its non-null fibres are smooth submanifolds of X, the tangent space at a point $x \in X$ to the fibre $f^{-1}f(x)$ through x being the kernel of Tf_x. A surjective submersion is a smooth projection.*

Example 21.63 below is an important example of a smooth projection. In this example a tangent vector at any non-zero point of a real linear space X will be said to be *radial* if it is of the form $(x + \lambda x, x)$, for some $\lambda \in \mathbf{R}$, or, equivalently, if it is of the form λx, when TX_x has been identified in the standard way with X.

Example 21.63 *For any positive integer n the map*

$$\pi : \mathbf{R}^{n+1} \rightarrowtail S^n;\ x \mapsto x/|x|$$

defined everywhere except at 0, is a smooth projection, the kernel of the tangent map at any point consisting of the radial tangent vectors there.

Proof Let $g : \mathbf{R}^{n+1} \rightarrowtail \mathbf{R}^{n+1}$ be the composite of the map π with the inclusion of S^n in $\mathbf{R}^{n+1}$. For any non-zero a and any $x \in \mathbf{R}^{n+1}$,

$$dga(x) = |a|^{-1}x - |a|^{-3}(x \cdot a)a = |a|^{-1}x',$$

where $x' = x - (x \cdot g(a))g(a)$ (cf. Proposition 21.19).

Moreover, for any non-zero $a \in \mathbf{R}^{n+1}$ and any $\lambda \in \mathbf{R}$,

$$x = \lambda a \quad \Rightarrow \quad x \cdot a = \lambda a \cdot a \quad \Rightarrow \quad x = (x \cdot g(a))g(a).$$

So $dga(x) = 0$ if and only if $x = \lambda a$, for some $\lambda \in \mathbf{R}$. That is the kernel of dga has dimension 1, implying, by Proposition 1.4, that $\operatorname{rk}(T\pi_a) = \operatorname{rk}(dga) = n$ and therefore that π is a submersion. Since π is surjective it follows, by the last part of Proposition 21.62, that π is a smooth projection. □

Exercises

21.1 Consider the map

$$f : L(X, X) \times L(X, Y) \rightarrowtail L(X, Y);\ (a, b) \mapsto b\,a^{-1},$$

where X and Y are finite-dimensional real linear spaces. Prove that f is differentiable, with $df(1_X, 0)(a', b') = b'$.

21.2 Prove that the map

$$]-1, 1[\rightarrow \mathbf{R};\ x \mapsto \frac{x}{1-x^2}$$

is a diffeomorphism.

21.3 Consider the map $f : \mathbf{R}(2) \rightarrow \mathbf{R};\ a \mapsto \det a$. Compute $dfa(b)$, for each $a, b \in \mathbf{R}(2)$, and show that $(df1)(b) = b_{00} + b_{11}$. Prove that $SL(2; \mathbf{R})$, the set of 2×2 real matrices with determinant 1, is a three-dimensional smooth submanifold of $\mathbf{R}(2)$.

21.4 Consider the map $f : \mathbf{R}(2) \rightarrow \mathbf{R}(2);\ t \mapsto t^- t$, where, for all $t = \begin{pmatrix} a & c \\ b & d \end{pmatrix} \in \mathbf{R}(2)$, $t^- = \begin{pmatrix} d & c \\ b & a \end{pmatrix}$. Verify that, for all $u, t \in \mathbf{R}(2)$, $dfu(t) = u^- t + t^- u$. Describe the matrices in the kernel and image of $df1$ and prove that $f^{-1}(\{1\})$ is a smooth submanifold of $\mathbf{R}(2)$.

21.5 Prove that, for any non-negative integer, the complex quasi-sphere $\mathscr{S}(\mathbf{C}^{n+1}) = \{x \in \mathbf{C}^{n+1} : x^{(2)} = 1\}$ is homeomorphic to TS^n.

21.6 Verify that the differential $d\mathbf{r}$ of the map

$$\mathbf{r} : \mathbf{R} \to \mathbf{R}^2;\ t \mapsto (t^2 - 1, t(t^2 - 1))$$

is continuous, with $d\mathbf{r}t$ injective everywhere, but that the map $\mathbf{r}$ itself is not injective.

Verify also that the differential dF of the map

$$F : \mathbf{R}^2 \to \mathbf{R};\ (x, y) \mapsto y^2 - (1 + x)x^2$$

is continuous, with $dF(x, y)$ everywhere surjective, except at the origin.

Note that $\operatorname{im} \mathbf{r} = F^{-1}(0)$.

21.7 Prove that the restriction of the map $\mathbf{r}$ of Exercise 21.6 to the interval $]-1, \infty[$ is an injective immersion that is not a topological embedding.

21.8 Let r be any irrational real number. Prove that the map

$$f : \mathbf{R} \to S^1 \times S^1;\ x \mapsto (e^{ix}, e^{irx})$$

is an injective immersion that is not a topological embedding.

(To see that f is not a topological embedding, it is convenient first to represent the torus as the quotient of $\mathbf{R}^2$ by the equivalence $(x + 2m\pi, y + 2n\pi) \sim (x, y)$, for any $(x, y) \in \mathbf{R}^2$ and any $(m, n) \in \mathbf{Z}^2$. Then f is the composite of the map $\mathbf{R} \to \mathbf{R}^2;\ x \mapsto (x, rx)$ with the partition induced by the equivalence.)

22

Lie groups

As we have remarked, there is a natural topology for a finite-dimensional real linear space X, that induced by any norm on X. It is a fair supposition that there should be more or less natural topologies also for the classical groups, Spin groups, Grassmannians and quadric Grassmannians, all of which are closely related to finite-dimensional linear spaces. It turns out that they also all have natural smooth structures as well, the groups being examples of *Lie groups*.

Important topological properties of the classical groups are their compactness or connectedness or otherwise.

Topological groups

A *topological group* consists of a group G and a topology for G such that the maps

$$G \times G \to G;\ (a, b) \mapsto ab \quad \text{and} \quad G \to G;\ a \mapsto a^{-1}$$

are continuous. An equivalent condition is that the map $G \times G;\ (a, b) \mapsto a^{-1}b$ is continuous.

Example 22.1 *Let X be a finite-dimensional real linear space. Then the group $GL(X)$ is a topological group.*

Topological group maps and *topological subgroups* are defined in the obvious ways.

Proposition 22.2 *Any subgroup of a topological group is a topological group.*

Corollary 22.3 *All the groups listed in Table 13.10 are topological groups.*

Proposition 22.4 *For any p, q the group $Spin(p, q)$, regarded as a subgroup of the Clifford algebra $\mathbf{R}_{p,q}$, is a topological group and the map*

$$Spin(p, q) \to SO(p, q); \; g \mapsto \rho_g,$$

defined in Proposition 16.14 is a topological group map.

The compactness, or otherwise, of the groups listed in Table 13.10 and of the Spin groups is easily settled.

Proposition 22.5 *For any n the topological groups $O(n), SO(n), U(n), SU(n)$ and $Sp(n)$ are compact.*

Proof By definition $O(n) = \{t \in \mathbf{R}(n) : t^\tau t = 1\}$, from which it follows that $O(n)$ is closed in $\mathbf{R}(n)$. Moreover if $\mathbf{R}^n$ is assigned the euclidean norm and $\mathbf{R}(n)$, identified with $L(\mathbf{R}^n, \mathbf{R}^n)$, the induced gradient norm, then, for any $t \in O(n)$, $|t| = 1$, from which it follows that $O(n)$ is bounded in $\mathbf{R}(n)$. So $O(n)$ is compact. Each of the other groups is isomorphic to a closed subgroup of $O(n), O(2n)$ or $O(4n)$, and is therefore compact, by Proposition 20.25. □

Proposition 22.6 *For any n the group $Spin(n)$ is compact.*

Proposition 22.7 *All the groups listed in Table 13.10, with the exception of those listed in Proposition 22.5, are non-compact (unless n or $pq = 0$).*

Show, for example, that each contains an unbounded copy of $\mathbf{R}^*$.

Proposition 22.8 *For any p, q with $pq > 0$, the group $Spin(p, q)$ is non-compact.*

Closely related to the concept of a topological group is the concept of a *homogeneous space*.

A Hausdorff topological space X is said to be a *homogeneous space* for a topological group G if there is a transitive continuous action of G on X, that is, a continuous map $G \times X \to X$; $(g, x) \mapsto g\,x$, such that

(i) for all $g, g' \in G$ and all $x \in X$,

$$(g'g)x = g'(g\,x), \quad \text{with} \quad 1\,x = x$$

(ii) (*transitivity*) for each $a, b \in X$, there is some $g \in G$ such that $b = g\,a$.

Proposition 22.9 *Let $G \times X \to X$; $(g, x) \mapsto gx$ be a continuous action of the topological group G on the topological space X. Then, for each $g \in G$, the map $X \to X$; $x \mapsto gx$ is a homeomorphism.*

Corollary 22.10 *Let X be a homogeneous space for a topological group G and let $a, b \in X$. Then there is a homeomorphism $h : X \to X$ such that $h(a) = b$.*

Example 22.11 *For any n, the sphere S^n is a homogeneous space for the group $O(n+1)$.*

The action that one has in mind is the obvious one, the map

$$O(n+1) \times S^n \to S^n; \ (t, x) \mapsto t(x),$$

which is well-defined, by Proposition 5.33.

Examples 22.12 *For any n, S^{2n+1} is a homogeneous space for $U(n+1)$ and S^{4n+3} is a homogeneous space for $Sp(n+1)$, while, for any positive n, S^n is a homogeneous space for $SO(n+1)$ and for $Spin(n+1)$, while S^{2n+1} is a homogeneous space for $SU(n+1)$*

The action in each case is the obvious analogue of the action of $O(n+1)$ on S^n described in Example 22.11.

The next few propositions explore the relationships between homogeneous spaces and coset space representations.

Proposition 22.13 *Let G be a topological group, let X be a homogeneous space for X and let $a \in X$. Then the map $a_R : G \to X$; $g \mapsto ga$ is surjective, the isotropy subgroup $G_a = \{g \in G : ga = a\}$ of the action of G at a is a closed subgroup of G and the fibres of a_R are the left cosets of G_a in G – in the terminology of Chapter 3 the sequence $G_a \xrightarrow{\iota} G \xrightarrow{a_R} X$ is left-coset exact.*

Proof This is straightforward, using parts of Exercise 3.1. □

Proposition 22.14 *Let F be a subgroup of a topological group G. Then the partition $\pi : G \to G/F$; $g \mapsto gF$ is open.*

Proposition 22.15 *Let F be a closed subgroup of a topological group G. Then the space of left cosets G/F is a homogeneous space of G with respect to the action*

$$G \times G/F \to G/F;\ (g, g'F) \mapsto gg'F.$$

Proof First, the space G/F is Hausdorff. For let gF, $g'F$ be distinct points of G/F, where $g, g' \in G$. Since F is closed and since $g^{-1}g' \notin F$ there exists an open neighbourhood A of $g^{-1}g'$ in the set complement $G \backslash F$. It then follows from the continuity of the map $G \times G \to G; (g, g') \mapsto g^{-1}g'$ that there exist open neighbourhoods B of g and C of g' in G such that, for all $b \in B$ and $c \in C$, $b^{-1}c \notin F$. Now define $U = \pi(B)$ and $V = \pi(C)$, where π is the partition $G \to G/F$. Then $U \cap V = \emptyset$, while, by Proposition 22.14, U is an open neighbourhood of gF and V is an open neighbourhood of $g'F$ in G/F.

Secondly, the action is continuous, for in the commutative diagram of maps

$$\begin{array}{ccc} G \times G & \xrightarrow{\text{group product}} & G \\ \downarrow{\scriptstyle 1\times\pi} & & \downarrow{\scriptstyle \pi} \\ G \times (G/F) & \xrightarrow{\text{action}} & G/F \end{array},$$

where, for each $(g, g') \in G \times G$, $(1 \times \pi)(g, g') = (g, \pi(g'))$, each of the maps except for the one labelled 'action' is continuous, while π, and therefore also $1 \times \pi$, is a projection. The continuity of the action then follows, by Proposition 20.14.

Finally, conditions (i) and (ii) are readily checked. □

Proposition 22.16 *Let X be a homogeneous space for a* compact *topological group G. Then, for any $a \in X$, the map $(a_R)_{bij};\ G/G_a \to X$ is a homeomorphism.*

Proposition 22.17 *Let F be a connected subgroup of a topological group G and suppose that G/F is connected. Then G is connected.*

Corollary 22.18 *For each positive integer n the groups $SO(n)$, $Spin(n)$, $U(n)$, $SU(n)$ and $Sp(n)$ are connected.*

Proposition 22.19 *For each positive integer n the group $O(n)$ is disconnected, with two components, namely $SO(n)$, the group of rotations of $\mathbf{R}^n$, and its coset, the set of anti-rotations of $\mathbf{R}^n$.*

Proof The map $O(n) \to S^0$; $t \mapsto \det t$, being the restriction of a multilinear map, is continuous, and for $n > 0$ it is surjective. □

The connectedness or otherwise of the non-compact groups does not come so easily. For this we require smoothness, as we shall see below.

We shall require to know the compactness or connectedness or otherwise of the various quasi-spheres. By Proposition 22.20 the ten cases reduce to four, namely $\mathscr{S}(\mathbf{R}^{p,q+1})$, $\mathscr{S}(\mathbf{C}^{n+1})$, $\mathscr{S}(\widetilde{\mathbf{H}}^{n+1})$, and $\mathscr{S}((^2\widetilde{\mathbf{H}}^{\sigma})^{n+1})$, for all p, q and all n. The symbol $\cong$ denotes homeomorphism.

Proposition 22.20

$$\begin{aligned}
\mathscr{S}((^2\mathbf{R}^{\sigma})^{n+1}) &= \{(a, b) \in (\mathbf{R}^{n+1})^2 : a^{\tau}b = 1\} \cong \mathscr{S}(\mathbf{R}_{\text{hb}}^{2n+2}) \\
&\cong \mathscr{S}(\mathbf{R}^{n+1,n+1}), \\
\mathscr{S}((^2\mathbf{C}^{\sigma})^{n+1}) &= \{(a, b) \in (\mathbf{C}^{n+1})^2 : a^{\tau}b = 1\} \cong \mathscr{S}(\mathbf{C}_{\text{hb}}^{2n+2}) \\
&\cong \mathscr{S}(\mathbf{C}^{n+1,n+1}), \\
\mathscr{S}(\mathbf{R}_{\text{sp}}^{2n+2}) &\cong \{(a, b) \in (\mathbf{R}^{2n+2})^2 : a^{\tau}b = 1\} \cong \mathscr{S}(\mathbf{R}_{\text{hb}}^{4n+4}) \\
&\cong \mathscr{S}(\mathbf{R}^{2n+2,2n+2}), \\
\mathscr{S}(\mathbf{C}_{\text{sp}}^{2n+2}) &\cong \{(a, b) \in (\mathbf{C}^{2n+2})^2 : a^{\tau}b = 1\} \cong \mathscr{S}(\mathbf{C}_{\text{hb}}^{4n+4}) \\
&\cong \mathscr{S}(\mathbf{C}^{2n+2,2n+2}), \\
\mathscr{S}(\widetilde{\mathbf{C}}^{p,q+1}) &\cong \mathscr{S}(\mathbf{R}^{2p,2q+2}), \\
\mathscr{S}(\widetilde{\mathbf{H}}^{p,q+1}) &\cong \mathscr{S}(\mathbf{R}^{4p,4q+4}).
\end{aligned}$$

The next four propositions cover the four outstanding cases.

Proposition 22.21 *For any p, q, $\mathscr{S}(\mathbf{R}^{p,q+1}) \cong \mathbf{R}^p \times S^q$, and so is connected for any positive q, but disconnected for $q = 0$, and non-compact, for any positive p, but compact for $p = 0$.*

Proof Use Exercise 5.4. It is not difficult to show that the bijection constructed in that exercise is a homeomorphism, by verifying that the map and its inverse are each continuous. □

Proposition 22.22 *The quasi-sphere $\mathscr{S}(\mathbf{C}^{n+1})$ is connected and non-compact, for any positive n.*

Proof By definition, $\mathscr{S}(\mathbf{C}^{n+1}) = \{z \in \mathbf{C}^{n+1} : z^{\tau}z = 1\}$. For any $z \in \mathbf{C}^{n+1}$, let $z = x + \mathrm{i}\,y$, where x and $y \in \mathbf{R}^{n+1}$, and let $\mathbf{R}^{n+1}$ have its standard positive-definite orthogonal structure. Then, since

$$z^{\tau}z = (x + \mathrm{i}\,y)^{\tau}(x + \mathrm{i}\,y) = x^{(2)} - y^{(2)} + 2\,\mathrm{i}\,x \cdot y,$$

it follows that $z \in \mathscr{S}(\mathbf{C}^{n+1})$ if and only if $x^{(2)} - y^{(2)} = 1$ and $x \cdot y = 0$. In particular, since $x^{(2)} = 1 + y^{(2)}$, $x \neq 0$.

Now S^n is a subset of $\mathscr{S}(\mathbf{C}^{n+1})$. Consider the continuous map $\pi : \mathscr{S}(\mathbf{C}^{n+1}) \to S^n;\ z \mapsto x/|x|$. It is surjective, with $\pi|S^n = 1_{S^n}$. For any $b \in S^n$, the fibre of π over b is the image of the continuous embedding

$$(\mathbf{R}\{b\})^{\perp} \to \mathscr{S}(\mathbf{C}^{n+1});\ y \mapsto (\sqrt{(}1 + y^{(2)})b,\ y),$$

where $(\mathbf{R}\{b\})^{\perp}$ denotes the orthogonal annihilator of $\mathbf{R}\{b\}$ in $\mathbf{R}^{n+1}$. This image is connected, since $(\mathbf{R}\{b\})^{\perp}$ is connected. It is also non-compact, since $(\mathbf{R}\{b\})^{\perp}$ is non-compact, n being positive. Since each fibre of π is connected, and since $\mathscr{S}^n$ is connected, for $n > 0$, it follows at once that $(\mathbf{C}^{n+1})$ is connected. Finally, since any fibre of π is a closed subset and is non-compact, $\mathscr{S}(\mathbf{C}^{n+1})$ is non-compact. □

Proposition 22.23 *The quasi-sphere $\mathscr{S}(\widetilde{\mathbf{H}}^{n+1})$ is connected and non-compact, for any positive n.*

Proof This follows the same pattern as the proof of Proposition 22.22. Here it is convenient to identify $\mathbf{C}^{n+1}$ with $\{a + \mathrm{j}\, b \in \mathbf{H}^{n+1} : a, b \in \mathbf{C}^{n+1}\}$, and to assign $\mathbf{C}^{n+1}$ its standard orthogonal structure, just as $\mathbf{R}^{n+1}$ was assigned its standard positive-definite orthogonal structure in the proof of Proposition 22.22.

By definition, $\mathscr{S}(\widetilde{\mathbf{H}}^{n+1}) = \{q \in \mathbf{H}^{n+1} : q^{\tau} q = 1\}$. For any $q \in \mathbf{H}^{n+1}$, let $q = x + \mathrm{i}\, y$, where x and $y \in \mathbf{C}^{n+1}$. Then since

$$\widetilde{q}^{\tau} q = (\overline{x} + \mathrm{i}\, \overline{y})^{\tau}(x + \mathrm{i}\, y) = \widetilde{x}^{\tau} x - \widetilde{y}^{\tau} y + \mathrm{i}(y^{\tau} x + x^{\tau} y),$$

and since $y^{\tau} x + x^{\tau} y = x \cdot y$, it follows that $q \in \mathscr{S}(\widetilde{\mathbf{H}}^{n+1})$ if and only if $\widetilde{x}^{\tau} x - \widetilde{y}^{\tau} y = 1$ and $x \cdot y = 0$. The rest of the proof consists of a consideration of the map $\pi : \mathscr{S}(\widetilde{\mathbf{H}}^{n+1}) \to S^{2n+1} : q \mapsto x/\sqrt{(\widetilde{x}^{\tau} x)}$ closely analogous to that given for the corresponding map in Proposition 22.22, the sphere S^{2n+1} being identified with $\mathscr{S}(\overline{\mathbf{C}}^{n+1})$ in this case. □

The final case is slightly trickier.

Proposition 22.24 *The quasi-sphere $\mathscr{S}(({}^2\widetilde{\mathbf{H}}^{\sigma})^{n+1})$ is connected and non-compact, for any n.*

Proof By definition, $\mathscr{S}(({}^2\widetilde{\mathbf{H}}^{\sigma})^{n+1}) = \{(q, r) \in (\mathbf{H}^{n+1})^2 : \widetilde{q}^{\tau} r = 1\}$. Let

$u = \hat{q}+r$, $v = \hat{q}-r$. Then it easily follows that $\mathscr{S}({}^2\widetilde{\mathbf{H}}^{n+1})$ is homeomorphic to

$$\mathscr{S}' \to S^{4n+3};\ (u, v) \mapsto u/\sqrt{(\bar{u}^\tau u)}.$$

This is handled just like the corresponding maps in Propositions 22.22 and 22.23. □

The various cases may be summarised as follows.

Theorem 22.25 *Let (X, ξ) be an irreducible, non-degenerate, symmetric or essentially skew, finite-dimensional correlated space over* $\mathbf{K}$ *or* ${}^2\mathbf{K}$, *where* $\mathbf{K} = \mathbf{R}$, $\mathbf{C}$ *or* $\mathbf{H}$. *Then, unless (X, ξ) is isomorphic to* $\mathbf{R}$, ${}^2\mathbf{R}^\sigma$ *or* $\mathbf{C}$, *the quasi-sphere $\mathscr{S}(X, \xi)$ is connected and, unless (X, ξ) is isomorphic to* $\mathbf{R}^n$, $\overline{\mathbf{C}}^n$ *or* $\overline{\mathbf{H}}^n$, *for any n, or to* $\mathbf{C}$ *or* $\widetilde{\mathbf{H}}$, *$\mathscr{S}(X, \xi)$ is non-compact.*

Lie groups

A *Lie group* is a topological group G with a specified smooth (C^1) structure, such that the maps

$$G \times G \to G;\ (a, b) \mapsto ab \quad \text{and} \quad g \mapsto g h$$

are smooth (C^1). For some purposes it is desirable to insist on a higher degree of smoothness than 1, but C^1 will do for the moment.

Elementary properties of Lie groups include the following.

Proposition 22.26 *Let G be a Lie group. Then, for any $a, b \in G$, the maps*

$$G \to G;\ g \mapsto ag \quad \textit{and} \quad g \mapsto gb$$

are smooth homeomorphisms.

Proof The map $g \mapsto ag$ is smooth, by Proposition 21.46, and its inverse, the map $G \to G$; $g \mapsto a^{-1}g'$, also is.

Similarly for the other map. □

A *Lie group map* is a smooth group map $G \to H$, where G and H are Lie groups, and a *Lie group isomorphism* is a bijective Lie group map whose inverse also is a Lie group map.

Proposition 22.27 *Let G be a Lie group. Then, for any $a \in G$, the map $G \to G$; $g \mapsto a\,g\,a^{-1}$ is a Lie group isomorphism.*

Examples of Lie groups include all the groups in Table 13.10.

Proposition 22.28 *Let (X, ξ) be a non-degenerate finite-dimensional irreducible A^{φ}-correlated space. Then the group of correlated automorphisms $O(X, \xi)$ is a smooth submanifold of* End Y *and is, with this smooth structure, a Lie group.*

Proof By Corollary 13.5, $O(X, \xi) = \{t \in \operatorname{End} X : t^{\xi}t = 1_X\}$. Now, by Proposition 13.2, the map $\operatorname{End} X \to \operatorname{End} X$; $t \mapsto t^{\xi}$ is real linear. It follows that the map

$$\pi : \operatorname{End} X \to \operatorname{End}_{+}(X, \xi);\ t \mapsto t^{\xi}\,t$$

is smooth.

For any $u \in \operatorname{End} X$,

$$d\pi u(t) = u\,t + t\,u.$$

From this it follows, as in Example 21.35, that, for any $u \in O(X, \xi), d\pi u$ is surjective and, by Proposition 21.62, that $O(X, \xi)$ is a smooth submanifold of End X of real dimension

$$\dim_{\mathbf{R}} \operatorname{End} X - \dim_{\mathbf{R}} \operatorname{End}_{+}(X, \xi) = \dim_{\mathbf{R}} \operatorname{End}_{-}(X, \xi).$$

□

The tangent space to $O(X, \xi)$ at 1_X, being the affine subspace of End X through 1_X parallel to the real linear subspace $\operatorname{End}_{-}(X, \xi)$, with 1_X chosen as origin, is commonly and tacitly identified with this real linear space.

Corollary 22.29 *The dimensions of the classical groups in the first list of Table 13.10 are as there stated.*

In the cases of the groups $GL(n;\mathbf{R})$, $GL(n;\mathbf{C})$ and $GL(n;\mathbf{H})$ it is simpler to observe that these are open subsets of $\mathbf{R}(n)$, $\mathbf{C}(n)$ and $\mathbf{H}(n)$, respectively.

Proposition 22.30 *For any non-negative integers p, q, n the maps*

$O(p, q) \to S^0$; $t \mapsto t \mapsto \det t$, $U(p,q) \to S^1$; $t \mapsto \det t$,
$GL(n;\mathbf{R}) \to \mathbf{R}^{\bullet}$; $t \mapsto \det t$, $GL(n;\mathbf{C}) \to \mathbf{C}^{\bullet}$; $t \mapsto \det t$, *and*
$GL(n;\mathbf{H}) \to \mathbf{R}^{>0}$; $t \mapsto \det t$ *are smooth projections.*

Proof Use Proposition 21.62, Proposition 21.58 and Proposition 21.15. □

Corollary 22.31 *The dimensions of the remainder of the groups listed in Table 13.10 are as there stated.*

Actions of Lie groups

There are many examples of a Lie group acting smoothly on a smooth manifold.

Proposition 22.32 *Let G be a Lie group, X a smooth manifold and $G \times X \to X$; $(g, x) \mapsto gx$ a smooth action of G on X. Then, for any $a \in G$, the map $X \mapsto X$; $x \mapsto ax$ is a diffeomorphism.*

Proposition 22.33 *Let G be a Lie group, X a smooth manifold and $G \times X \to X$; $(g, x) \mapsto gx$ a smooth action of G on X. Then, for any $b \in X$, the map $\pi : G \to X : g \mapsto gb$ is a submersion if and only if $T\pi_1$ is surjective, where $1 = 1_{(G)}$.*

Proof For any $a \in G$ the map π admits the decomposition

$$G \to G \xrightarrow{\pi} X \to X;\ g \mapsto a^{-1}g \mapsto a^{-1}g\,b \mapsto g\,b.$$

From this, and Proposition 22.26 and Proposition 21.59, it follows that if $T\pi_1$ is surjective then $T\pi_a$ is surjective. This proves $\Leftarrow$. The proof of $\Rightarrow$ is trivial. □

The quasi-spheres of Theorem 13.13 are all smooth manifolds, the appropriate correlated group for each quasi-sphere acting smoothly on it.

Proposition 22.34 *Let (X, ξ) be a symmetric non-degenerate finite-dimensional irreducible $\mathbf{A}^{\psi}$-correlated space. Then the quasi-sphere $\mathscr{S}((X, \xi) \times \mathbf{A}^{\psi})$ is a smooth submanifold of $X \times \mathbf{A}$ with tangent space at $(0, 1)$ the linear subspace*

$$\{(c, d) \in X \times \mathbf{A} : d^{\psi} + d = 0\}$$

or, more strictly, its parallel through $(0, 1)$ with that point chosen as 0.

Proof The quasi-sphere is the fibre over 1 of the map

$$X \times \mathbf{A} \to \{\lambda \in \mathbf{A} : \lambda^{\varphi} = \lambda\};\ (c, d) \mapsto c^{\xi}c + d^{\xi}d$$

is a smooth submersion, with tangent map at (0, 1) the map

$$X \times \mathbf{A} \to \{\lambda \in \mathbf{A} : \lambda^{\varphi} = \lambda\};\ (c, d) \mapsto (O, d^{\varphi} + d).$$

□

There is an analogue of Proposition 22.34 for the essentially skew cases. The reader is invited to formulate the analogue and prove it.

Proposition 22.35 *Let* (X, ξ) *be a symmetric non-degenerate finite-dimensional irreducible* $\mathbf{A}^{\varphi}$*-correlated space and let* G *and* S *be the group of correlated automorphisms and the unit quasi-sphere, respectively, of the* $\mathbf{A}^{\varphi}$*-correlated space* $(X, \xi) \times \mathbf{A}^{\varphi}$. *Then the map*

$$G \times X \to S;\ (g, x) \mapsto g(x)$$

is smooth.

Proof This map is a restriction of the linear map

$$\mathrm{End}(X \times \mathbf{A}) \times (X \times \mathbf{A}) \to X \times A;\ (t, x) \mapsto t(x).$$

□

Proposition 22.36 *Let* G *and* S *be as in Proposition 22.35. Then the map* $\pi : \mathcal{G} \to S;\ g \mapsto g(0, 1)$ *is a smooth projection.*

Proof By Theorem 13.13 the map π is surjective with fibres the left cosets in G of the group $O(X, \xi)$ regarded as a subgroup of G in the obvious way. By Proposition 22.35 and Proposition 21.46 the maps π and $S \to S;\ x \mapsto u(x)$, for any $u \in G$, are smooth. By Proposition 22.33 it remains to prove that $T\pi_1$ is surjective.

Now $\begin{pmatrix} a & c \\ b & d \end{pmatrix} \in TG_1$ if and only if $a^{\xi} + a = 0$, $b = c^{\xi}$ and $d^{\xi} + d = 0$, and $(c, d) \in TS_{(0,1)}$ if and only if $d^{\varphi} + d = 0$, from which the surjectivity of the linear map

$$T\pi_1 : TG_1 \to TS_{(0,1)};\ g \mapsto g(0, 1)$$

is evident. □

Corollary 22.37 *For any p, q, n the maps*

$$\begin{array}{rcl} O(p, q+1) & \longrightarrow & \mathcal{S}(\mathbf{R}^{p,q+1}), \\ O(n+1; \mathbf{C}) & \longrightarrow & \mathcal{S}(\mathbf{C}^{n+1}), \\ O(n+1; \mathbf{H}) & \longrightarrow & \mathcal{S}(\mathbf{H}^{n+1}), \\ U(p, q+1) & \longrightarrow & \mathcal{S}(\overline{\mathbf{C}}^{p,q+1}), \\ Sp(p, q+1) & \longrightarrow & \mathcal{S}(\overline{\mathbf{H}}^{p,q+1}), \\ GL(n+1; \mathbf{R}) & \longrightarrow & \mathcal{S}({}^2\mathbf{R}^{n+1}), \\ GL(n+1; \mathbf{C}) & \longrightarrow & \mathcal{S}({}^2\mathbf{C}^{n+1}), \\ GL(n+1; \mathbf{H}) & \longrightarrow & \mathcal{S}({}^2\mathbf{H}^{n+1}) \end{array}$$

defined in Theorem 13.13 are open continuous surjections.

Corollary 22.38 *For any p, q, n the groups*

$$O(n;\mathbf{H}),\ U(p, q),\ Sp(p, q),\ GL(n;\mathbf{C}) \text{ and } GL(n;\mathbf{H})$$

are each connected.

Proof Add the information in Corollary 22.37 to Theorem 22.25 and apply Proposition 20.45. □

Similar methods prove the following.

Proposition 22.39 *For any p, q, n the maps*

$$\begin{array}{rcl} SO(p, q+1) & \longrightarrow & \mathcal{S}(\mathbf{R}^{p,q+1}), \\ SO(n+1; \mathbf{C}) & \longrightarrow & \mathcal{S}(\mathbf{C}^{n+1}), \\ SU(p, q+1) & \longrightarrow & \mathcal{S}(\overline{\mathbf{C}}^{p,q+1}), \\ SL(n+1; \mathbf{R}) & \longrightarrow & \mathcal{S}({}^2\mathbf{R}^{n+1}) \end{array}$$

are open continuous surjections

Corollary 22.40 *For any p, q, n the groups $SO(n; \mathbf{C})$, $SU(p, q)$ and $SL(n; \mathbf{R})$ are connected.*

The groups $SO(p, q)$, by contrast, are not connected unless p or $q = 0$. See Proposition 22.47 below.

Once again there is an analogue for the essentially skew cases. The conclusion is as follows.

Proposition 22.41 *For any n the groups $Sp(2n;\mathbf{R})$ and $Sp(2n;\mathbf{C})$ are connected.*

Further examples of smooth manifolds and maps are provided by the Grassmannians and quadric Grassmannians studied in Chapter 14.

Proposition 22.42 *Let (X, ξ) be any non-degenerate finite-dimensional irreducible symmetric or skew $\mathbf{A}^{\varphi}$-correlated space. Then, for any $k \leq \dim X$, the quadric Grassmannian $\mathcal{N}_{(}X, \xi)$ is a smooth submanifold of $\mathcal{G}_k(X)$. The parabolic atlas is a smooth atlas for $\mathcal{N}_{(}X, \xi)$ and determines the same smooth structure.*

Proposition 22.43 *Let*

$$G = \left\{ \begin{pmatrix} a & \bar{b} \\ b & \bar{a} \end{pmatrix} \in \mathbf{C}(2n) : \begin{pmatrix} a & \bar{b} \\ b & \bar{a} \end{pmatrix}^{\xi} \begin{pmatrix} a & \bar{b} \\ b & \bar{a} \end{pmatrix} = 1 \right\},$$

where, for any $\begin{pmatrix} a & \bar{b} \\ b & \bar{a} \end{pmatrix} \in \mathbf{C}(2n), \begin{pmatrix} a & \bar{b} \\ b & \bar{a} \end{pmatrix}^{\xi} = \begin{pmatrix} a & \bar{b} \\ b & \bar{a} \end{pmatrix}$. *Then G is a Lie group, with tangent space at 1 the real linear subspace of* $\mathbf{C}(2n)$,

$$\left\{ \begin{pmatrix} a & \bar{b} \\ b & \bar{a} \end{pmatrix} \in \mathbf{C}(2n) : a \in \operatorname{End}_-(\overline{\mathbf{C}}^n) \times \operatorname{End}_-(\mathbf{C}^n) \right\},$$

isomorphic in an obvious way with $\operatorname{End}_-(\overline{\mathbf{C}}^n) \times \operatorname{End}_-(\mathbf{C}^n)$. *Moreover the map*

$$O(2n) \to \mathbf{C}(2n);\ t \mapsto c^{-1} t\, c,$$

with $c = \frac{1}{\sqrt{2}} \begin{pmatrix} 1 & \mathrm{i} \\ \mathrm{i} & 1 \end{pmatrix}$, *is a smooth embedding, with image G inducing a Lie group isomorphism between $O(2n)$ and G.*

Proposition 22.44 *For any n the map*

$$f : O(2n) \to \mathcal{N}_n({}^2\mathbf{C}^{2n}_{hb}) : \begin{pmatrix} a & \bar{b} \\ b & \bar{a} \end{pmatrix} \mapsto \operatorname{im} \begin{pmatrix} a \\ b \end{pmatrix}$$

(cf. the proof of Proposition 14.9) *is a smooth projection.*

Proof In this instance $O(2n)$ is embedded in $\mathbf{C}(2n)$, as in Proposition 22.43. It is enough to prove that the map is smooth at 1, with surjective differential there, the surjectivity of f having been already proved in Chapter 14.

The image of 1 by f is $\operatorname{im}\begin{pmatrix}1\\0\end{pmatrix}$ and near this point of $\mathcal{N}_n(\mathbf{C}^{2n}_{\mathrm{hb}})$ one has the chart

$$\operatorname{End}_-(\mathbf{C}^n) \to \mathcal{N}_n(\mathbf{C}^{2n}_{\mathrm{hb}});\ b' \mapsto \operatorname{im}\begin{pmatrix}1\\b'\end{pmatrix},$$

with inverse

$$\mathcal{N}_n(\mathbf{C}^{2n}_{\mathrm{hb}}) \to \operatorname{End}_-(\mathbf{C}^{2n});\ \operatorname{im}\begin{pmatrix}a\\b\end{pmatrix} \mapsto b\,a^{-1},$$

sending $\operatorname{im}\begin{pmatrix}1\\0\end{pmatrix}$, in particular, to 0.

Near 1, therefore, the map f is representable by the map

$$O(2n) \to \operatorname{End}_-(\mathbf{C}^n);\ \begin{pmatrix}a & \overline{b}\\ b & \overline{a}\end{pmatrix} \mapsto b\,a^{-1},$$

which is smooth, with tangent map at 1

$$\operatorname{End}_-(\overline{\mathbf{C}}^n) \times \operatorname{End}_-(\mathbf{C}^n) \to \operatorname{End}_-(\mathbf{C}^n);\ (a, b) \mapsto b.$$

(Cf. Exercise 21.1.) This is clearly surjective. □

There are nine other examples like this one, and the reader is invited to formulate and to discuss them. (Cf. Proposition 14.10.)

It remains to consider several examples involving Spin groups.

Proposition 22.45 *For any p, q the group $Spin(p, q)$ is a smooth submanifold of $\mathbf{R}^0_{0,n}$ and is, with this structure, a Lie group.*

Proof The Pfaffian charts of $Spin(p, q)$ are smooth embeddings, and the group operations are restrictions of maps that are known to be smooth. □

Proposition 22.46 *The group surjection $\rho : Spin(p, q) \to SO(p, q)$ is a smooth projection, the fibres being pairs of points.*

Proof Use Pfaffian and Cayley charts. □

Proposition 22.47 *The groups $Spin^+(p, q)$ and $SO^+(p, q)$ are Lie groups. All of these are connected, except for $Spin^+(0, 0)$, $Spin^+(0, 1)$, $Spin^+(1, 0)$ and $Spin^+(1, 1)$, homeomorphic to S^0, S^0, S^0 and $S^0 \times \mathbf{R}$, respectively.*

Proposition 22.48 *The Lorentz group $SO^+(p, q)$ consists of the rotations of $\mathbf{R}^{p,q}$ preserving the semi-orientations of $\mathbf{R}^{p,q}$.*

Proof Since $SO^+(p, q)$ is connected, by Proposition 22.47, the continuous map $SO^+(p, q) \to \mathbf{R}^{\bullet}; \begin{pmatrix} a & c \\ b & d \end{pmatrix} \mapsto \det a$ is of constant sign and, since its value at 1 is 1, it is always of positive sign. Similarly $\det d$ is positive on $SO^+(p, q)$. By a similar argument $\det a$ and $\det d$ are negative on the coset $SO^-(p, q)$ of $SO^+(p, q)$ in $SO(p, q)$. □

Lie algebras

In all the examples of Lie groups given above the standard atlases or embeddings defining the smooth structure and the group operations have been not only C^1, but also C^2, C^∞ and even C^ω. Here we shall assume that all groups are C^2 at least. This is no restriction, since it can be shown (Pontrjagin (1946)) that any C^1 Lie group admits a unique C^2, C^∞ or even C^ω Lie group structure compatible with the given C^1 Lie group structure.

Proposition 22.49 *Let G be a C^2 Lie group. Then the map*

$$G \times G \to G;\ (a, g) \mapsto a g a^{-1}$$

is C^2.

In particular, the group map

$$\rho_a : G \to G;\ g \mapsto a g a^{-1}$$

is C^2, for any $a \in G$. The map

$$\mathrm{Ad}_G : G \to \operatorname{Aut} TG_1;\ a \mapsto (T\rho_a)_1,$$

where $1 = 1_{(G)}$, is called the *adjoint representation* of the Lie group G.

Proposition 22.50 *Let G be a C^2 Lie group. Then Ad_G is a C^1 group map.*

Proof By Proposition 21.59, Ad_G is a group map. To prove that it is C^1 it is enough to prove that it is C^1 at 1.

Let $L = TG_1$ and let $h : L \rightarrowtail G$ be any C^2 chart on G with $h(0) = 1$ and $Th_0 = 1_L$, the identity map on L. Let $f : L \times L \rightarrowtail L$ be the map defined, for any $(x, y) \in L$ sufficiently near to 0, by the formula

$$h(f(x, y)) = h(x)h(y)(h(x))^{-1}.$$

Then, for any $a \in G$ sufficiently near 1,

$$(T\rho_a)_1 = d_1 f(x, 0), \quad \text{where} \quad h(x) = a,$$

so that $(\mathrm{Ad}_G)h = d_1 f(-, 0)$, which is C^1 at 1. Therefore Ad_G is C^1 at 1. □

The adjoint representation of a Lie group need not be injective.

Example 22.51 *Let* $G = S^1$. *Then* Ad_G *is the constant map with value* 1.

Example 22.52 *Let* G *be any abelian Lie group. Then* Ad_G *is the constant map with value* 1.

Example 22.53 *Let* $G = S^3$. *Then* Ad_G *has image* $SO(3)$, *the surjective map* $S^3 \to SO(3)$ *being the familiar double covering.*

The map

$$\mathrm{ad}_G = T(\mathrm{Ad}_G)_1 : TG_1 \to \mathrm{End}\, TG_1$$

is called the *adjoint representation* of TG_1.

Proposition 22.54 *For any* C^2 *Lie group* G, *the map*

$$TG_1 \times TG_1 \to TG_1;\ (x, y) \mapsto [x, y] = \mathrm{ad}_G(x)(y)$$

is bilinear.

The product defined in Proposition 22.54 is known as the *Lie bracket*, and the linear space TG_1 with this product is known as the *Lie algebra* of G. The Lie bracket is normally neither commutative nor associative. (See Theorems 22.58 and 22.62 below.)

Proposition 22.55 *Let* G *be a* C^2 *Lie group and let* h *and* f *be defined as in the proof of Proposition 22.50. Then, for any* $x, y \in TG_1$,

$$[x, y] = d_0 d_1 f(0, 0)(x)(y).$$

Theorem 22.56 *Let* $t : G \to H$ *be a* C^2 *Lie group map, where* G *and* H *are* C^2 *Lie groups. Then* Tt_1 *is a Lie algebra map; that is, for all* $x, y \in TG_1$,

$$Tt_1([x, y]) = [Tt_1(x), Tt_1(y)].$$

Proof For any $a, g \in G$,

$$t(\rho_a(g)) = t(a\,g\,a^{-1}) = \rho_{t(a)}t(g),$$

and therefore, for any $a \in G$, the diagram of maps

$$\begin{array}{ccc} G & \xrightarrow{\rho_a} & G \\ \downarrow{\scriptstyle t} & & \downarrow{\scriptstyle t} \\ H & \xrightarrow{\rho_{t(a)}} & H \end{array}$$

is commutative. The induced diagram of tangent maps is

$$\begin{array}{ccc} TG_1 & \xrightarrow{\mathrm{Ad}_G\,a} & TG_1 \\ \downarrow{\scriptstyle Tt_1} & & \downarrow{\scriptstyle Tt_1} \\ TH_1 & \xrightarrow{\mathrm{Ad}_H\,t(a)} & TH_1\,, \end{array}$$

leading, for any $y \in TG_1$, to the commutative diagram

$$\begin{array}{ccccc} G & \xrightarrow{\mathrm{Ad}_G} & \mathrm{Aut}\,TG_1 & \xrightarrow{\text{evaluation at } y} & TG_1 \\ \downarrow{\scriptstyle t} & & & & \downarrow{\scriptstyle Tt_1} \\ H & \xrightarrow{\mathrm{Ad}_H} & \mathrm{Aut}\,TH_1 & \xrightarrow{\text{evaluation at } Tt_1(y)} & TH_1\,. \end{array}$$

The induced tangent map diagram this time is

$$\begin{array}{ccccc} TG_1 & \xrightarrow{\mathrm{ad}_G} & \mathrm{End}\,TG_1 & \xrightarrow{\text{evaluation at } y} & TG_1 \\ \downarrow{\scriptstyle Tt_1} & & & & \downarrow{\scriptstyle Tt_1} \\ TH_1 & \xrightarrow{\mathrm{ad}_H} & \mathrm{End}\,TH_1 & \xrightarrow{\text{evaluation at } Tt_1(y)} & TH_1\,. \end{array}$$

This also is commutative. That is, for all $x, y \in TG_1$,

$$Tt_1[x, y] = [Tt_1(x), Tt_1(y)],$$

which is what had to be proved. □

Proposition 22.57 *Let G be a C^2 Lie group and let L and h be defined as in the proof of Proposition 22.50. For all $x, y \in L$, let $\phi(x, y) = x \cdot y$ be defined by the formula*

$$h(x \cdot y) = h(x)h(y)$$

whenever $h(x)\,h(y) \in \operatorname{im} h$, and let $\chi(x) = x^{(-1)}$ be defined by the formula

$$h(x^{(-1)}) = h(x)^{-1},$$

whenever $h(x)^{-1} \in \operatorname{im} h$. *Then* ϕ *and* χ *are* C^2 *maps with non-null open domains,*

$$d_0\phi(0,0) = 1_L, \quad d_1\chi(0,0) = 1_l$$

and $d\chi 0 = -1_L$.

(Note that, for any $x \in L$ sufficiently near 0, $\phi(x, 0) = x$, $\phi(0, x) = x$ and $\phi(x, \chi(x)) = 0$.)

Theorem 22.58 *Let* G *be a* C^2 *Lie group and let* $L = TG_1$. *Then, for all* $x, y \in L$,

$$[y, x] = [x, y].$$

Proof Let h and f be defined as in the proof of Proposition 22.50 and let ϕ and χ be defined as in Proposition 22.57. Then, for any $x \in \operatorname{dom}\chi$, since the map $f(x, -)$ admits the decomposition

$$\begin{array}{ccccc} L & \rightarrowtail & L & \rightarrowtail & L \\ y & \mapsto & x \cdot y = w & \mapsto & w \cdot x^{(-1)}, \end{array}$$

it follows that

$$d_1 f(x, 0) = d_0\phi(0, x^{(-1)})\, d_1\phi(x, 0).$$

From this, and from Proposition 22.57, it follows that, for any $x \in L$,

$$\begin{aligned} d_0 d_1 f(0,0)(x) &= (d_1 d_0\phi(0,0))(d\chi 0(x)) d_1\phi(0,0) \\ &\quad + (d_0\phi(0,0))(d_0 d_1\phi(0,0))(x) \\ &= d_0 d_1\phi(0,0)(x) - d_1 d_0\phi(0,0)(x), \end{aligned}$$

implying, by Proposition 22.55 and Corollary 21.26, that, for any $x, y \in L$,

$$[x, y] = d_0 d_1\phi(0,0)(x)(y) - d_0 d_1\phi(0,0)(y)(x),$$

and therefore that $[y, x] = [x, y]$. □

Note that, though $d_0 d_1 f(0, 0)$ is independent of the choice of the chart h, this is not so for $d_0 d_1\phi(0, 0)$. Consider, for example, $G = \mathbf{R}^{\bullet}$. Then, if h is the chart $\mathbf{R} \rightarrowtail \mathbf{R}^{\bullet};\ x \mapsto 1 + x$, ϕ is a restriction of the map

$$\mathbf{R} \times \mathbf{R} \to \mathbf{R};\ (x, y) \mapsto x + y + xy$$

and $d_0 d_1\phi(0,0)(x)(y) = xy$, while, if h is the chart $\mathbf{R} \to \mathbf{R}^{\bullet};\ x \mapsto e^x$, ϕ is the map

$$\mathbf{R} \times \mathbf{R} \to \mathbf{R};\ (x, y) \mapsto x + y$$

and $d_0 d_1\phi(0,0)(x)(y) = 0$.

Corollary 22.59 *Let X be a finite-dimensional real linear space. Then, for any $u, v \in T(\operatorname{Aut} X)_1 = \operatorname{End} X$,*

$$[u, v] = u\,v - v\,u.$$

Proof Let h be the chart

$$\operatorname{End} X \rightarrowtail \operatorname{Aut} X;\ t \mapsto t\,.$$

Then, for any $u, v \in \operatorname{End} X$, since $\phi(u, v) = u\,v$,

$$d_0 d_1 \phi(0, 0)(u)(v) = u\,v.$$

□

Corollary 22.60 *Let G be a Lie subgroup of* $\operatorname{Aut} x$, *where X is a finite-dimensional real linear space. Then, for any $u, v \in TG_1$,*

$$[u, v] = u\,v - v\,u.$$

Example 22.61 *For any $x, y \in (TS^3)_1$, the space of pure quaternions,*

$$[x, y] = x\,y - y\,x = 2\,x \times y,$$

where $\times$ denotes the vector product.

Theorem 22.62 *Let G be a C^2 Lie group and let $L = TG_1$. Then, for all $x, y, z \in L$,*

$$[[x, y], z] = [x, [y, z]] - [y, [x, z]].$$

Proof By Theorem 22.56 applied to the C^1 group map Ad_G, ad_G is a Lie algebra map. Therefore, for all $x, y \in L$, by Corollary 22.59,

$$\operatorname{ad}_G[x, y] = (\operatorname{ad}_G x)(\operatorname{ad}_G y) - (\operatorname{ad}_G y)(\operatorname{ad}_G x),$$

and so, for all $x, y, z \in L$, $[[x, y, z] = [x, [y, z]] - [y, [x, z]]$. □

The equation proved in Theorem 22.62 is known as the *Jacobi identity* for the Lie algebra L. By Theorem 22.58 this can also be written in the more symmetrical form

$$[x, [y, z]] + [y, [z, x]] + [z, [x, y]] = 0.$$

A *Lie algebra* over a commutative field K is an algebra L over K such that, for any $x, y \in L$,

$$[y, x] = -[x, y]$$

and, for any $x, y, z \in L$,

$$x, [y, z]] + [y, [z, x]] + [z, [x, y]] = 0,$$

where $L \times L \to L$; $(x, y) \mapsto [x, y]$ is the algebra product.

By Theorem 22.58 and Theorem 22.62 the Lie algebra of a Lie group is a Lie algebra in this more general sense.

For a good survey article on Lie algebras see Kaplansky (1963).

The theory of Lie groups is developed in many places. See, for example, Weyl (1939), Chevalley (1946), Pontrjagin (1946), Helgason (1962) and Gilmore (1974).

Exercises

22.1 Construct the following homeomorphisms:

$$O(1) \cong \mathscr{S}(\mathbf{R}^{0,1}) \cong S^0, \quad \mathscr{S}(\mathbf{R}^{0,2}) \cong S^1, \quad \mathscr{S}(\mathbf{R}1, 1) \cong \mathbf{R} \times S^0,$$
$$\mathbf{R}^* \cong \mathscr{S}({}^2\mathbf{R}) \cong \mathbf{R} \times S^0, \quad \mathscr{S}({}^2\mathbf{R})^2 \cong \mathbf{R}^2 \times S^1,$$
$$Sp(2, \mathbf{R}) \cong \mathscr{S}(\mathbf{R}^2_{\mathrm{sp}}) \cong \mathbf{R}^2 \times S^1, \quad \mathscr{S}(\mathbf{R}^4_{\mathrm{sp}}) \cong \mathbf{R}^4 \times S^3,$$
$$Sp(2, \mathbf{C}) \cong \mathscr{S}(\mathbf{C}^2_{\mathrm{sp}}) \cong \mathbf{C}^2 \times S^1, \quad \mathscr{S}(\mathbf{C}^4_{\mathrm{sp}}) \cong \mathbf{R}^7 \times S^7,$$
$$Sp(1) \cong \mathscr{S}(\overline{\mathbf{H}}^{0,1}) \cong S^3, \quad \mathscr{S}(\overline{\mathbf{H}}^{0,2}) \cong S^7, \quad \mathscr{S}(\overline{\mathbf{H}}^{1,1}) \cong \mathbf{R}^4 \times S^3,$$
$$\mathbf{H}^* \cong \mathscr{S}({}^2\tilde{\mathbf{H}}) \cong \mathbf{R} \times S^3, \quad \mathscr{S}({}^2\tilde{\mathbf{H}})^2 \cong \mathbf{R}^5 \times S^7,$$
$$O(1, \mathbf{H}) \times \mathscr{S}(\tilde{\mathbf{H}}^1) \cong S^1, \quad \mathscr{S}(\tilde{\mathbf{H}}^2) \cong \mathbf{R}^2 \times S^3,$$
$$O(1, \mathbf{C}) \times \mathscr{S}(\mathbf{C}^1) \cong S^0, \quad \mathscr{S}(\mathbf{C}^2) \cong \mathbf{R} \times S^1,$$
$$U(1) \cong \mathscr{S}(\overline{\mathbf{C}}^{0,1}) \cong S^1, \quad \mathscr{S}(\overline{\mathbf{C}}^{0,2}) \cong S^3, \quad \mathscr{S}(\overline{\mathbf{C}}^{1,1}) \cong \mathbf{R}^2 \times S^1,$$
$$\mathbf{C}^* \cong \mathscr{S}({}^2\mathbf{C}) \cong \mathbf{R} \times S^1, \quad \mathscr{S}({}^2\mathbf{C})^2 \cong \mathbf{R}^3 \times S^3.$$

Exercise 9.4 and Exercises 19.8 and 19.9 may be of assistance in constructing several of the harder ones.

22.2 Prove that $\mathbf{R}P^1$, $\mathbf{C}P^1$ and $\mathbf{H}P^1$ are homeomorphic, respectively, to S^1, S^2 and S^4.

Prove also that $\mathbf{O}P^1$ is homeomorphic to S^8.

22.3 Prove that $SO(2) \cong \mathbf{R}P^1$, that $SO(3) \cong \mathbf{R}P^3$ and that $SO(4) \cong \mathscr{N}_1(\mathbf{R}^{4,4}) \cong \mathscr{N}_1(\mathbf{R}^8_{\mathrm{hb}})$. (Cf. Proposition 14.18.)

22.4 Prove that $\mathscr{N}_n(\mathbf{R}^{2n}_{\mathrm{hb}})$ and $\mathscr{N}_n(\mathbf{C}^{2n}_{\mathrm{hb}})$ are each the union of two disjoint connected components, and that either component of $\mathscr{N}_4(\mathbf{R}^8_{\mathrm{hb}})$ is homeomorphic to $\mathscr{N}_1(\mathbf{R}^8_{\mathrm{hb}})$.

Is either component of $\mathscr{N}_4(\mathbf{C}^8_{\mathrm{hb}})$ homeomorphic to $\mathscr{N}_1(\mathbf{C}^8_{\mathrm{hb}})$? (We give the answer eventually at the end of Chapter 24.)

22.5 Let X be a real linear space of finite dimension n say, let $k \leq n$, and let X be assigned a positive-definite quadratic form. Let $O(\mathbf{R}^k, X)$ denote the set of orthogonal maps $\mathbf{R}^k \to X$, the images of the standard basis vectors in $\mathbf{R}^k$ forming an orthonormal k-frame in X. Prove that $O(\mathbf{R}^k, X)$ is a compact submanifold of $L(\mathbf{R}^k, X)$ of dimension $\frac{1}{2}k(2n-k-1)$ and that the map

$$\pi : O(\mathbf{R}^k, X) \to \mathscr{G}_k(X);\ t \mapsto \operatorname{im} t$$

is a smooth projection, inferring, in particular, that $\mathscr{G}_k(X)$ is compact.

The manifolds $O(\mathbf{R}^k, X)$ are known as *Stiefel manifolds.*

23
Conformal groups

Our concern in this chapter is with the group $Conf(X)$ of conformal transformations of a non-degenerate real quadratic space X of finite dimension n and signature (p, q), and with a description of such groups that involves Clifford algebras. In doing so we shall draw heavily on Chapter 18 which was concerned with the study of 2×2 Clifford matrices.

Let X and Y be finite-dimensional quadratic spaces and $f : X \rightarrowtail Y$ a smooth map. Then f is said to be *conformal* if the differential dfx of f at any point x is of the form $\rho(x)t$, where $\rho(x)$ is a non-zero real number and $t : X \to Y$ is an orthogonal map, and so is such that, for any $u, v \in X$, $dfx(u)\cdot dfx(v) = (\rho(x))^2 u \cdot v$; that is it is conformal if it *preserves angles.* More generally, let X and Y be finite-dimensional smooth manifolds and $f : X \rightarrowtail Y$ a smooth map. Then f is said to be *conformal* if the differential dfx of f at any point x of X is a non-zero real multiple of an orthogonal map.

It is well-known that any holomorphic map $f : \mathbf{C} \rightarrowtail \mathbf{C}$, with $\mathbf{C}$ identified as a quadratic space with $\mathbf{R}^2$ with its standard scalar product, is conformal. Conformal transformations of quadratic spaces of dimension greater than 2 are more restricted, as follows, in the positive-definite case at least, from a theorem of Liouville (1850). It turns out that in studying such maps it is appropriate to compactify the quadratic spaces in question in a particular way that is known as the *conformal compactification*, this being quite distinct from the possibly more familiar projective compactification.

Liouville's theorem

The theorem of Liouville, as originally stated, concerns smooth maps $f : \mathbf{R}^3 \rightarrowtail \mathbf{R}^3$, the vector space $\mathbf{R}^3$ being assigned its standard euclidean metric. It follows from the standard theory of the curvature of a smooth surface in $\mathbf{R}^3$ that the parallels to such a surface together with the families of surfaces generated by the normals to either set of lines of curvature form a triply orthogonal set of families of surfaces. Dupin showed that the converse is true, in the sense that if one has such a triply orthogonal system then the surfaces of any two of the families cut out lines of curvature on the surfaces of the third family. Clearly a conformal transformation of $\mathbf{R}^3$ sends any such triply orthogonal system to another, so maps lines of curvature on any smooth surface to lines of curvature on the image surface. In particular it maps umbilical points of such a surface to umbilical points of the image surface. It then follows, by a theorem of Meusnier (1785), that the image by any conformal map of a sphere or plane is a sphere or plane. Such a map, by a theorem of Möbius, is representable as the composite of a finite number of orthogonal maps, translations or inversions of $\mathbf{R}^3$ in spheres, the simplest such inversion being inversion in the sphere, centre the origin, with unit radius, namely the map $\mathbf{R}^3 \rightarrowtail \mathbf{R}^3$; $x \mapsto x/|x|^2$. Moreover all such Möbius maps are conformal. This is the *theorem of Liouville.*

The obvious analogue of this theorem then holds for positive-definite quadratic spaces of any finite dimension greater than 3. The analogous statement for indefinite quadratic spaces also is true by a theorem of Haantjes (1937).

The projective compactification

Let X be a finite-dimensional real linear space. Then any norm on X induces the same topology. That is any subset of X that is open in X with respect to any particular norm is also open with respect to any other norm.

Consider the map $X \to X \times \mathbf{R}$; $x \mapsto (x, 1)$. This map is clearly injective but also induces an injective map

$$X \to \mathscr{G}_1(X \times \mathbf{R});\; x \mapsto [x, 1] = \mathbf{R}\{x, 1\}$$

of X to the *projective space* of lines through the origin in $X \times \mathbf{R}$. The projective space inherits the quotient topology from the topology of the linear space $X \times \mathbf{R}$ and is compact. The complement in the projective space of the image of the original space X is by definition the *hyperplane at infinity* of X.

The conformal compactification

For a real non-degenerate quadratic space X of signature (p, q) there is an alternative compactification that often offers advantages. This is the conformal compactification defined as follows (cf. É. Cartan (1947), (1949), Kuiper (1949), Hermann (1979)).

For simplicity let the scalar product on X be denoted by $\cdot$, and consider the injective map

$$\imath : X \to X'' = X \times \mathbf{R} \times \mathbf{R};\ x \mapsto (x, 1, x \cdot x) = (x, \mu, \nu).$$

This is not linear, but the image of X is a subset of the quadric cone Q in X'' with equation

$$x \cdot x - \mu \nu = 0.$$

The map then induces an injective map to the quadric in the projective space $\mathscr{G}_1(X'')$ with this homogeneous equation. This quadric is compact, being a closed subset of a compact space, and is defined to be the *conformal compactification* $\widehat{X}$ of the quadratic space X.

By Proposition 14.18 the quadric $\widehat{X}$ is homeomorphic to $(S^p \times S^q)/S^0$, where $S^0 = \{1, -1\}$ acts on $S^p \times S^q$ by $(-1)(x, y) = (-x, -y)$. In particular, in the case that $p = 0$, $q = n$ the quadric is homeomorphic to S^n and in that case is a *one-point compactification* of $\mathbf{R}^n$.

Let X'' be assigned the quadratic form $(x, \mu, \nu) \mapsto x \cdot x - \mu \nu$, this being of signature $(p + 1, q + 1)$. The central result is then the corollary of the following.

Theorem 23.1 *Let X, X'' and Q be as above. Then*

(i) *the map $\imath : X \to X''$; $x \mapsto (x, x \cdot x, 1)$, with image a subset of Q, is an isometry,*
(ii) *the map $\pi : Q \rightarrowtail X$; $(x, \mu, \nu) \mapsto x/\nu$, defined where $\nu \neq 0$, is conformal.*

Proof

(i) The differential of $\imath$ at x is the linear map

$$dx \mapsto (dx, 2x \cdot dx, 0),$$

and $dx \cdot dx - (2x \cdot dx)(0) = dx \cdot dx$.

(ii) The differential of π at (x, μ, ν) is the linear map

$$(dx, d\mu, d\nu) \mapsto \nu^{-2}(\nu\, dx - x\, d\nu),$$

with $x \cdot x = \mu\, \nu$ and $2x \cdot dx = \mu\, dv + \nu\, d\mu$, implying that

$$\begin{aligned}(\nu\, dx - x\, d\nu) \cdot (\nu\, dx - x\, d\nu) &= \nu^2 dx \cdot dx - \nu\, d\nu(\mu\, d\nu + \nu\, d\mu) + \mu\nu\, d\nu^2 \\ &= \nu^2(dx \cdot dx - d\mu\, d\nu),\end{aligned}$$

so that $\nu^{-2}(\nu\, dx - x\, d\nu) \cdot \nu^{-2}(\nu\, dx - x\, d\nu) = \nu^{-2}(dx \cdot dx - d\mu\, d\nu)$. □

Corollary 23.2 *Let $t : X'' \to X''$ be any orthogonal transformation of X''. Then the map $f = \pi\, t\, \iota : X \rightarrowtail X$ is conformal.*

From their form it is clear that such maps map *conformal spheres* (that is, quasi-spheres or hyperplanes) to conformal spheres, a *quasi-sphere* in the quadratic space X being a submanifold of X defined by an equation of the form $a\, x \cdot x + b \cdot x + c = 0$, where $a, c \in \mathbf{R}$ and $b \in X$, a, b and c not all being zero, this being a genuine sphere in the case that the quadratic form on X is positive-definite and $a \neq 0$ and a plane in the case that $a = 0$. It is a consequence of Liouville's theorem that, for $\dim X \geq 3$, all conformal maps $X \rightarrowtail X$ are so induced. Clearly any such map f extends to a map $\widehat{f} : \widehat{X} \to \widehat{X}$, with domain the whole of $\widehat{X}$.

It is clear in Corollary 23.2 that both t and $-t$ induce the same conformal transformation of X. The *conformal group* $Conf(X)$ is accordingly defined to be the quotient group $O(X'')/S^0$. Since the signature $(p+1, q+1)$ of X'' is indefinite, the group $O(X'')$ has four components. So $Conf(X)$ has four or two, according to where the element $-1_{X''}$ lies. If it lies in the connected component of the identity, which is the case when p and q are both odd, implying that $p+1$ and $q+1$ are both even, then $Conf(X)$ has four components, but, if not, it has only two. The connected component of the identity is known as the *Möbius group* of X, denoted by $M(X)$.

It should be noted that $x \cdot x - \mu\nu = x \cdot x + (\frac{1}{2}(\mu - \nu))^2 - (\frac{1}{2}(\mu + \nu))^2$. With this in mind the most usual chart to employ on the projective space $\mathcal{G}_1(X'')$ is the map

$$[x, \mu, \nu] \mapsto \left(\frac{2x}{\mu + \nu}, \frac{\mu - \nu}{\mu + \nu}\right).$$

Then the composite of the embedding of X in X'' with the projection

$$X'' \rightarrowtail X \times \mathbf{R};\ (x, \mu, \nu) \mapsto \left(\frac{2x}{\mu + \nu}, \frac{\mu - \nu}{\mu + \nu}\right)$$

is conformal, the product space $X \times \mathbf{R}$ being assigned the quadratic form

$(x, w) \mapsto x \cdot x + w^2$, since its differential is

$$dx \mapsto \left(\frac{2\,dx}{x \cdot x + 1} - \frac{4x\, x \cdot dx}{(x \cdot x + 1)^2}, \frac{2x \cdot dx}{x \cdot x + 1} - \frac{2x \cdot dx(x \cdot x - 1)}{(x \cdot x + 1)^2}\right),$$

the quadratic norm of the image being

$$\frac{4}{(x \cdot x + 1)^4}\left(((x \cdot x + 1)dx - 2x\, x \cdot dx) \cdot ((x \cdot x + 1)dx - 2x\, x \cdot dx) + 4(x \cdot dx)^2\right)$$

$$= \frac{4}{(x \cdot x + 1)^2} dx \cdot dx.$$

To clarify all this let us look at some simple examples.

Example 23.3 *Let $X = \mathbf{R}^2$ with its standard positive-definite scalar product. That is $X = \mathbf{R}^{0,2}$. Then the image of the map*

$$\mathbf{R}^2 \to \mathscr{G}_1(\mathbf{R}^4);\ (x, y) \mapsto [x, y, 1, x^2 + y^2]$$

lies in the quadric with equation $x^2 + y^2 - \mu\, v = 0$, this quadric being the conformal compactification of X.

Suppose that we make a change of variables to express the equation of the quadric as a sum of squares, an appropriate such choice being $x = x$, $y = y$, $z = \frac{1}{2}(\mu - v)$, $t = \frac{1}{2}(\mu + v)$. Then the equation reduces to $x^2 + y^2 + z^2 - t^2$ and the image lies entirely in the affine chart given by $t = 1$, the map to this chart being the map

$$\mathbf{R}^2 \rightarrowtail \mathbf{R}^3;\ (x, y) \mapsto \left(\frac{2x}{1 + x^2 + y^2}, \frac{2y}{1 + x^2 + y^2}, \frac{1 - x^2 - y^2}{1 + x^2 + y^2}\right),$$

with image a subset of the unit sphere S^2 in R^3. Indeed the image is the whole of this sphere with the exception of one point, the *South pole*, $(0, 0, -1)$.

This compactification, being a one-point compactification, is quite distinct from the projective compactification of the previous section. It is often presented in inverse form as the *stereographic projection* of the sphere, minus its South pole, to its equatorial plane. Indeed the three points

$$\left(\frac{2x}{1 + x^2 + y^2}, \frac{2y}{1 + x^2 + y^2}, \frac{1 - x^2 - y^2}{1 + x^2 + y^2}\right)$$

of the sphere, $(x, y, 0)$ of the equatorial plane and $(0, 0, -1)$, the South Pole, are collinear, as is readily verified.

The inverse map is the map

$$S^2\backslash(0,0,-1) \to \mathbf{R}^2;\ (x',y',z') \mapsto \left(\frac{x'}{1+z'}, \frac{y'}{1+z'}\right).$$

Example 23.4 *Let $X = \mathbf{R}^{1,1}$. Then the image of the map*

$$\mathbf{R}^2 \to \mathscr{G}_1(\mathbf{R}^4);\ (x,y) \mapsto [x, y, 1, -x^2+y^2]$$

lies in the quadric with equation $-x^2+y^2-\mu\nu = 0$, this quadric being the conformal compactification of X.

Suppose that we make the same change of variables as in Example 23.3 to express the equation of the quadric as a sum of squares, that is $x = x$, $y = y$, $z = \frac{1}{2}(\mu - \nu)$, $t = \frac{1}{2}(\mu + \nu)$. Then the equation reduces to $-x^2+y^2+z^2-t^2$. The image no longer lies entirely in any affine chart, but the map to the chart with $t = 1$ is the map

$$\mathbf{R}^2 \rightarrowtail \mathbf{R}^3;\ (x,y) \mapsto \left(\frac{2x}{1-x^2+y^2}, \frac{2y}{1-x^2+y^2}, \frac{1+x^2-y^2}{1-x^2+y^2}\right),$$

with domain the complement of the hyperbola with equation $x^2 - y^2 = 1$ and image a subset of the hyperboloid of one sheet with equation $-X^2+Y^2+Z^2 = 1$.

Möbius transformations of $\mathbf{R}^2$

As we saw in Chapter 16 the appropriate Clifford algebra in which to study rotations of the euclidean plane $\mathbf{R}^2$ is the algebra $\mathbf{R}_{0,2}^0$ of complex numbers $\mathbf{C}$. The plane itself is represented by the plane of paravectors which in this case is the whole of $\mathbf{C}$. Now the conformal compactification sits naturally inside the Clifford algebra $\mathbf{R}_{1,3}^0$ which is representable by the matrix algebra $\mathbf{C}(2)$, the complex number z then being represented by the matrix

$$\begin{pmatrix} z & z\bar{z} \\ 1 & \bar{z} \end{pmatrix},$$

that is explicitly by the element

$$x\begin{pmatrix} 1 & 0 \\ 0 & 1 \end{pmatrix} + iy\begin{pmatrix} 1 & 0 \\ 0 & -1 \end{pmatrix} + \frac{1}{2}(1+z\bar{z})\begin{pmatrix} 0 & 1 \\ 1 & 0 \end{pmatrix} + \frac{1}{2}(1-z\bar{z})\begin{pmatrix} 0 & -1 \\ 1 & 0 \end{pmatrix}.$$

More generally any element of $\mathbf{R}^{1,3}$ sitting as a paravector in $\mathbf{R}^0_{1,3}$ is represented by an element of $\mathbf{C}(2)$ of the form

$$\begin{pmatrix} z & v \\ \mu & \bar{z} \end{pmatrix},$$

where $z \in \mathbf{C}$ and μ and v are real.

Consider now an element of $Spin^+(1,3)$ represented by an element of $\mathbf{C}(2)$ of the form

$$\begin{pmatrix} a & c \\ b & d \end{pmatrix}.$$

This maps the above paravector to

$$\begin{pmatrix} a & c \\ b & d \end{pmatrix}\begin{pmatrix} z & v \\ \mu & \bar{z} \end{pmatrix}\begin{pmatrix} a & c \\ b & d \end{pmatrix}^{\sim} = \begin{pmatrix} a & c \\ b & d \end{pmatrix}\begin{pmatrix} z & v \\ \mu & \bar{z} \end{pmatrix}\begin{pmatrix} \bar{d} & \bar{c} \\ \bar{b} & \bar{a} \end{pmatrix}.$$

In particular it maps the paravector

$$\begin{pmatrix} z & z\bar{z} \\ 1 & \bar{z} \end{pmatrix} = \begin{pmatrix} z \\ 1 \end{pmatrix}\begin{pmatrix} 1 & \bar{z} \end{pmatrix}$$

representing the complex number z to

$$\begin{pmatrix} a & c \\ b & d \end{pmatrix}\begin{pmatrix} z & z\bar{z} \\ 1 & \bar{z} \end{pmatrix}\begin{pmatrix} \bar{d} & \bar{c} \\ \bar{b} & \bar{a} \end{pmatrix} = \begin{pmatrix} \lambda z' & \lambda z'\overline{z'} \\ \lambda & \lambda\overline{z'} \end{pmatrix} = \lambda\begin{pmatrix} z' & z'\overline{z'} \\ 1 & \overline{z'} \end{pmatrix},$$

where $z' = \dfrac{az+c}{bz+d}$ and λ is the real number $|bz+d|^2$.

Now the conformal compactification of $\mathbf{C}$ is a one-point compactification and it is natural to denote the additional point by ∞. One easily verifies that this point is represented in $\mathbf{C}(2)$ by the matrix

$$\begin{pmatrix} 0 & 1 \\ 0 & 0 \end{pmatrix}$$

and that the image of this matrix by the above element of $Spin^+(1,3)$ is the matrix

$$\begin{pmatrix} a\bar{b} & a\bar{a} \\ b\bar{b} & b\bar{a} \end{pmatrix}$$

that represents the complex number a/b, when $b \neq 0$ and ∞ otherwise.

A map

$$\mathbf{C} \cup \{\infty\} \to \mathbf{C} \cup \{\infty\};\ z \mapsto \frac{az+c}{bz+d},$$

where $a, b, c, d \in \mathbf{C}$ and $ad - bc = 1$, is known as a *Möbius map*. It represents a *special conformal transformation* of the conformal compactification of $\mathbf{R}^{0,2}$, namely one that respects the orientations of $\mathbf{R}^{0,2}$ and its compactification.

For such a Möbius map the induced map of $\mathbf{C}^2$ defined by the matrix

$$\begin{pmatrix} a & c \\ b & d \end{pmatrix}$$

will restrict to a rotation of this sphere if and only if this matrix lies in the copy of $Spin^+(0,3) = Spin\ 3$ naturally included in $Spin^+(1,3)$. Such matrices, as we know from earlier work, are those that represent unit quaternions, namely those of the form

$$\begin{pmatrix} a & -\bar{b} \\ b & \bar{a} \end{pmatrix}.$$

Example 23.5 *There is a unique Möbius map of* $\mathbf{C}$ *to* $\mathbf{C}$ *that sends* 0 *to* 1, 1 *to* i *and* i *to* 0. *The induced rotation of the Riemann sphere is rotation of the sphere through an angle* $\frac{2\pi}{3}$ *about the line with equations* $x = y = z$.

Proof Let the Möbius map be

$$z \mapsto \frac{az + c}{bz + d}.$$

Then $\frac{c}{d} = 1$, that is $c = d$, $\frac{a+c}{b+d} = \mathrm{i}$, that is $a + c = (b + d)\,\mathrm{i}$, and $\frac{a\mathrm{i}+c}{b\mathrm{i}+d} = 0$, that is $a\mathrm{i} + c = 0$. Try $a = 1$. Then $c = d = -\mathrm{i}$ and $1 - \mathrm{i} = b\,\mathrm{i} + 1$ so that $b = -1$. This gives the map

$$z \mapsto \frac{z - \mathrm{i}}{-z - \mathrm{i}}.$$

However, then $ad - bc = 1(-\mathrm{i}) - (-1)(-\mathrm{i}) = -2\mathrm{i} = (1 - \mathrm{i})^2$, and the inverse of $1 - \mathrm{i}$ is $\frac{1}{2}(1 + \mathrm{i})$. So finally the required map is

$$z \mapsto \frac{\frac{1}{2}(1 + \mathrm{i})z + \frac{1}{2}(1 - \mathrm{i})}{-\frac{1}{2}(1 + \mathrm{i})z + \frac{1}{2}(1 - \mathrm{i})}.$$

The fixed points of this map are given by $z - \mathrm{i} = (-z - \mathrm{i})z$, that is by the quadratic equation $z^2 + (1 + \mathrm{i})z - \mathrm{i} = 0$, with roots $z = \frac{1}{2}(-1 - \mathrm{i} \pm \sqrt{6\mathrm{i}}\,)$. □

Möbius transformations of $\mathbf{R}^{p,q}$

The general case can be handled in exactly the same way as the case of $\mathbf{R}^2$ has been treated in the previous section. The appropriate Clifford algebra in which to study rotations of the quadratic space $X = \mathbf{R}^{p,q}$ is the algebra $\mathbf{R}^0_{p,q}$, isomorphic to the Clifford algebra $\mathbf{R}_{p,q-1}$. The quadratic space X itself is then represented by, and will be identified with, the space of paravectors in this Clifford algebra, with $x \cdot x = x x^-$, for any $x \in X$. The conformal compactification sits naturally inside the Clifford algebra $\mathbf{R}^0_{p+1,q+1}$ which is isomorphic to the algebra $\mathbf{R}_{p+1,q}$ of 2×2 matrices with entries in $\mathbf{R}_{p,q-1}$.

The vector x is then represented by the matrix

$$\begin{pmatrix} x & x x^- \\ 1 & x^- \end{pmatrix},$$

that is explicitly by the element

$$\begin{pmatrix} x & 0 \\ 0 & x^- \end{pmatrix} + \frac{1}{2}(1 + x x^-)\begin{pmatrix} 0 & 1 \\ 1 & 0 \end{pmatrix} + \frac{1}{2}(1 - x x^-)\begin{pmatrix} 0 & -1 \\ 1 & 0 \end{pmatrix}.$$

More generally any element of $\mathbf{R}^{p+1,q+1}$ is represented by a paravector in $\mathbf{R}_{p+1,q} \cong \mathbf{R}_{p,q-1}(2)$ of the form

$$\begin{pmatrix} x & \nu \\ \mu & x^- \end{pmatrix},$$

where x is a paravector in $\mathbf{R}_{p,q-1}$ and μ and ν are real.

Consider now an element of $Spin^+(p+1, q)$ represented by an element of $\mathbf{R}_{p,q-1}(2)$ of the form

$$\begin{pmatrix} a & c \\ b & d \end{pmatrix}.$$

This maps the above paravector to

$$\begin{pmatrix} a & c \\ b & d \end{pmatrix}\begin{pmatrix} x & \upsilon \\ u & x^- \end{pmatrix}\begin{pmatrix} a & c \\ b & d \end{pmatrix}^{\sim} = \begin{pmatrix} a & c \\ b & d \end{pmatrix}\begin{pmatrix} x & \nu \\ \mu & x^- \end{pmatrix}\begin{pmatrix} d^- & c^- \\ b^- & a^- \end{pmatrix}$$

by Proposition 18.1. In particular it maps the paravector

$$\begin{pmatrix} x & x x^- \\ 1 & x^- \end{pmatrix} = \begin{pmatrix} x \\ 1 \end{pmatrix}\begin{pmatrix} 1 & x^- \end{pmatrix}$$

representing the vector x to

$$\begin{pmatrix} a & c \\ b & d \end{pmatrix}\begin{pmatrix} x & x x^- \\ 1 & x^- \end{pmatrix}\begin{pmatrix} d^- & c^- \\ b^- & a^- \end{pmatrix} = \lambda\begin{pmatrix} x' & x' x'^- \\ 1 & x' \end{pmatrix},$$

where $x' = (ax+c)(bx+d)^{-1}$ and λ is the real number $(bx+d)(bx+d)^{-}$.

The matrix

$$\begin{pmatrix} a & c \\ b & d \end{pmatrix}$$

representing the induced *special conformal transformation* of the conformal compactification of $\mathbf{R}^{p,q}$ is known as a *Vahlen representation* (see Vahlen, (1902)), a *special* transformation being one that respects the orientation (and, in the case that p and q are both odd, the semi-orientations) of the compactification, this representation being unique up to sign in the case that pq is even, but with a four-fold ambiguity in its definition in the case that pq is odd.

For example the *translation* $x \mapsto x + c$ is represented by the matrix

$$\begin{pmatrix} 1 & c \\ 0 & 1 \end{pmatrix}$$

and *inflation* by the positive scalar ρ by the matrix

$$\begin{pmatrix} \sqrt{\rho} & 0 \\ 0 & \sqrt{\rho}^{-1} \end{pmatrix},$$

while inversion in the unit quasi-sphere composed with the hyperplane reflection $x \mapsto -x^{-}$ is represented by the matrix

$$\begin{pmatrix} 0 & -1 \\ 1 & 0 \end{pmatrix}.$$

An important special case in which pq is odd is that of $\mathbf{R}^{1,3}$ in which case the component of the identity of the conformal group is covered four times rather than twice by the group $SU(2,2)$, the identity transformation of the space being represented by each of the matrices

$$\begin{pmatrix} 1 & 0 \\ 0 & 1 \end{pmatrix}, \begin{pmatrix} -1 & 0 \\ 0 & -1 \end{pmatrix}, \begin{pmatrix} \mathrm{I} & 0 \\ 0 & \mathrm{I} \end{pmatrix} \text{ and } \begin{pmatrix} -\mathrm{I} & 0 \\ 0 & -\mathrm{I} \end{pmatrix},$$

where I is the 2×2 complex matrix $\begin{pmatrix} \mathrm{i} & 0 \\ 0 & \mathrm{i} \end{pmatrix}$.

Vahlen's characterisation (1902) of matrices

$$\begin{pmatrix} a & c \\ b & d \end{pmatrix}$$

that represent special conformal transformations of $\mathbf{R}^{0,n}$ was rediscovered by Ahlfors ten years ago. See, for example, Ahlfors (1985). That the appropriate setting for this in the case of the indefinite quadratic space

$\mathbf{R}^{p,q}$ is the study of the Clifford group $\Gamma(p,q)(2)$ of the Clifford algebra $\mathbf{R}_{p+1,q+1} \cong \mathbf{R}_{p,q}(2)$ and the determination of Vahlen type conditions on the entries in the matrix is the work of several people since that time, in particular Elstrodt, Grunewald and Mennicke (1987), Maks (1989), Fillmore and Springer (1990) and Cnops (1994). See Chapter 18 for the details. For further references consult Ryan (1995).

The complete set of Möbius groups, for $p+q \leq 4$, is given in the following theorem.

Theorem 23.6 *Let $M_{p,q}$ denote the Möbius group $M(\mathbf{R}^{p,q})$. Then*

$$\begin{aligned}
M_{0,1} &\cong M_{1,0} \cong Sp(2;\mathbf{R})/S^0,\\
M_{0,2} &\cong M_{2,0} \cong Sp(2;\mathbf{C})/S^0,\\
M_{1,1} &\cong (Sp(2;\mathbf{R})/S^0) \times (Sp(2;\mathbf{R})/S^0) \quad (pq \text{ odd}),\\
M_{0,3} &\cong M_{3,0} \cong Sp(1,1)/S^0,\\
M_{1,2} &\cong M_{2,1} \cong Sp(4;\mathbf{R})/S^0,\\
M_{0,4} &\cong M_{4,0} \cong SL(2;\mathbf{H})/S^0,\\
M_{1,3} &\cong M_{3,1} \cong SU(2,2))/\mathbf{C}(4) \quad (pq \text{ odd}),\\
M_{2,2} &\cong SL(4;\mathbf{R})/S^0.
\end{aligned}$$

Proof These results follow directly from Table 14 of Theorem 17.9. □

Exercise

23.1 Let X be a finite-dimensional quadratic space, let $X \times \mathbf{R}$ be assigned the quadratic form $x \cdot x - t^2 = 0$, for all $(x,t) \in X \times \mathbf{R}$, and let Q be the quadric cone $\{(x,t) \in X \times \mathbf{R} : x \cdot x - t^2 = 0\}$. Prove that the projection $Q \rightarrowtail X : x \mapsto t^{-1}x$, defined wherever $x \cdot x \neq 0$, is conformal.

24
Triality

At the beginning of Chapter 19 we remarked that the Cayley division algebra **O** can ultimately be held 'responsible' for a rich variety of exceptional phenomena. Among these is the triality which we study in this chapter – an automorphism of order 3 of $Spin(8)$ that does not project to an automorphism of $SO(8)$. As a byproduct we make acquaintance with the fourteen-dimensional Lie group G_2, the group of automorphisms of the Cayley algebra **O**.

Triality has something of interest to say about the projective quadrics $\mathcal{N}_1(\mathbf{C}^8)$ and $\mathcal{N}_1(\mathbf{R}^{4,4})$. This quadric triality seems first to have been noted by Study (1903), (1913), though the word 'triality' is due to Élie Cartan (1925), who placed the phenomenon in its proper Lie group context.

Transitive actions on spheres

To put the group $Spin(8)$ in context we begin by looking at all the groups $Spin(n)$, with $n < 10$. By virtue of earlier work we know that

$$\begin{array}{llll}
Spin(1) \cong O(1) & \subset \mathbf{R}^0_{0,1} \cong \mathbf{R}, & & \\
Spin(2) \cong U(1) & \subset \mathbf{R}^0_{0,2} \cong \mathbf{C} & \subset & \mathbf{R}(2), \\
Spin(3) \cong Sp(1) & \subset \mathbf{R}^0_{0,3} \cong \mathbf{H} & \subset & \mathbf{R}(4), \\
Spin(4) \cong Sp(1) \times Sp(1) & \subset \mathbf{R}^0_{0,4} \cong {}^2\mathbf{H} & \subset & {}^2\mathbf{R}(4), \\
Spin(5) \cong Sp(2) & \subset \mathbf{R}^0_{0,5} \cong \mathbf{H}(2) & \subset & \mathbf{R}(8), \\
Spin(6) \cong SU(4) \subset U(4) & \subset \mathbf{R}^0_{0,6} \cong \mathbf{C}(4) & \subset & \mathbf{R}(8), \\
Spin(7) \subset O(8) & \subset \mathbf{R}^0_{0,7} \cong \mathbf{R}(8), & & \\
Spin(8) \subset O(8) \times O(8) & \subset \mathbf{R}^0_{0,8} \cong {}^2\mathbf{R}(8), & & \\
Spin(9) \subset O(16) & \subset \mathbf{R}^0_{0,9} \cong \mathbf{R}(16), & & \\
Spin(10) \subset U(16) & \subset \mathbf{R}^0_{0,10} \cong \mathbf{C}(16), & &
\end{array}$$

and so on. The induced *Clifford* or *spinor* actions of $Spin(1)$ on S^0, $Spin(2)$ on S^1, $Spin(3)$ and $Spin(4)$ (in two ways) on S^3, $Spin(5)$, $Spin(6)$, $Spin(7)$ and $Spin(8)$ (in two ways) on S^7 and $Spin(9)$ on S^{15} are, moreover, all transitive, although the Clifford action of $Spin(10)$ on S^{31} is not, as we shall see – a good reason for stopping at this point!

Apart from these Clifford actions of the groups $Spin(n)$ on spheres there are the standard orthogonal actions.

In studying the standard orthogonal action of $Spin(n+1)$ on S^n, for a positive integer n, it is appropriate to work in the Clifford algebra $\mathbf{R}_{0,n} \cong \mathbf{R}^0_{0,n+1}$, identifying $\mathbf{R}^{n+1}$ with $Y \cong \mathbf{R} \oplus \mathbf{R}^n$, $\mathbf{R}$ and $\mathbf{R}^n$ being embedded in $\mathbf{R}_{0,n}$ in the standard way. Then, for any $y \in Y$, $\widehat{y} = y^-$,

$$S^n = \{y \in Y : y^- y = 1\} \text{ and } Spin(n) = \{g \in Spin(n+1) : \widehat{g} = g\}.$$

It is worth a passing mention that Y is closed under the operation of squaring and therefore can be assigned the bilinear product

$$Y^2 \to Y;\ (y_0, y_1) \mapsto y_0\, y_1 + y_1\, y_0.$$

This gives Y the structure of a *Jordan algebra* (Paige (1961)). Moreover the squaring map $Y \to Y;\ y \mapsto y^2$ is surjective, since any element of Y with non-zero real part (and there are such, since $n \geq 1$) generates a subalgebra of $\mathbf{R}_{0,n}$ isomorphic to $\mathbf{C}$. The standard orthogonal action of $Spin(n+1)$ on Y is

$$Spin(n+1) \times Y \to Y;\ (h, y) \mapsto h\, y\, \widehat{h}^{-1},$$

the map $Y \to Y;\ y \mapsto h\, y\, \widehat{h}^{-1}$ being a rotation of Y, for each $h \in Spin(n+1)$.

Proposition 24.1 *Any element of* $Spin(n+1)$ *is expressible in the form* $z\,g$, *where* $z \in S^n$, $g \in Spin(n)$.

Proof Let $h \in Spin(n+1)$. Since $1 \in S^n$ so also $h\widehat{h}^{-1} \in S^n$. Let $z \in Y$ be such that $z^2 = h\widehat{h}^{-1}$ and let $g = z^{-1}\, h = \widehat{z}\, h$. Such z exists, since squaring on Y is surjective, while, since $h\widehat{h}^{-1} \in S^n$, so also $z \in S^n$. Moreover, $g\,\widehat{g}^{-1} = z^{-1}\, h\widehat{h}^{-1}\, z^{-1} = 1$. So $g = \widehat{g}$, implying that $g \in Spin(n)$. □

It is easy to verify that, for any $n > 0$, the sequence

$$Spin(n) \xrightarrow{\ \iota\ } Spin(n+1) \longrightarrow S^n$$
$$h \mapsto h\widehat{h}^{-1}$$

is left-coset exact, and projects to the left-coset exact sequence

$$SO(n) \longrightarrow SO(n+1) \longrightarrow S^n$$

studied in Chapter 14. Thus $Spin(n+1)$ acts transitively on S^n, each isotropy subgroup of the action being isomorphic to $Spin(n)$.

All these transitive actions of the groups $Spin(n)$ on spheres bear closer study, not only independently, but in relation to each other. Of particular interest are the isotropy groups of the various Clifford actions.

The story is summarised in the following sequence of commutative diagrams:

Diagram 24.1

$$\begin{array}{ccccc} Spin(1) & \longrightarrow & Spin(2) & \longrightarrow & S^1 \\ \cong & & \cong & & = \\ S^0 & \longrightarrow & S^1 & \xrightarrow{h_{\mathbf{R}}} & S^1 \end{array} \quad \text{in } \mathbf{R}_{0,1} \cong \mathbf{C},$$

involving the Hopf map $h_{\mathbf{R}}$, the restriction to S^1 of the Hopf map $\mathbf{R}^2 \to \mathbf{R}P^1$, composed with a stereographic projection;

Diagram 24.2

$$\begin{array}{ccccc} Spin(2) & \longrightarrow & Spin(3) & \longrightarrow & S^2 \\ \cong & & \cong & & = \\ S^1 & \longrightarrow & S^3 & \xrightarrow{h_{\mathbf{C}}} & S^2 \end{array} \quad \text{in } \mathbf{R}_{0,2} \cong \mathbf{H},$$

involving the Hopf map $h_{\mathbf{C}}$, the restriction to S^3 of the Hopf map $\mathbf{C}^2 \to \mathbf{C}P^1$, composed with a stereographic projection;

Diagram 24.3

$$\begin{array}{ccccc} & & S^3 & \longrightarrow & S^3 \\ & & \downarrow & & = \\ Spin(3) & \longrightarrow & Spin(4) & \longrightarrow & S^3 \\ \downarrow & & \downarrow & & \\ S^3 & = & S^3 & & \end{array} \quad \text{in } \mathbf{R}_{0,3} \cong {}^2\mathbf{H},$$

involving various transitive actions of $Spin(4)$ on S^3, $Spin(4)$ being isomorphic to $Sp(1) \times Sp(1) \cong S^3 \times S^3$;

Diagram 24.4

$$\begin{array}{ccccc} S^3 & = & Sp(1) & & \\ \downarrow & & \downarrow & & \\ Spin(4) & \longrightarrow & Spin(5) & \longrightarrow & S^4 \quad \text{in } \mathbf{R}_{0,4} \cong \mathbf{H}(2), \\ \downarrow & & \downarrow & & = \\ S^3 & \longrightarrow & S^7 & \xrightarrow{h_H} & S^4 \end{array}$$

involving the Hopf map h_H, the restriction to S^7 of the Hopf map $\mathbf{H}^2 \to \mathbf{H}P^1$, composed with a stereographic projection, and the isomorphism $Spin(5) \cong Sp(2)$;

Diagram 24.5

$$\begin{array}{ccccc} Sp(1) & \longrightarrow & SU(3) & \longrightarrow & S^5 \\ \downarrow & & \downarrow & & = \\ Spin(5) & \longrightarrow & Spin(6) & \longrightarrow & S^5 \quad \text{in } \mathbf{R}_{0,5} \cong \mathbf{C}(4), \\ \downarrow & & \downarrow & & \\ S^7 & = & S^7 & & \end{array}$$

involving the isomorphisms $Sp(1) \cong SU(2)$, $Spin(5) \cong Sp(2)$ and $Spin(6) \cong SU(4)$;

Diagram 24.6

$$\begin{array}{ccccc} SU(3) & \longrightarrow & G_2 & \longrightarrow & S^6 \\ \downarrow & & \downarrow & & = \\ Spin(6) & \longrightarrow & Spin(7) & \longrightarrow & S^6 \quad \textit{in } \mathbf{R}_{0,6} \cong \mathbf{R}(8), \\ \downarrow & & \downarrow & & \\ S^7 & = & S^7 & & \end{array}$$

introducing G_2, the automorphism group of the Cayley algebra $\mathbf{O}$, and involving the transitive action of G_2 on S^6;

Diagram 24.7

$$\begin{array}{ccccc} G_2 & \longrightarrow & Spin(7) & \longrightarrow & S^7 \\ \downarrow & & \downarrow & & = \\ Spin(7) & \longrightarrow & Spin(8) & \longrightarrow & S^7 \quad \textit{in } \mathbf{R}_{0,7} \cong {}^2\mathbf{R}(8), \\ \downarrow & & \downarrow & & \\ S^7 & = & S^7 & & \end{array}$$

involving various transitive actions of $Spin(8)$ on S^7 and the associated triality automorphism of $Spin(8)$ of order 3; and finally,

Diagram 24.8

$$\begin{array}{ccccc} Spin(7) & = & Spin(7) & & \\ \downarrow & & \downarrow & & \\ Spin(8) & \longrightarrow & Spin(9) & \longrightarrow & S^8 \quad \textit{in } \mathbf{R}_{0,8} \cong \mathbf{R}(16), \\ \downarrow & & \downarrow & & \| \\ S^7 & \longrightarrow & S^{15} & \xrightarrow{h_{\mathbf{O}}} & S^8 \end{array}$$

involving the Hopf map $h_{\mathbf{O}}$, the restriction to S^{15} of the Hopf map $\mathbf{O}^2 \to \mathbf{O}P^1$, composed with a stereographic projection.

The first few diagrams

Diagrams 24.1 to 24.5 can be dealt with fairly rapidly, for much of the detail has appeared above, in Chapter 16, or even earlier.

Diagram 24.1

We work in $\mathbf{R}_{0,2} = \mathbf{C}$. The map $h_R : S^1 \to S^1$ is the restriction to S^1 of the map $\mathbf{C} \to \mathbf{C};\ z \mapsto z^2$, or, equivalently, the map $\mathbf{R}^2 \to \mathbf{R}^2$; $(x, y) \mapsto (x^2 - y^2, 2yx)$, which admits the factorisation

$$\begin{array}{ccccc} S^1 & \longrightarrow & \mathbf{R}P^1 & \longrightarrow & S^1 \\ (x, y) & \mapsto & [x, y] & \mapsto & (2x^1 - 1, 2yx) \\ \text{with } x^2 + y^2 = 1 & & = & & = \\ & & [2x^2, 2yx] & & (x^2 - y^2, 2yx). \\ & & \text{(at least when } x \neq 0) & & \end{array}$$

The map $\mathbf{R}P^1 \to S^1$ may be interpreted as stereographic projection from $(-1, 0)$ in $\mathbf{R}^2$.

Diagram 24.2

The Clifford algebra in which we work is $\mathbf{R}_{0,2} \cong \mathbf{H}$, with the real linear space $\mathbf{R}^3 \cong \mathbf{R} \oplus \mathbf{C}$ embedded in $\mathbf{H} \subset \mathbf{C}(2)$ by the real linear map

$$\mathbf{R} \oplus \mathbf{C} \to \mathbf{C}(2);\ (\lambda, z) \mapsto \begin{pmatrix} \lambda & -\bar{z} \\ z & \lambda \end{pmatrix},$$

S^2 in $\mathbf{R}^3$ being represented by those (λ, z) such that $\lambda^2 + z\overline{z} = 1$. The map $h_C : S^3 \to S^3$ is the restriction to S^3 of the map

$$\mathbf{H} \to \mathbf{H};\ q = \begin{pmatrix} w & -\overline{z} \\ z & \overline{w} \end{pmatrix} \mapsto q\widetilde{q} = \begin{pmatrix} w & -\overline{z} \\ z & \overline{w} \end{pmatrix} \begin{pmatrix} \overline{w} & -\overline{z} \\ z & w \end{pmatrix}$$

$$= \begin{pmatrix} w\overline{w} - \overline{z}z & -2w\overline{z} \\ 2z\overline{w} & \overline{w}w = z\overline{z} \end{pmatrix},$$

or, equivalently, the map

$$\mathbf{C}^2 \to \mathbf{C}^2;\ (w, z) \mapsto (w\overline{w} - z\overline{z},\ 2z\overline{w}),$$

which admits the factorisation

$$\begin{array}{ccccc} S^3 & \longrightarrow & \mathbf{C}P^1 & \longrightarrow & S^2 \\ (w, z) & \mapsto & [w, z] & \mapsto & (2w\overline{w} - 1,\ 2z\overline{w}) \\ \text{with } w\overline{w} + z\overline{z} = 1 & & = & & = \\ & & [2w\overline{w},\ 2z\overline{w}] & & (w\overline{w} - z\overline{z}, 2z\overline{w}), \\ & & \text{at least when } w \neq 0 & & \end{array}$$

the last of these maps being stereographic projection from $(-1, 0) \in \mathbf{C}^2$.

Diagram 24.3

The Clifford algebra in which we work is $\mathbf{R}_{0,3} \cong {}^2\mathbf{H}$, with the real linear space $\mathbf{R}^4 \cong \mathbf{H}$ embedded in ${}^2\mathbf{H} \subset \mathbf{H}(2)$ by the real linear map

$$\mathbf{H} \to \mathbf{H}(2);\ q \mapsto \begin{pmatrix} q & 0 \\ 0 & \widetilde{q} \end{pmatrix},$$

S^3 in $\mathbf{H}$ being represented by those q such that $q\overline{q} = 1$. With this choice $\mathbf{R}_{0,3}^0 = \mathbf{H}$ is embedded in ${}^2\mathbf{H}$ by the real linear map

$$\mathbf{H} \to \mathbf{H}(2);\ q \mapsto \begin{pmatrix} q & 0 \\ 0 & \widehat{q} \end{pmatrix}.$$

The diagram is

$$\begin{array}{ccccc} & & S^3 & \longrightarrow & S^3 \\ & & \downarrow & & \| \\ Spin(3) & \longrightarrow & Spin(4) & \longrightarrow & S^3 \\ \downarrow & & \downarrow & & \\ S^3 & = & S^3 & & \end{array}$$

where the horizontal maps are

$$S^3 \to S^3;\ r \mapsto \overline{r},\ \text{where } r\,\overline{r} = 1,$$

$$Spin(3) \to Spin(4) q \mapsto \begin{pmatrix} q & 0 \\ 0 & \widehat{q} \end{pmatrix},\ \text{where } q\,\overline{q} = 1,$$

and

$$Spin(4) \to S^3;\ \begin{pmatrix} q & 0 \\ 0 & \widehat{r} \end{pmatrix} \mapsto \begin{pmatrix} q & 0 \\ 0 & \widehat{r} \end{pmatrix} \begin{pmatrix} \overline{r} & 0 \\ 0 & \tilde{q} \end{pmatrix} = \begin{pmatrix} q\overline{r} & 0 \\ 0 & \widehat{r}\tilde{q} \end{pmatrix},$$

where $q\,\overline{q} = r\,\overline{r} = 1$.

The central vertical maps are, simply,

$$S^3 \to Spin(4);\ r \mapsto \begin{pmatrix} 1 & 0 \\ 0 & \widehat{r} \end{pmatrix},$$

and

$$Spin(4) \to S^3;\ \begin{pmatrix} q & 0 \\ 0 & \widehat{r} \end{pmatrix} \mapsto q.$$

The diagram relates one of the Clifford actions of $Spin(4)$ on S^3 to the standard orthogonal action and in so doing delates two distinct product structures on $Spin(4)$, the group isomorphism $Spin(4) \cong Sp(1) \times Sp(1)$ and the diffeomorphism $Spin(4) \cong Spin(3) \times S^3$ with (q, r) corresponding to $(q, q\,\overline{r})$. A similar dagram relates the other Clifford action of $Spin(4)$ on S^3 to the standard orthogonal action.

One way in which the 'vertical' embeddings of $S^3 = Spin(3)$ in $Spin(4)$ differ from the 'horizontal' one is that they do not project to embeddings of $SO(3)$ in $SO(4)$. We refer to these embeddings in the sequel as the *Clifford* embeddings of $Spin(3)$ in $Spin(4)$. It is, in a sense, fortuitous that the *Clifford* homogeneous space $Spin(4)/Spin(3)$ is homeomorphic to the standard one, the real Stiefel manifold (cf. Exercise 22.5)

$$O(\mathbf{R}^3, \mathbf{R}^4) = O(4)/O(3) = SO(4)/SO(3) = Spin(4)/Spin(3).$$

The force of this remark will become more evident in the sequel.

Diagram 24.4

The Clifford algebra in which we work is $\mathbf{R}_{0,4} \cong \mathbf{H}(2)$, with the real linear space $\mathbf{R}^5 \cong \mathbf{R} \oplus \mathbf{R}^4 \cong \mathbf{R} \oplus \mathbf{H}$ embedded in $\mathbf{R}_{0,4}$ by the real linear map

$$\mathbf{R} \oplus \mathbf{H} \to \mathbf{H}(2);\ (\lambda, q) \mapsto \begin{pmatrix} \lambda & -\overline{q} \\ q & \lambda \end{pmatrix},$$

S^4 in $\mathbf{R}^5$ being represented by those (λ, q) such that $\lambda^2 + q\bar{q} = 1$. With this choice $\mathbf{R}_{0,4}^0$ is the standard copy of ${}^2\mathbf{H}$ in $\mathbf{H}(2)$, namely the subalgebra of diagonal matrices. The diagram

$$\begin{array}{ccccc} S^3 & = & Sp(1) & & \\ \downarrow & & \downarrow & & \\ Spin(4) & \longrightarrow & Spin(5) & \longrightarrow & S^4 \\ \downarrow & & \downarrow & & = \\ S^3 & \longrightarrow & S^7 & \xrightarrow{h_H} & S^4 \end{array}$$

relates one of the Clifford actions of $Spin(4)$ on S^3 and the Clifford action of $Spin(5)$ on S^7 to the standard orthogonal action of $Spin(5)$ on S^4. The horizontal maps are

$$Spin(4) \cong Sp(1) \times Sp(1) \to Spin(5) \cong Sp((2); \begin{pmatrix} a & 0 \\ 0 & d \end{pmatrix} \mapsto \begin{pmatrix} a & 0 \\ 0 & d \end{pmatrix},$$

$$Spin(5) \cong Sp(2) \to S^4; \begin{pmatrix} a & c \\ b & d \end{pmatrix} \mapsto \begin{pmatrix} a & c \\ b & d \end{pmatrix} \begin{pmatrix} \bar{a} & -\bar{b} \\ -\bar{c} & \bar{d} \end{pmatrix}$$

$$= \begin{pmatrix} 2a\bar{a} - 1 & -2a\bar{b} \\ 2b\bar{a} & 2a\bar{a} - 1 \end{pmatrix},$$

$$\text{for } \begin{pmatrix} a & c \\ b & d \end{pmatrix} \begin{pmatrix} \bar{a} & b \\ \bar{c} & \bar{d} \end{pmatrix} = \begin{pmatrix} 1 & 0 \\ 0 & 1 \end{pmatrix} = \begin{pmatrix} \bar{a} & b \\ \bar{c} & \bar{d} \end{pmatrix} \begin{pmatrix} a & c \\ b & d \end{pmatrix},$$

$$S^3 \subset \mathbf{H} \to S^7 \subset \mathbf{H}^2; \; a \mapsto (a, 0), \text{ with } a\bar{a} = 1,$$

and

$$\begin{array}{ccccc} S^7 \subset \mathbf{H}^2 & \longrightarrow & \mathbf{H}P^2 & \longrightarrow & S^4 \\ (a, b) & \mapsto & [a, b] & \mapsto & (2a\bar{a} - 1, 2b\bar{a}). \\ \text{with } a\bar{a} + b\bar{b} = 1 & & = & & = \\ & & [2a\bar{a}, 2b\bar{a}] & & (a\bar{a} - b\bar{b}, 2b\bar{a}) \\ & & \text{at least when } a \neq 0 & & \end{array}$$

The vertical maps are, simply,

$$S^3 \cong Sp(1) \to Spin(4) \cong Sp(1) \times Sp(1); \; d \mapsto \begin{pmatrix} 1 & 0 \\ 0 & d \end{pmatrix},$$

and this composed with the inclusion $Spin(4) \to Spin(5)$,

$$Spin(4) = Sp(1) \times Sp(1) \to S^3, \begin{pmatrix} a & 0 \\ 0 & d \end{pmatrix} \mapsto a;$$

and

$$Spin(5) \cong Sp(2) \to S^7;\ \begin{pmatrix} a & c \\ b & d \end{pmatrix} \mapsto (a, b).$$

The vertical embedding of $Sp(1)$ in $Sp(2)$ is a standard one, but the induced embedding of $Spin(3)$ in $Spin(5)$ factors through one of the Clifford embeddings of $Spin(3)$ in $Spin(4)$ and is not standard. We refer to it as a *Clifford* embedding of $Spin(3)$ in $Spin(5)$. The *Clifford* homogeneous space $Spin(5)/Spin(3)$ is homeomorphic to $Sp(2)/Sp(1) \cong S^7$. On the other hand it can be shown, by methods of algebraic topology (see, for example, Steenrod and Epstein (1962), Theorem 4.5 or James (1976), that the standard homogeneous space $Spin(5)/Spin(3)$, homeomorphic to the real Stiefel manifold $O(\mathbf{R}^3, \mathbf{R}^5) \cong SO(5)/SO(3)$, is not homeomorphic to S^7, nor to the product $S^3 \times S^4$.

There is, of course, an analogous diagram involving the other Clifford action of $Spin(4)$ on S^3.

Diagram 24.5

We have already met this diagram in Proposition 17.3 where we proved that $Spin(6) \cong SU(4)$. We give below a slight variant of that proof.

The Clifford algebra in which we work is $\mathbf{R}_{0,5} \cong \mathbf{C}(4)$, with the real linear space $\mathbf{R} \oplus \mathbf{R}^5 \cong \mathbf{R}^6 \cong \mathbf{C}^3$ embedded in it by the real linear injection

$$\mathbf{C}^3 \to \mathbf{C}(4);\ (z_0, z_1, z_2) \mapsto \begin{pmatrix} \overline{z_2} & 0 & z_0 & \overline{z_1} \\ 0 & \overline{z_2} & z_1 & -\overline{z_0} \\ -\overline{z_0} & -\overline{z_1} & z_2 & 0 \\ -z_1 & z_0 & 0 & z_2 \end{pmatrix}.$$

With this choice, $\mathbf{R}_{0,5}^0$ is the standard copy of $\mathbf{H}(2)$ in $\mathbf{C}(4)$. The sphere S^5 in $\mathbf{R}^6$ is represented by those (z_0, z_1, z_2) such that $z_0\overline{z_0} + z_1\overline{z_1} + z_2\overline{z_2} = 1$. The determinant of the matrix representing (z_0, z_1, z_2) is easily computed to be $(z_0\overline{z_0} + z_1\overline{z_1} + z_2\overline{z_2})^2$, which is equal to 1 when $(z_0, z_1, z_2) \in S^5$. Since, by Proposition 24.1, any element of $Spin(6) \subset U(4)$ is of the form $z\,g$, where $z \in S^5$ and $g \in Spin(5) \cong Sp(2)$, and since, by what we have just proved and by Corollary 10.14, both z and g have determinant equal to 1 (as elements of $\mathbf{C}(4)$), it follows that $Spin(6) \subset SU(4)$. Since both these groups are connected and of the same dimension, namely 15, it follows that they coincide.

Filling out the rest of the detail of the diagram is then straightforward. For any $t \in \mathbf{C}(4)$, $t^{-} = \widehat{t}^{\,-}$ is given as in Exercise 10.2. A direct computation (in which Exercise 10.1 is relevant) shows that, for any $u \in SU(3)$,

$$\begin{pmatrix} u & 0 \\ 0 & 1 \end{pmatrix} \begin{pmatrix} u & 0 \\ 0 & 1 \end{pmatrix}^{\sim} = \begin{pmatrix} \overline{u_{22}} & 0 & u_{02} & \overline{u_{12}} \\ 0 & \overline{u_{22}} & u_{12} & -\overline{u_{02}} \\ -\overline{u_{02}} & -\overline{u_{12}} & u_{22} & 0 \\ -u_{12} & u_{02} & 0 & u_{22} \end{pmatrix},$$

which is the identity matrix if and only if $u = \begin{pmatrix} v & 0 \\ 0 & 1 \end{pmatrix}$, with $v \in SU(2) \cong Sp(1)$.

So, finally, we obtain the commutative diagram

$$\begin{array}{ccccc} Sp(1) \cong SU(2) & \longrightarrow & SU(3) & \longrightarrow & S^5 \\ \downarrow & & \downarrow & & = \\ Sp(2) \cong Spin(5) & \longrightarrow & Spin(6) \cong SU(4) & \longrightarrow & S^5 \\ \downarrow & & \downarrow & & \\ S^7 & = & S^7 & & \end{array}$$

where the maps not explicitly described above are all standard ones, each row and each column being left-coset exact.

The embedding of $SU(2)$ in $SU(4)$ here is the standard one. It is therefore a corollary of the diagram that the complex Stiefel manifold $U(\mathbf{C}^2, \mathbf{C}^4) \cong SU(4)/SU(2)$ is homeomorphic to $S^3 \times S^7$. Since $SU(4) \cong Spin(6)$ and $SU(2) \cong Spin(3)$, this complex Stiefel manifold may also be regarded as a Clifford homogeneous space $Spin(6)/Spin(3)$, the embedding of $Spin(3)$ in $Spin(6)$ being a Clifford one, as it factors through a Clifford embedding of $Spin(3)$ in $Spin(4)$. By contrast, it can be shown, by methods of algebraic topology, that the standard homogeneous space $Spin(6)/Spin(3)$, homeomorphic to the real Stiefel manifold $O(\mathbf{R}^3, \mathbf{R}^6) \cong SO(6)/SO(3)$, is not homeomorphic to $S^5 \times S^7$.

Getting further

To get any further it is appropriate to jump a stage and to take a look first at $Spin(8)$. Any linear automorphism of $\mathbf{R}^n$ induces an automorphism of $Spin(n)$, which projects to an automorphism of $SO(n)$, since the original automorphism of $\mathbf{R}^n$ commutes with $-{}^n1$. Of all the groups $Spin(n)$ the group $Spin(8)$ is unique in that it possesses automorphisms of order 3 that do *not* project to automorphisms of $SO(8)$.

To construct such an automorphism we begin with $Spin(8)$ as a subgroup of $O(8) \times O(8)$, or rather, since $Spin(8)$ is connected, as a subgroup of $SO(8) \times SO(8)$. The Clifford algebra in which $Spin(8)$ lies is $\mathbf{R}^0_{0,8} \cong \mathbf{R}_{0,7} \cong {}^2\mathbf{R}(8)$, where we may suppose that $\mathbf{R}^8$ is embedded by the injection

$$\mathbf{R}^8 \to {}^2\mathbf{R}(8);\ a \mapsto \begin{pmatrix} \upsilon(a) & 0 \\ 0 & \upsilon(a)^\tau \end{pmatrix},$$

υ(upsilon) : $\mathbf{O} = \mathbf{R}^8 \to \mathbf{R}(8)$ being the injection with image $\mathbf{Y}$, with which we became familiar in our discussion of the Cayley algebra $\mathbf{O}$ early in Chapter 19. Here, as on that occasion, the product on $\mathbf{R}(8)$ and the product on $\mathbf{O}$ will both be denoted by juxtaposition, as will be the action of $\mathbf{R}(8)$ on $\mathbf{O}$, the unit element in $\mathbf{O}$ being denoted by e. One technical detail is worth isolating as a lemma.

Lemma 24.2 *Let $x \in \mathbf{Y}$, let $g \in \mathbf{R}(8)$ and suppose that $g\,y\,\mathrm{e} = x\,y\,\mathrm{e}$, for all $y \in \mathbf{Y}$. Then $g = x$ and $g\,y\,\mathrm{e} = (g\,\mathrm{e})(y\,\mathrm{e})$.*

Our purpose in singling this out is to emphasise that it is incorrect to contract $(g\,\mathrm{e})(y\,\mathrm{e})$ to $g\,y\,\mathrm{e}$ or to expand $g\,y\,\mathrm{e}$ to $(g\,\mathrm{e})(y\,\mathrm{e})$, unless we know that $g \in \mathbf{Y}$.

The companion involution

Conjugation on the Cayley algebra $\mathbf{O}$ is associated not only with the conjugation anti-involution of the Clifford algebra $\mathbf{R}^0_{0,7} \cong \mathbf{R}(8)$, namely transposition, but also with an involution of $\mathbf{R}(8)$, which we term the *companion* involution of $\mathbf{R}(8)$, and which restricts to an involution of $SO(8)$. This involution is defined by means of the element of $O(8)$ which induces conjugation on $\mathbf{O}$ (by left multiplication), namely the symmetric anti-rotation $\begin{pmatrix} 1 & 0 \\ 0 & -{}^7 1 \end{pmatrix}$.

Proposition 24.3 *The map*

$$\mathbf{R}(8) \to \mathbf{R}(8);\ g \mapsto \check{g} = \begin{pmatrix} 1 & 0 \\ 0 & -{}^7 1 \end{pmatrix} g \begin{pmatrix} 1 & 0 \\ 0 & -{}^7 1 \end{pmatrix}$$

is a linear involution of the algebra $\mathbf{R}(8)$ which commutes with transposition and restricts to a group involution of $SO(8)$ and is such that

$$\check{g}\,y\,\mathrm{e} = \overline{g\,\overline{y\mathrm{e}}},\ \textit{for all } g \in \mathbf{R}(8) \textit{ and all } y \in \mathbf{Y},$$

or, equivalently,

$$\check{g}\,b = \overline{g\,\overline{b}},\ \textit{for all } g \in \mathbf{R}(8) \textit{ and all } b \in \mathbf{O}.$$

In particular, by setting $y = 1$, *or* $b = \mathrm{e}$,

$$\check{g}\,\mathrm{e} = \overline{g\,\mathrm{e}}, \textit{ for all } g \in \mathbf{R}(8).$$

Moreover, for all $g \in SO(8)$,

$$\check{g} = g \Leftrightarrow g\,\mathrm{e} = \mathrm{e} \Leftrightarrow \check{g}\,\mathrm{e} = \mathrm{e},$$

g, *in such a case, being of the form* $\begin{pmatrix} 1 & 0 \\ 0 & h \end{pmatrix}$, *where* $h \in SO(7)$.

The element $\check{g}$ will be called the *companion* of the element g.

The triality automorphism

Consider now an element $\begin{pmatrix} g_0 & 0 \\ 0 & \check{g}_1 \end{pmatrix}$ of $Spin(8)$, g_0 and g_1 being elements of $SO(8)$ and $\check{g}_1$ being the companion of g_1. Its action on $\mathbf{R}^8$ and in particular on S^7 is given by

$$\begin{pmatrix} y & 0 \\ 0 & y^\tau \end{pmatrix} \mapsto \begin{pmatrix} g_0 & 0 \\ 0 & \check{g}_1 \end{pmatrix} \begin{pmatrix} y & 0 \\ 0 & y^\tau \end{pmatrix} \begin{pmatrix} \check{g}_1^{\,\tau} & 0 \\ 0 & g_0^{\,\tau} \end{pmatrix},$$

where $y \in \mathbf{Y}$ or S^7, that is by $y \mapsto g_0\, y\, \check{g}_1^{-1}$, since $\check{g}_1^{\dagger} = \check{g}_1^{\tau} = \check{g}_1^{-1}$; the corresponding action on $\mathbf{O}$ being given by $y\,\mathrm{e} \mapsto g_0\, y\, \check{g}_1^{-1}\,\mathrm{e}$. In this way the pair (g_0, g_1) of elements of $SO(8)$ defines a third element $g_2 \in SO(8)$ by

$$g_0\, y\, \check{g}_1^{-1}\,\mathrm{e} = \check{g}_2\, y\,\mathrm{e}, \text{ for all } y \in \mathbf{Y}.$$

An ordered triple (g_0, g_1, g_2) of elements of $SO(8)$ such that

$$\begin{pmatrix} g_0 & 0 \\ 0 & \check{g}_1 \end{pmatrix} \in Spin(8),$$

or, equivalently, such that $g_0\, y\, \check{g}_1^{-1} \in \mathbf{Y}$ for all $y \in \mathbf{Y}$, and such that $g_0\, y\, \check{g}_1^{-1}\,\mathrm{e} = \check{g}_2\, y\,\mathrm{e}$, for all $y \in \mathbf{Y}$, will be called a *θ-triad* of $SO(8)$, θ, the *triality* automorphism of $Spin(8)$, being the automorphism of order 3

$$\theta : Spin(8) \to Spin(8); \begin{pmatrix} g_0 & 0 \\ 0 & \check{g}_1 \end{pmatrix} \mapsto \begin{pmatrix} g_1 & 0 \\ 0 & \check{g}_2 \end{pmatrix},$$

which exists by virtue of the following theorem.

Theorem 24.4 *Let* (g_0, g_1, g_2) *be a* θ*-triad of* $SO(8)$*. Then* (g_1, g_2, g_0) *and* (g_2, g_0, g_1) *are* θ*-triads, as also are* $(g_0^{-1}, g_1^{-1}, g_2^{-1})$, $(g_1^{-1}, g_2^{-1}, g_0^{-1})$ *and* $(g_2^{-1}, g_2^{-1}, g_1^{-1})$*. Moreover*

$$\theta : Spin(8) \to Spin(8); \begin{pmatrix} g_0 & 0 \\ 0 & \check{g}_1 \end{pmatrix} \mapsto \begin{pmatrix} g_1 & 0 \\ 0 & \check{g}_2 \end{pmatrix}$$

is an automorphism of $Spin(8)$ *of order 3.*

Proof The key to the proof is the scalar triple product on $\mathbf{O}$ with which we made acquaintance in Proposition 19.5. What we need to recall is that, for all $a, b, c \in \mathbf{O}$,

$$\bar{a} \cdot bc = \bar{b} \cdot ca = \bar{c} \cdot ab,$$

where $\cdot$ denotes the standard scalar product on $\mathbf{R}^8$.

So let (g_0, g_1, g_2) be a θ-triad. Then

$$g_0\, y\, \check{g}_1{}^{-1} \in \mathbf{Y}, \text{ with } g_0\, y\, \check{g}_1{}^{-1}\,\mathrm{e} = \check{g}_2\, y\,\mathrm{e}, \text{ for all } y \in \mathbf{Y}.$$

Then, by Lemma 24.2,

$$g_0\, y\, \check{g}_1{}^{-1}\, z\,\mathrm{e} = (g_0\, y\, \check{g}_1{}^{-1}\,\mathrm{e})(z\,\mathrm{e}) = (\check{g}_2\, y\,\mathrm{e})(z\,\mathrm{e}), \text{ for all } y \in \mathbf{Y}.$$

Since g_0 is orthogonal it follows, by Proposition 24.3, that

$$\begin{aligned} \overline{x\mathrm{e}} \cdot (y\,\mathrm{e})(\check{g}_1{}^{-1}\, z\,\mathrm{e}) &= g_0\, \overline{x\mathrm{e}} \cdot g_0\, y\, \check{g}_1{}^{-1}\, z\,\mathrm{e} \\ &= \overline{\check{g}_0\, x\,\mathrm{e}} \cdot \overline{g_2\, \overline{y\mathrm{e}}}(z\,\mathrm{e}), \text{ for all } x, y, z \in \mathbf{Y}, \end{aligned}$$

and so, by Proposition 19.5, as we promised above,

$$\begin{aligned} \overline{y\mathrm{e}} \cdot (\check{g}_1{}^{-1}\, z\,\mathrm{e})(x\,\mathrm{e}) &= g_2\, \overline{y\mathrm{e}} \cdot (z\,\mathrm{e})(\check{g}_0\, x\,\mathrm{e}) \\ &= \overline{y\mathrm{e}} \cdot g_2^{-1}\, z\, \check{g}_0\, x\,\mathrm{e}, \text{ for all } x, y, z \in \mathbf{Y}, \end{aligned}$$

g_2 being orthogonal. Therefore

$$(\check{g}_1{}^{-1}\, z\,\mathrm{e})(x\,\mathrm{e}) = g_2^{-1}\, z\, \check{g}_0\, x\,\mathrm{e}, \text{ for all } x, z \in \mathbf{Y}.$$

So, by Lemma 24.2 yet again,

$$g_2^{-1}\, z\, \check{g}_0 \in \mathbf{Y}, \text{ with } g_2^{-1}\, z\, \check{g}_0\,\mathrm{e} = \check{g}_1{}^{-1}\, z\,\mathrm{e}, \text{ for all } z \in \mathbf{Y}.$$

That is, $(g_2^{-1}, g_0^{-1}, g_1^{-1})$ is a θ-triad. Repeating the whole argument with this θ-triad as starting point one deduces at once that (g_1, g_2, g_0) is a θ-triad.

The rest of the proof, including the proof that θ is a group homomorphism of order 3, is obvious. □

With this we have the companion theorem:

Theorem 24.5 *Let* (g_0, g_1, g_2) *be a* θ*-triad of* $SO(8)$. *Then so also is* $(\check{g}_1, \check{g}_0, \check{g}_2)$. Note the change of order!

Proof Let (g_0, g_1, g_2) be a θ-triad of $SO(8)$. Then

$$g_0\, y\, \check{g}_1^{-1} \in \mathbf{Y}, \text{ with } g_0\, y\, \check{g}_1^{-1}\mathrm{e} = \check{g}_2\, y\, ee, \text{ for all } y \in \mathbf{Y}.$$

Then $\check{g}_1\, y^-\, g_0^{-1} = (g_0\, y\, \check{g}_1^{-1})^- \in \mathbf{Y}$, with

$$\check{g}_1\, y^-\, g_0^{-1}\mathrm{e} = (g_0\, y\, \check{g}_1^{-1})^-\mathrm{e} = \overline{g_0\, y \check{g}_1^{-1}\mathrm{e}} = \overline{\check{g}_2\, y\, \mathrm{e}} = g_2\, y^-\mathrm{e},$$

for all $y^- \in \mathbf{Y}$. That is, $(\check{g}_1, \check{g}_0, \check{g}_2)$ is a θ-triad. □

It is therefore appropriate to call $\begin{pmatrix} \check{g}_1 & 0 \\ 0 & g_0 \end{pmatrix}$ the *companion* of $\begin{pmatrix} g_0 & 0 \\ 0 & \check{g}_1 \end{pmatrix}$ in $Spin(8)$.

Corollary 24.6 *Suppose that* $(g, \check{g}, \check{g})$ *is a* θ*-triad of* $SO(8)$. *Then* $\check{g} = g$ *so that* (g, g, g) *is a* θ*-triad of* $SO(8)$.

Proof Since $(g, \check{g}, \check{g})$ is a θ-triad, so is $(\check{\check{g}}, \check{g}, g) = (g, \check{g}, g)$. So $\check{g} = g$. □

Theorem 24.7 *The triality automorphism* θ *of* $Spin(8)$ *does not project to an automorphism of* $SO(8)$. *However, it does project to an automorphism of* $SO(8)/S^0$.

Proof Under the projection $Spin(8) \to SO(8)$ the elements $\begin{pmatrix} g_0 & 0 \\ 0 & \check{g}_1 \end{pmatrix}$ and $\begin{pmatrix} -g_0 & 0 \\ 0 & -\check{g}_1 \end{pmatrix}$ project to the same element $\check{g}_2$ of $SO(8)$. However, $\theta\begin{pmatrix} 1 & 0 \\ 0 & 1 \end{pmatrix} = \begin{pmatrix} 1 & 0 \\ 0 & 1 \end{pmatrix}$, while $\theta\begin{pmatrix} -1 & 0 \\ 0 & -1 \end{pmatrix} = \begin{pmatrix} -1 & 0 \\ 0 & 1 \end{pmatrix}$, since $(1, 1, 1)$, and therefore $(-1, -1, 1)$, is a θ-triad of $SO(8)$. From the fact that $\begin{pmatrix} 1 & 0 \\ 0 & 1 \end{pmatrix}$ and $\begin{pmatrix} -1 & 0 \\ 0 & 1 \end{pmatrix}$ project to distinct elements of $SO(8)$, namely 1 and -1, it follows that θ does not project to an automorphism of $SO(8)$.

Under the projection $Spin(8) \to SO(8)/S^0$, however, the four elements $\begin{pmatrix} \pm g_0 & 0 \\ 0 & \pm\check{g}_1 \end{pmatrix}$ project to the same element $\pm\check{g}_2$ of $SO(8)/S^0$, while under

θ they map to the four elements $\begin{pmatrix} \pm g_1 & 0 \\ 0 & \pm \check{g}_2 \end{pmatrix}$, which then project to the same element $\pm\check{g}_0$ of $SO(8)/S^0$. The automorphism θ therefore projects to the automorphism

$$SO(8)/S^0 \to SO(8)/S^0;\ \pm\check{g}_0 \mapsto \pm\check{g}_0 \text{ (and } \pm g_0 \mapsto \pm g_2),$$

where (g_0, g_1, g_2) is a θ-triad of $SO(8)$. □

It is incorrect to say that θ and θ^{-1} are the only automorphisms of order 3 of $Spin(8)$ that do not project to automorphisms of $SO(8)$, for if ϕ is the automorphism of $Spin(8)$ induced by a change of coordinates on $\mathbf{R}^8$ then $\phi\,\theta\,\phi^{-1}$ will also be an automorphism of order 3 that does not project to an automorphism of $SO(8)$, and not all such ϕ commute with θ. Essentially, however, θ is unique. The proof that $Spin(8)$ is the only one of the groups $Spin(n)$ to admit a triality automorphism depends on a much deeper analysis of the structure of the groups $Spin(n)$ and their Lie algebras than it is possible to give here. See, for example, Loos (1969).

The group G_2

Let G be any group and let $\psi : G \to G$ be an automorphism of G. Then the subset $\{g \in G : \psi(g) = g\}$ of elements of G left untouched by ψ is clearly a subgroup of G.

Consider, in particular, the group $Spin(8)$ and its triality automorphism θ. The subgroup of $Spin(8)$ left untouched by θ consists of those elements $\begin{pmatrix} g_0 & 0 \\ 0 & \check{g}_1 \end{pmatrix}$ of $Spin(8)$ such that

$$\theta\begin{pmatrix} g_0 & 0 \\ 0 & \check{g}_1 \end{pmatrix} = \begin{pmatrix} g_1 & 0 \\ 0 & \check{g}_2 \end{pmatrix} = \begin{pmatrix} g_0 & 0 \\ 0 & \check{g}_1 \end{pmatrix},$$

that is, those $\begin{pmatrix} g & 0 \\ 0 & \check{g} \end{pmatrix} \in Spin(8)$ such that (g, g, g) is a θ-triad of $SO(8)$. Clearly this group is isomorphic to the subgroup of $SO(8)$ consisting of those $g \in SO(8)$ such that (g, g, g) is a θ-triad of $SO(8)$. This group we define to be the group G_2. (The name derives from the classification of Lie algebras. There is no group G_1 !)

Theorem 24.8 *Let* $g \in G_2$. *Then* $g = \check{g}$ *and* $g\,\mathrm{e} = \mathrm{e}$.

Proof Let $g \in G_2$. Then

$$g\,y\,\breve{g}^{-1} \in \mathbf{Y}, \text{ with } g\,y\,\breve{g}^{-1}\,\mathrm{e} = \breve{g}\,y\,\mathrm{e}, \text{ for all } y \in \mathbf{Y}.$$

In particular, by setting $y = 1$, $g\,\breve{g}^{-1} \in \mathbf{Y}$ and $g\,\breve{g}^{-1}\,\mathrm{e} = \breve{g}\,\mathrm{e}$, from which it follows that $\breve{g}^{-1}\,g\,\breve{g}^{-1}\,\mathrm{e} = \mathrm{e}$, so that, by the last part of Proposition 24.3, $g^{-1}\breve{g}\,g^{-1} = \breve{g}^{-1}\,g\,\breve{g}^{-1}$, implying that $(g\,\breve{g}^{-1})^3 = 1$. Let $x = g\,\breve{g}^{-1}$. Then $x \in G_2$ and $x\,y\,\breve{x}^{-1}\,\mathrm{e} = \breve{x}\,y\,\mathrm{e}$, for all $y \in \mathbf{Y}$. But $\breve{x} = x^{-1}$ and $x^3 = 1$. So $y\,x\,\mathrm{e} = x\,y\,\mathrm{e}$, for all $y \in \mathbf{Y}$; that is $(y\,\mathrm{e})(x\,\mathrm{e}) = (x\,\mathrm{e})(y\,\mathrm{e})$, for all $y \in \mathbf{Y}$. So $x\,\mathrm{e} = \pm\mathrm{e}$; that is $x = \pm 1$. But $(-1, -1, -1)$ is not a θ-triad of $SO(8)$. So $x = 1$. That is $g = \breve{g}$. Then $\breve{g}\,\mathrm{e} = \mathrm{e}$. So $g\,\mathrm{e} = \mathrm{e}$. □

The next theorem characterises G_2.

Theorem 24.9 *G_2 is the group of automorphisms of the Cayley division algebra* $\mathbf{O}$.

Proof Suppose first that $g \in G_2$, acting on $\mathbf{O}$ by left multiplication. Then, for all $b = y\,e$, $c = z\,e \in \mathbf{O}$,

$$\begin{aligned} g(b\,c) &= g\,y\,z\,\mathrm{e} = g\,y\,g^{-1}\,g\,z\,\mathrm{e} \\ &= g\,y\,g^{-1}\,\mathrm{e}\,g\,z\,\mathrm{e}, \text{ since } g = \breve{g}, \\ &= g\,y\,\mathrm{e}\,g\,z\,\mathrm{e}, \text{ again since } g = \breve{g}, \\ &= (g\,b)(g\,c), \end{aligned}$$

with, in particular, $g\,\mathrm{e} = \mathrm{e}$. Thus g is an automorphism of $\mathbf{O}$.

Conversely, by the argument of Proposition 8.18, applied to $\mathbf{O}$ rather than to $\mathbf{H}$, any automorphism or anti-automorphism g of $\mathbf{O}$ is of the form $\begin{pmatrix} 1 & 0 \\ 0 & t \end{pmatrix} : \mathbf{R} \oplus \mathbf{R}^7 \to \mathbf{R} \oplus \mathbf{R}^7;\ a \mapsto \mathrm{re}\,a + t(\mathrm{pu}\,a)$, where t is an orthogonal automorphism of $\mathbf{O}$. Then

$$\begin{aligned} & g(b\,c) = (g\,b)(g\,c), \text{ for all } b = y\,\mathrm{e},\ c = z\,\mathrm{e} \in \mathbf{O}, \\ \text{that is } & g\,y\,z\,\mathrm{e} = g\,y\,\mathrm{e}\,g\,z\,\mathrm{e}, \text{ for all } y, z \in \mathbf{Y}, \\ \text{that is } & g\,y\,\breve{g}^{-1}\,g\,z\,\mathrm{e} = \breve{g}\,y\,\mathrm{e}\,g\,z\,\mathrm{e}, \text{ for all } y, z \in \mathbf{Y}, \text{ since } \breve{g} = g, \\ \text{that is } & g\,y\,\breve{g}^{-1}\,\mathrm{e} = \breve{g}\,y\,\mathrm{e}, \text{ for all } y \in \mathbf{Y}, \\ \text{that is } & g \in G_2. \end{aligned}$$

□

Since $\breve{g} = g$, for all $G \in G_2$, it follows, from the last part of Proposition 24.3, that G_2 actually is a subgroup of $SO(7)$. We shall prove shortly that G_2 is in fact a Lie group of dimension 14 ($SO(7)$ and $SO(8)$ being

Lie groups of dimension 21 and 28, respectively). This we can do after we have established Diagrams 24.6 and 24.7.

Before turning to these we prove one further result about the way that the group G_2 lies in $Spin(8)$. In doing so it is helpful to think of $Spin(8)$ as the group of θ-triads of $SO(8)$ themselves, under the group multiplication

$$(g_0, g_1, g_2)(g_0', g_1', g_2') = (g_0 g_0', g_1 g_1', g_2 g_2'),$$

with G_2 the subgroup of triads of the form (g, g, g). Now, for any θ-triad (g_0, g_1, g_2),

$$g_0 \check{g}_1^{-1} = 1 \Leftrightarrow g_2\,\mathrm{e} = \mathrm{e} \Leftrightarrow \check{g}_2\,\mathrm{e} = \mathrm{e}.$$

Bearing this in mind, we define, for each $i \in \mathbf{3}$,

$$H_i = \{(g_0, g_1, g_2) \in Spin(8) : g_i\,\mathrm{e} = \mathrm{e}\}.$$

Theorem 24.10 *For each $i \in \mathbf{3}$, H_i is a subgroup of $Spin(8)$ isomorphic to $Spin(7)$ the three subgroups being permuted cyclically by θ, namely*

$$\theta(H_0) = H_1,\ \theta(H_1) = H_2 \quad \textit{and} \quad \theta(H_2) = H_0.$$

Moreover,

$$H_1 \cap H_2 = H_2 \cap H_0 = H_0 \cap H_1 = G_2\,.$$

Proof It is clear that H_2 is the isotropy subgroup at 1 of the standard orthogonal action of $Spin(8)$ on S^7, this subgroup being isomorphic to $Spin(7)$. It is clear also that the three subgroups are isomorphic and that they are permuted cyclically by θ.

To prove the last part, suppose that $(g_0, g_1, g_2) \in H_1 \cap H_2$. Then $g_2 = \check{g}_0$ and $g_0 = \check{g}_1$, implying that $(g_0, g_1, g_2) = (g_0, \check{g}_0, \check{g}_0)$ and therefore, by Corollary 24.6, that $g_0 = g_1 = g_2$. That is $H_1 \cap H_2 \subset G_2$. Conversely it is clear that G_2 is a subgroup of each of the H_i. So $H_1 \cap H_2 = G_2$. Likewise $H_2 \cap H_0 = H_0 \cap H_1 = G_2$. □

We are at last in a position to appreciate Diagrams 24.6 and 24.7. Paradoxically it is convenient to consider Diagram 24.7 first.

Diagram 24.7

The diagram is

$$\begin{array}{ccccc} G_2 & \longrightarrow & Spin(7) = H_0 & \longrightarrow & S^7 \\ \downarrow & & \downarrow & & = \\ Spin(7) = H_2 & \longrightarrow & Spin(8) & \longrightarrow & S^7, \\ \downarrow & & \downarrow & & \\ S^7 & = & S^7 & & \end{array}$$

where, as has been explicit throughout the preceding discussion, except momentarily in Theorem 24.10, $Spin(8)$ lies in the Clifford algebra $\mathbf{R}_{0,7} = {}^2\mathbf{R}(8)$, any element $\begin{pmatrix} g_0 & 0 \\ 0 & \check{g}_1 \end{pmatrix}$ of $Spin(8)$ being such that (g_0, g_1, g_2) is a θ-triad of $SO(8)$.

The diagram relates two of the three actions of $Spin(8)$ on S^7, the standard orthogonal action and one or other of the two Clifford actions of $Spin(8)$ on S^7. Suppose that we choose the action

$$Spin(8) \times S^7 \to S^7$$

$$\left(\begin{pmatrix} g_0 & 0 \\ 0 & \check{g}_1 \end{pmatrix}, y\right) \mapsto g_0\, y$$

with isotropy subgroup at 1 the subgroup H_0 defined above. The central vertical sequence of maps is then the corresponding left-coset exact sequence, while the central horizontal sequence involves the standard orthogonal action with isotropy subgroup at 1 the subgroup H_2.

In view of Theorem 24.10 the whole structure of the diagram should now be clear.

An analogous diagram relates the standard orthogonal action to the other Clifford action of $Spin(8)$ on S^7.

It is the case that, under the standard projection $\rho : Spin(8) \to SO(8)$, with $\rho\begin{pmatrix} 1 & 0 \\ 0 & 1 \end{pmatrix} = \rho\begin{pmatrix} -1 & 0 \\ 0 & -1 \end{pmatrix} = 1$, $\rho(H_2) \cong SO(7)$, while $\rho(H_0) \cong \rho(H_1) \cong Spin(7)$.

In line with our practice above we refer to the horizontal embedding of $Spin(7)$ in $Spin(8)$, with image H_2, as the standard embedding, the vertical embeddings, with image H_0 or H_1, being the *Clifford* embeddings of $Spin(7)$ in $Spin(8)$.

It is a corollary of Diagram 24.7 that $Spin(8)/G_2 \cong S^7 \times S^7$.

Diagram 24.6

The details of Diagram 24.6 can now be inferred.

From Diagrams 24.5 and 24.7 and the standard left-coset exact sequence

$$Spin(6) \to Spin(7) \to S^6$$

we have the diagram

$$\begin{array}{ccccc} SU(3) & \longrightarrow & G_2 & \longrightarrow & S^6 \\ \downarrow & & \downarrow & & \| \\ Spin(6) & \longrightarrow & Spin(7) & \longrightarrow & S^6, \\ \downarrow & & \downarrow & & \\ S^7 & = & S^7 & & \end{array}$$

where the elements of $Spin(6)$ are those of $Spin(7)$ which lie in $\mathbf{C}(4)$, regarded as a subspace of $\mathbf{R}(8)$ in the standard way. It follows at once that the group $SU(3)$ coincides with the subgroup of G_2 consisting of all those automorphisms of $\mathbf{O}$ which belong to $\mathbf{C}(4)$ rather than to $\mathbf{R}(8)$. Moreover the sequence

$$SU(3) \to G_2 \to S^6$$

is left-coset exact, the sphere S^6 being thus representable as the homogeneous space $G_2/SU(3)$.

Diagram 24.8

Diagram 24.8, concerning $Spin(9)$, is now easy to establish. The Clifford algebra in which we work is $\mathbf{R}_{0,8} \cong \mathbf{R}(16)$, with the real linear space $\mathbf{R}^9 = \mathbf{R} \oplus \mathbf{R}^8$ embedded by the real linear map

$$\mathbf{R} \oplus \mathbf{R}^8 \longrightarrow \mathbf{R}(16)$$

$$(\lambda, b) \mapsto \begin{pmatrix} \lambda & -\upsilon(b)^\tau \\ \upsilon(b) & \lambda \end{pmatrix},$$

where $\upsilon : \mathbf{O} \cong \mathbf{R}^8 \to \mathbf{Y} \subset \mathbf{R}(8)$ is the standard embedding of $\mathbf{O}$ in $\mathbf{R}(8)$, the sphere S^8 being represented in $\mathbf{R}(16)$ by the matrices $\begin{pmatrix} \lambda & -y^\tau \\ y & \lambda \end{pmatrix}$, where $\lambda \in \mathbf{R}$, $y \in \mathbf{Y}$ and $\lambda^2 + y\,y^\tau = 1$.

With this choice $\mathbf{R}_{0,8}^{0}$ is the standard copy of ${}^2\mathbf{R}(8)$ in $\mathbf{R}(16)$. Any element of $Spin(9)$ is of the form

$$\begin{pmatrix} \lambda & -y^{\tau} \\ y & \lambda \end{pmatrix}\begin{pmatrix} g_0 & 0 \\ 0 & \check{g}_1 \end{pmatrix}$$

with $\begin{pmatrix} g_0 & 0 \\ 0 & \check{g}_1 \end{pmatrix} \in Spin(8)$, and $\lambda^2 + y\,y^{\tau} = 1$. The detail of the diagram, namely

$$\begin{array}{ccccc} Spin(7) & = & Spin(7) & & \\ \downarrow & & \downarrow & & \\ Spin(8) & \longrightarrow & Spin(9) & \longrightarrow & S^8, \\ \downarrow & & \downarrow & & \| \\ S^7 & \longrightarrow & S^{15} & \xrightarrow{h_{\mathbf{O}}} & S^8 \end{array}$$

is then very similar to that of Diagram 24.4, with $\mathbf{O}$ replacing $\mathbf{H}$. The map

$$Spin\,9 \to S^8;\ \begin{pmatrix} \lambda & -y^{\tau} \\ y & \lambda \end{pmatrix}\begin{pmatrix} g_0 & 0 \\ 0 & \check{g}_1 \end{pmatrix} \mapsto \begin{pmatrix} \lambda & -y^{\tau} \\ y & \lambda \end{pmatrix}^2,$$

with isotropy subgroup at 1 the subgroup $Spin(8)$, determines the central horizontal exact sequence. The lower vertical maps are

$$Spin(8) \to S^7;\ \begin{pmatrix} g_0 & 0 \\ 0 & \check{g}_1 \end{pmatrix} \mapsto g_0\,e,$$

$$Spin(9) \to S^{15};\ \begin{pmatrix} \lambda & -y^{\tau} \\ y & \lambda \end{pmatrix}\begin{pmatrix} g_0 & 0 \\ 0 & \check{g}_1 \end{pmatrix} \mapsto (\lambda\, g_0\, e,\ y\, g_0\, e)$$

and the restriction of the latter to S^8, while the lower horizontal maps are $S^7 \subset \mathbf{O} \to S^{15} \subset \mathbf{O}^2$; $g_0\,\mathrm{e} \mapsto (g_0\,\mathrm{e}, 0)$ and

$$\begin{array}{ccccc} S^{15} \subset \mathbf{O}^2 & \longrightarrow & \mathbf{O}P^2 & \longrightarrow & S^9 \\ (\lambda\, g_0\,\mathrm{e},\ y\, g_0\,\mathrm{e}) & \mapsto & [\lambda\, g_0\,\mathrm{e},\ y\, g_0\,\mathrm{e}] & \mapsto & (2\lambda^2 - 1,\ 2\lambda\, y\,\mathrm{e}) \end{array},$$

where $\lambda^2 + (y\,\mathrm{e})(y\,\mathrm{e})^{\tau} = 1$,

$$[\lambda\, g_0\,\mathrm{e},\ y\, g_0\,\mathrm{e}] = [2(\lambda\, g_0\,\mathrm{e})(\overline{\lambda\, g_0\,\mathrm{e}}),\ 2(y\, g_0\,\mathrm{e})(\overline{\lambda\, g_0\,\mathrm{e}})]\ ,$$

at least when $\lambda \neq 0$, and

$$\begin{aligned}[2(\lambda g_0 e)(\overline{\lambda g_0 e}), 2(y g_0 e)(\overline{\lambda g_0 e})] &= (2\lambda^2 - 1, 2\lambda y e) \\ &= ((\lambda^2 - y y^\tau) e, 2\lambda y e).\end{aligned}$$

Here we have assumed, for the sake of definiteness, that the left-hand column of the diagram corresponds to the Clifford action of $Spin(8)$ on S^7 with isotropy group at 1 equal to H_0, one of the two Clifford $Spin(7)$'s in $Spin(8)$. There is of course an analogous diagram involving the other Clifford action of $Spin(8)$ on S^7.

It is a corollary of the diagram that the Clifford homogeneous space $Spin(9)/Spin(7) = Spin(9)/H_0$ is homeomorphic to S^{15}. On the other hand, $Spin(9)/H_2$, which is homeomorphic to the real Stiefel manifold $O(\mathbf{R}^7, \mathbf{R}^9)$, is not homeomorphic to S^{15} (by Steenrod and Epstein (1962), Theorem 4.5).

The action of $Spin(10)$ *on* S^{31}

All the Clifford actions on spheres discussed up until now have been transitive. By contrast, the Clifford action of $Spin(10)$ on S^{31} is not, for the isotropy subgroup at 1 at least contains a Clifford copy of $Spin(7)$ as a subgroup, from which it follows that the dimension of the orbit of 1 is at most equal to

$$\dim Spin(10) - \dim Spin(7) = 45 - 21 = 24.$$

In fact the space of orbits, assigned the quotient topology, can be shown to be homeomorphic to a closed interval of the real line, one end-point of which represents an orbit A_{21} of dimension 21, homeomorphic both to $Spin(9)/Spin(6)$ and to $Spin(10)/SU(5)$, the embedding of $Spin(6)$ in $Spin(9)$ in the former case being a Clifford one, while the other end-point represents an orbit B_{24} of dimension 24, homeomorphic to $Spin(10)/Spin(7)$, the embedding of $Spin(7)$ in $Spin(10)$ being Clifford, $Spin(7) = H_0$ being indeed the isotropy subgroup at 1. Each of the interior points of the interval represents an orbit of dimension 30, homeomorphic to

$$C_{30} = Spin(10)/Spin(6) \cong A_{21} \times S^9$$

(the embedding of $Spin(6)$ in $Spin(10)$ being Clifford).

More information about $Spin(10)$, the orbit A_{21} and various relationships between A_{21}, C_{30} and B_{24} will be found in Exercises 24.2 to 24.4. The standard text on differentiable group actions is Bredon (1972), though the above example is not to be found there. I am grateful to Christopher Spurgeon and to Dr Hugh Morton for establishing many of the details of the action as part of an M.Sc. project in 1978.

G_2 as a Lie group

In our treatment of the groups *Spin n* so far we have regarded them certainly as topological groups and not just as groups, but, apart from one brief argument when discussing Diagram 24.5, we have disregarded the fact that they are Lie groups and that the various maps between them and the homogeneous spaces formed from them are not only continuous but smooth (in fact C^∞). What about G_2?

Theorem 24.11 *The group G_2 is a compact, connected Lie group of dimension* 14.

Proof Consider Diagrams 24.5 and 24.6. It is enough to prove that the vertical map, the surjection $Spin(7) \to S^7$; $g \mapsto g\,e$ of Diagram 24.6, is a submersion and to prove this it is enough to prove that the tangent map at 1, namely $T(Spin(7))_1 \to T(S^7)_1;\ \gamma \mapsto \gamma\,e$, is surjective. However, this map composed with the injection $T(Spin(6))_1 \to T(Spin(7))_1$ is the tangent map at 1 of the standard submersion $SU(4) \to S^7$ of Diagram 24.5.

Hence G_2 is a smooth submanifold of the Lie group $Spin(7)$, of dimension $\dim Spin(7) - \dim S^7 = 21 - 7 = 14$. The group is clearly closed in the compact group $Spin(7)$, so is compact. Finally, since $SU(3)$ and S^6 are connected, so is G_2.

The group G_2 is one of a clutch of five compact exceptional simple Lie groups all associated in one way or another with the Cayley algebra **O**, the other four being known as F_4, E_6, E_7 and E_8, of dimensions 52, 78, 133 and 248, respectively. For the definitions of *semi-simple* and *simple* Lie algebras and Lie groups the reader must refer to one of the standard texts on Lie groups, such as Helgason (1962), Gilmore (1974) or Jacobson (1962). Most treatments construct the exceptional groups by first constructing their Lie algebras. An elementary account of them, as groups, is hard to find. □

Other aspects of triality

For any positive integer n the Lie group surjection $\rho : Spin(n) \to SO(n)$, with kernel S^0, induces an isomorphism $T\rho_1$ between the Lie algebras

$T(Spin(n))_1$ and $T(SO(n))_1$, the latter normally being identified with $\mathrm{End}_-(\mathbf{R}^n)$, by the remark following Proposition 22.28. What about θ?

Proposition 24.12 *The triality automorphism $\theta : Spin(8) \to Spin(8)$ is smooth (indeed C^∞) and induces a Lie algebra automorphism*

$$T\theta_1 : T(Spin(8))_1 \longrightarrow T(Spin(8))_1$$

of order 3. Although θ does not project globally to $SO(8)$ its restriction to a suitably small neighbourhood U of 1 in $Spin(8)$ does project to a smooth map $\theta_V : V \to V$, where $V = \rho(U)$, the diagram of Lie algebra maps

$$\begin{array}{ccc} T(Spin(8))_1 & \xrightarrow{T\theta_1} & T(Spin(8))_1 \\ T\rho_1 \big\downarrow \cong & & T\rho_1 \big\downarrow \cong \\ T(SO(8))_1 & \xrightarrow{T(\theta_V)_1} & T(SO(8))_1 \end{array}$$

being commutative.

Triality may be formulated entirely in terms of the action of $SO(8)$ on the Cayley division algebra $\mathbf{O}$ as follows.

Theorem 24.13 *The triple (g_0, g_1, g_2) is a θ-triad of $SO(8)$ if and only if $\check{g}_0(a\,b) = (g_1 a)(g_2 b)$, for all $a, b \in \mathbf{O}$.*

Proof (g_0, g_1, g_2) is a θ-triad of $SO(8)$

$\Leftrightarrow$ for all $y \in \mathbf{Y}$, $\quad g_0\, y\, \check{g}_1^{-1} \in \mathbf{Y}$ and $g_0\, y\, \check{g}_1^{-1}\mathrm{e} = \check{g}_2\, y\, \mathrm{e}$,
$\Leftrightarrow$ for all $y, z \in \mathbf{Y}$, $g_0\, y\, \check{g}_1^{-1} z\, \mathrm{e} = (\check{g}_2\, y\, \mathrm{e})(z\, \mathrm{e})$, by Lemma 24.2,
$\Leftrightarrow$ for all $x, y \in \mathbf{Y}$, $g_0\, y\, x\, \mathrm{e} = (\check{g}_2\, y\, \mathrm{e})(\check{g}_1\, x\, \mathrm{e})$, setting
$x\,\mathrm{e} = \check{g}_1^{-1} z\, \mathrm{e}$, $z\,\mathrm{e} = \check{g}_1\, x\, \mathrm{e}$,
$\Leftrightarrow$ for all $x, y \in \mathbf{Y}$, $\check{g}_0(\overline{x\,\mathrm{e}}\,\overline{y\,\mathrm{e}}) = (g_1\, \overline{x\,\mathrm{e}})(g_2\, \overline{y\,\mathrm{e}})$, conjugating both sides,
$\Leftrightarrow$ for all $a, b \in \mathbf{O}$, $\check{g}_0(a\,b) = g\, a)(g\, b)$, setting
$\overline{x\,\mathrm{e}} = a$, $\overline{y\,\mathrm{e}} = b$.

□

Theorem 24.13 is due to Élie Cartan (1925) (page 370). The Lie algebra version is known as Freudenthal's principle of triality (1951):

Theorem 24.14 *For any $\gamma_0 \in T(SO(8))_1$ ($\cong \mathrm{End}_-(\mathbf{R}^8)$), there exist unique $\gamma_1, \gamma_2 \in T(SO(8))_1$ such that*

$$\check{\gamma}_0(a\,b) = (\gamma_1\, a)b + a(\gamma_2\, b), \text{ for all } a \in \mathbf{O}.$$

Proof For existence let V be as in Proposition 24.12 and take tangents at 1 of each side of the equation

$$\check{g}_0(a\,b) = ((\theta_V{}^2 g_0)a)((\theta_V g_0)b)$$

for each $g_0 \in V$ and each $a, b \in \mathbf{O}$. Then

$$\check{\gamma}_0(a\,b) = (((T\theta_v{}^2)_1\gamma_0)a)b + a(((T\theta_V)_1\gamma_0)b),$$

for each $\gamma_0 \in T(SO(8))_1$ and each $a, b \in \mathbf{O}$; for Cayley multiplication is bilinear, while the companion involution and evaluation at a or at b are restrictions of linear maps.

So take $\gamma_1 = (T\theta_V{}^2)_1\gamma_0$ and $\gamma_2 = (T\theta_V)_1\gamma_)$.

For uniqueness we have to prove that if $\gamma_1, \gamma_2, \gamma_1', \gamma_2' \in T(SO(8))_1$ are such that $(\gamma_1\,b)c + b(\gamma_2\,c) = (\gamma_1'b)c + b(\gamma_2'c)$, for all $b, c \in \mathbf{O}$, then $\gamma_1 = \gamma_1'$ and $\gamma_2 = \gamma_2'$. It is clearly enough to prove that if

$$(\gamma_1\,b)c + b(\gamma_2\,c) - 0,$$

for all $b, c \in \mathbf{O}$, then $\gamma_1 = \gamma_2 = 0$. Let $a = \gamma_1\,\mathrm{e}$. Then

$$\mathrm{e}\cdot a = \mathrm{e}\cdot\gamma_1\,\mathrm{e} = -\gamma_1\,\mathrm{e}\cdot\mathrm{e}\ (\text{since } \gamma_1 \text{ is skew}) \ = -a\cdot\mathrm{e}.$$

So a is a pure Cayley number. However, we have

$$0 = a\,c + \gamma_2\,c, \text{ for all } c \in \mathbf{O}.$$

So $\quad 0 = (\gamma_1\,b)c - b(a\,c)$, for all $b, c \in \mathbf{O}$.

So $\quad 0 = \gamma_1\,b \ - b\,a$, for all $b \in \mathbf{O}$.

So $\quad 0 = (b\,a)c - b(a\,c)$, for all $b, c \in \mathbf{O}$,

which is not the case, unless $a = 0$. So $\gamma_1 b = 0$, for all $b \in \mathbf{O}$ and $\gamma_2 = 0$, for all $c \in \mathbf{O}$. So $\gamma_1 = \gamma_2 = 0$. □

It is more usual to start the entire discussion of triality by first proving Theorem 24.13 directly and defining $(\gamma_0, \gamma_1, \gamma_2)$ to be a triality triad of $T(SO(8))_1$ if

$$\check{\gamma}_0\,b\,c = (\gamma_1\,b) \ + b(\gamma_2\,c), \text{ for all } b, c \in \mathbf{O}.$$

See, for example Loos (1969), Vol.II.

Quadric triality

In Exercise 22.4 we noted that either component of $\mathcal{N}_4(\mathbf{R}^8_{\mathrm{hb}})$ is homeomorphic to $\mathcal{N}_1(\mathbf{R}^8_{\mathrm{hb}})$, each being homeomorphic to $SO(4)$ by Exercise 22.3, and we asked the question whether or not either component of

$\mathcal{N}_4(\mathbf{C}^8_{\text{hb}})$ is homeomorphic to $\mathcal{N}_1(\mathbf{C}^8_{\text{hb}})$. Now, back in Theorem 14.17 we have represented each of these quadric Grassmannians as follows:

$$\begin{array}{lllll} \mathcal{N}_4(\mathbf{R}^{4,4}) & \cong & \mathcal{N}_4(\mathbf{R}^8_{\text{hb}}) & \cong & (O(4)\times O(4))/O(4), \\ \mathcal{N}_1(\mathbf{R}^{4,4}) & \cong & \mathcal{N}_1(\mathbf{R}^8_{\text{hb}}) & \cong & (O(4)\times O(4))/(O(1)\times O(3)\times O(3)), \\ \mathcal{N}_4(\mathbf{C}^8) & \cong & \mathcal{N}_4(\mathbf{C}^8_{\text{hb}}) & \cong & O(8)\times U(4). \\ \mathcal{N}_1(\mathbf{C}^8) & \cong & \mathcal{N}_1(\mathbf{C}^8_{\text{hb}}) & \cong & O(8)\times U(1)\times O(6). \end{array}$$

Hence one component of $\mathcal{N}_4(\mathbf{R}^8_{\text{hb}})$ is homeomorphic to the homogeneous space $(SO(4)\times SO(4))/SO(4)$, clearly homeomorphic to $SO(4)$, while $\mathcal{N}_1(\mathbf{R}^8_{\text{hb}})$ is homeomorphic to

$$\begin{aligned}(SO(4)\times SO(4))/(S^0\times SO(3)\times SO(3)) &\cong (S^3\times S^3)/S^0 \\ &\cong Spin(4)/S^0 \cong SO(4).\end{aligned}$$

Likewise one component of $\mathcal{N}_4(\mathbf{C}^8_{\text{hb}})$ is homeomorphic to $SO(8)/U(4)$, while $\mathcal{N}_1(\mathbf{C}^8_{\text{hb}})$ is homeomorphic to $SO(8)/(U(1)\times SO(6))$.

It looks at first sight as though the isomorphism of $U(4)$ to $U(1)\times SO(6)$ is a necessary condition for the homeomorphism of $\mathcal{N}_4(\mathbf{C}^8_{\text{hb}})$ to $\mathcal{N}_1(\mathbf{C}^8_{\text{hb}})$ – yet it can be shown by methods of algebraic topology that these groups are *not* homeomorphic! However, $SO(4)$ is not homeomorphic to $S^0\times SO(3)\times SO(3)$ – though $SO(4)/S^0$ is homeomorphic to $SO(3)\times SO(3)$ by an obvious isomorphism, $Spin(4)$ being isomorphic to $Spin(3)\times Spin(3)$. A better question therefore is:

Are $U(4)/S^0$ and $(U(1)\times SO(6))/S^0$ homeomorphic?

Triality provides an affirmative answer.

Once again we work in $\mathbf{R}(8)\cong\mathbf{R}_{0,6}$, with $\mathbf{R}^6$ embedded in this Clifford algebra in such a way that the even Clifford algebra $\mathbf{R}^0_{0,6}$ is the standard copy of $\mathbf{C}(4)$ in $\mathbf{R}(8)$. It is easily verified that the product of the basis elements for $\mathbf{R}^6$ is then either of the diagonal elements $\pm\mathrm{i}$ of $\mathbf{C}(4)$ and we so order them that the product is in fact i. The elements 1, i and the six basis elements of $\mathbf{R}^6$ then span the copy of $\mathbf{R}^8$ in $\mathbf{R}(8)$ that we have found it convenient in Chapter 19 and in this chapter to denote by $\mathbf{Y}$. With these notational conventions we can now state

Theorem 24.15 *Let $g\in U(4)\subset SO(8)$, let z be the inverse of either of the square roots of the determinant of g, regarded as an element of $\mathbf{C}(4)$, so that $z^{\frac{1}{2}}g$, defined up to sign, is an element of $SU(4)$, and let $\rho: Spin(6)\cong SU(4)\to SO(6)$ be the standard projection. Then*

$$\left(g,\ \bar{z}\check{g},\ \begin{pmatrix} z & 0 \\ 0 & \rho(z^{\frac{1}{2}}g)\end{pmatrix}\right) \quad \textit{is a } \theta\textit{-triad of } SO(8).$$

In particular, the projection of θ to $SO(8)/S^0$ maps the subgroup $U(4)/S^0$ of $SO(8)/S^0$ to the subgroup $(U(1) \times SO(6))/S^0$ by the isomorphism

$$\pm g \mapsto \pm \begin{pmatrix} z & 0 \\ 0 & \rho(z^{\frac{1}{2}}g) \end{pmatrix}.$$

Corollary 24.16 *The triality automorphism of $SO(8)/S^0$ permutes cyclically the two components of $\mathscr{N}_4(\mathbf{C}^8_{\text{hb}})$ with the projective quadric $\mathscr{N}_1(\mathbf{C}^8_{\text{hb}})$ itself.*

In fact triality also is involved in the case of the real quadric $\mathscr{N}_1(\mathbf{R}^8_{\text{hb}})$. We note first the following.

Proposition 24.17 *Let $SO(4)$ be embedded in $SU(4) \cong Spin(6)$ in the obvious way. Then $\rho(SO(4))$ is a copy of $SO(3) \times SO(3)$ in $SO(6)$.*

With an obvious reordering of basis elements where necessary for sense we then have the following.

Theorem 24.18 *The triality automorphism θ of $Spin(8)$ restricts to an automorphism of $\rho^{-1}(SO(4) \times SO(4))$, the induced automorphism of the group $SO(8)/S^0$ likewise restricting to an automorphism of $(SO(4) \times SO(4))/S^0$. Moreover, for any $g \subset SO(4) \subset SU(4)$,*

$$\left(g, \pm\check{g}, \begin{pmatrix} \pm^2 1 & 0 \\ 0 & \rho((\pm 1)^{\frac{1}{2}}g) \end{pmatrix}\right) \quad \text{is a } \theta\text{-triad of } SO(4) \times SO(4).$$

In particular, the projection of θ to $(SO(4) \times SO(4))/S^0$ maps the subgroup $SO(4)/S^0$ of $(SO(4) \times SO(4))/S^0$ to the subgroup $SO(3) \times SO(3)$ by the isomorphism $\pm g \mapsto \rho(\pm g)$.

Corollary 24.19 *The triality automorphism of $(SO(4) \times SO(4))/S^0$ permutes cyclically the two components of $\mathscr{N}_4(\mathbf{R}^8_{\text{hb}})$ with the projective quadric $\mathscr{N}_1(\mathbf{R}^8_{\text{hb}})$ itself.*

We do not wish to deny the reader the fun of filling in the details of the proofs of these last few theorems for him- or herself.

We have noted several times that the real projective quadric $\mathscr{N}_1(\mathbf{R}^8_{\text{hb}})$ is homeomorphic to $SO(4)$. Study's interest in this quadric first arose in (1891) in connection with the problem of representing the group of rigid motions of $\mathbf{R}^3$. As we saw towards the end of Chapter 18 such rigid motions can be represented, uniquely up to non-zero real multiples, by pairs of quaternions (α, β) with $\alpha \cdot \beta = 0$ but with $\alpha \neq 0$, so that the group is representable by the quadric $\mathcal{N}_1(\mathbf{R}^8_{hb})$ with one of its isotropic half spaces removed (the group product is $(\alpha, \beta)(\gamma, \delta) = (\alpha\gamma, \alpha\delta + \beta\gamma)$, and the unit element is $(1, 0)$). The relationship of this representation to the representation of the group $SO(4)$ by the whole quadric, which we have explored in a wider setting in the section on groups of rigid motions in Chapter 18, is hinted at in Study (1903) and stated quite explicitly in Study (1913). The same passage in Study (1913) contains a clear statement of what is now known as Study's *principle of triality*, but which he called the *Reziprozitätsgezetz* or *reciprocity law*, namely the existence of an analytic homeomorphism between the quadric $\mathcal{N}_1(\mathbf{R}^8_{\mathrm{hb}})$ and either component of the quadric Grassmannian $\mathcal{N}_4(\mathbf{R}^8_{\mathrm{hb}})$ and also between the quadric $\mathcal{N}_1(\mathbf{C}^8)$ and either component of $\mathcal{N}_4(\mathbf{C}^8)$. For an exhaustive treatment of quadric triality in a general setting see Tits (1959).

Exercises

24.1 Verify that the set of copies of the algebra $\mathbf{C}$ in the Cayley algebra $\mathbf{O}$ can be represented as the homogeneous space $G_2/SU(3)$, while the set of copies of the algebra $\mathbf{H}$ in $\mathbf{O}$ can be represented as the homogeneous space $G_2/SO(4)$.

24.2 Prove that the map

$$\begin{aligned} \mathbf{R} \oplus \mathbf{R}^9 \cong \mathbf{C} \oplus \mathbf{O} &\rightarrow \mathbf{C}(16) \\ (\xi, x) = (\zeta, c) &\mapsto \begin{pmatrix} \zeta & \iota z^\tau \\ \iota z & \zeta^\tau \end{pmatrix}, \end{aligned}$$

where $\xi, \eta \in \mathbf{R}$, $\zeta = \xi + i\eta$, $c \in \mathbf{O} \cong \mathbf{R}^8$, $x = (\eta, c)$, $z = \upsilon(c) \in \mathbf{Y}$ and ι (iota) denotes the square root of -1 in the coefficients of the elements of $\mathbf{C}(16)$, is a real linear embedding of $\mathbf{R} \oplus \mathbf{R}^9$ in $\mathbf{R}_{0,9} = \mathbf{C}(16)$ such that $\mathbf{R}^0_{0,9}$ is the standard copy of $\mathbf{R}(16)$ in $\mathbf{C}(16)$.

Hence prove that any element of $Spin(10) \subset \mathbf{C}(16)$ is expressible in the form

$$\begin{pmatrix} \zeta & \iota z^\tau \\ \iota z & \zeta^\tau \end{pmatrix} \begin{pmatrix} \lambda & -y^\tau \\ y & \lambda \end{pmatrix} \begin{pmatrix} g_0 & 0 \\ 0 & \check{g}_1 \end{pmatrix},$$

where

$$\begin{pmatrix} g_0 & 0 \\ 0 & \check{g}_1 \end{pmatrix} \in Spin(8), \text{ with } g_0, g_1 \in SO(8),$$

$$\begin{pmatrix} \lambda & -y^\tau \\ y & \lambda \end{pmatrix} \in S^8, \text{ with } \lambda \in \mathbf{R}, y \in \mathbf{Y} \text{ and } \lambda^2 + y\,y^\tau = 1,$$

and

$$\begin{pmatrix} \zeta & \imath\, z^\tau \\ \imath\, z & \zeta^\tau \end{pmatrix} \in S^0, \text{ with } \zeta \in \mathbf{C}, z \in \mathbf{Y} \text{ and } \zeta\,\zeta^\tau + z\,z^\tau = 1,$$

the image of such an element by the standard projection $Spin(10) \to S^9$ being $\begin{pmatrix} \zeta & \imath\, z^\tau \\ \imath\, z & \zeta^\tau \end{pmatrix}^2$.

24.3 With the elements of $Spin(10)$ represented as in Exercise 24.2 acting on $\mathbf{R}^{10} \cong \mathbf{C} \oplus \mathbf{O} \subset \mathbf{C}(16)$ by left multiplication, prove that the isotropy group at 1 is isomorphic to $Spin(7)$ (in fact to the Clifford subgroup H_0 of $Spin(8)$ – cf. Theorem 24.10) and that the isotropy group at $\begin{pmatrix} 1+\imath\,\mathrm{i} & 0 \\ 0 & 1-\imath\,\mathrm{i} \end{pmatrix}$ consists of those elements of $Spin(10)$ for which

$$g_0 = \begin{pmatrix} \cos\theta & -\sin\theta & 0 \\ \sin\theta & \cos\theta & 0 \\ 0 & 0 & g_0' \end{pmatrix}, \text{ with } g_0' \in SO(6),$$

with $\zeta = \lambda(\cos\theta + \mathrm{i}\sin\theta)$ and $z = y(\sin\theta - \mathrm{i}\cos\theta)$, the latter group having dimension equal to

$$\dim SO(6) + \dim S^9 = 15 + 9 = 24,$$

and acting transitively on S^9 with isotropy group at 1 isomorphic to $Spin(6) \cong SU(4)$. (By a theorem (Montgomery and Samelson (1943))that lists all compact Lie groups acting transitively on spheres it follows that this latter group must be isomorphic to $SU(5)$.)

24.4 Establish the following commutative diagrams for the orbits A_{21}, B_{24} and C_{30} of the Clifford action of $Spin(10)$ on S^{31}:

$$\begin{array}{ccccc} Spin(6) & = & Spin(6) & & \\ \downarrow & & \downarrow & & \\ Spin(7) & \to & Spin(9) & \to & S^{15} \\ \downarrow & & \downarrow & & \| \\ S^6 & \to & A_{21} & \to & S^{15} \end{array},$$

$$\begin{array}{ccccc}
Spin(7) & = & Spin(7) & & \\
\downarrow & & \downarrow & & \\
Spin(9) & \to & Spin(10) & \to & S^9 \\
\downarrow & & \downarrow & & = \\
S^{15} & \to & B_{24} & \to & S^9
\end{array},$$

$$\begin{array}{ccccc}
Spin\,6 \cong SU(4) & \to & SU(5) & \to & S^9 \\
\downarrow & & \downarrow & & = \\
Spin(9) & \to & Spin(10) & \to & S^9 \\
\downarrow & & \downarrow & & \\
A_{21} & \to & A_{21} & &
\end{array},$$

implying that $C_{30} \cong Spin\,(10)/Spin\,(6) \cong A_{21} \times S^9$,

$$\begin{array}{ccccc}
Spin(6) & = & Spin(6) & & \\
\downarrow & & \downarrow & & \\
Spin(9) & \to & Spin(10) & \to & S^9 \\
\downarrow & & \downarrow & & = \\
A_{21} & \to & C_{30} & \to & S^9
\end{array},$$

$$\begin{array}{ccccc}
Spin(6) \cong SU(4) & = & Spin(6) & & \\
\downarrow & & \downarrow & & \\
SU(5) & \to & Spin(10) & \to & A_{21} \\
\downarrow & & \downarrow & & = \\
S^9 & \to & C_{30} & \to & A_{21}
\end{array}$$

and

$$\begin{array}{ccccc}
Spin(6) & = & Spin(6) & & \\
\downarrow & & \downarrow & & \\
Spin(7) & \to & Spin(10) & \to & B_{24} \\
\downarrow & & \downarrow & & = \\
S^6 & \to & C_{30} & \to & B_{24}
\end{array}.$$

References

Adams, J.F. (1958) 'On the nonexistence of elements of Hopf invariant one', *Bull. Am. math. Soc.* 64, 279-282.

Adams, J.F. (1960) 'On the nonexistence of elements of Hopf invariant one', *Ann. Math* 72, 20-104.

Ahlfors, L. (1985) 'Möbius transformations and Clifford numbers', in I. Chavel, H.M. Parkas (eds) *Differential Geometry and Complex Analysis*, (dedicated to H.E. Rauch) Springer-Verlag, Berlin, 65-73.

Ahlfors, L. (1986) 'Möbius transformations in $\mathbf{R}^n$ expressed through 2×2 matrices of Clifford numbers', *Complex Variables* 5, 215-224.

Ahlfors, L. and Lounesto, P. (1989) 'Some remarks on Clifford algebras', *Complex Variables* 12, 201-209.

Arnol'd, V.I. (1974) *Mathematical Methods of Classical Mechanics*, Izdat. 'Nauka', Moscow. English translation by K. Vogtmann and A. Weinstein, Graduate Texts in Mathematics, 60, Springer Verlag, New York – Heidelberg (1978).

Arnol'd, V.I. (1995) 'The geometry of spherical curves and the algebra of quaternions', *Uspekhi Mat. Nauk* 20, 3–68 = *Russian Math. Surveys* 50.

Artin, E. (1957) *Geometric Algebra*, Interscience, New York.

Atiyah, M.F., Bott, R. and Shapiro, A. (1964) 'Clifford modules', *Topology* 3 (Supp. 1), 3-38.

Benn, I.M. and Tucker. R.W. (1987) *An Introducion to Spinors and Geometry wih Applications to Physics*, Hilger, Bristol.

Brackx, F. Delanghe, R and Serras, H. (1993) *Clifford Algebras and their Applications in Mathematical Physics*, Kluwer Academic Publishers, Dordrecht, Netherlands.

Bredon, G.E. (1972) *Introduction to Compact Transformation Groups*, Academic Press, New York – London.

Bruck, R.H. (1955) 'Recent advances in the foundations of Euclidean plane geometry', *Am. Math. Mon.* 62, 2-17.

Cartan, É. (1925) 'Le principe de dualité et la théorie des groupes simples et semi-simples', *Bull. Sci. Math.* 49, 361-374.

Cartan, É. (1947) 'Sur l'espace anallagmatique réel à n dimensions', *Ann. Polon. Math.* 20, 266-278.

Cartan, É. (1949) 'Deux théorèmes de géométrie anallagmatique réelle à n dimensions', *Ann. Mat. Pura Appl.* (4) 28, 1-12.

Cayley, A. (1845) 'On Jacobi's elliptic functions, in reply to Rev. Brice Brownin and on quaternions', *Phil. Mag.* 3, 210-213.

Chevalley, C. (1946) *Theory of Lie groups*, Princeton University Press, Princeton, N.J.

Chisholm, J.S.R. and Common, A.K. (1986) *Clifford Algebras and their applications in Mathematical Physics*, Reidel, Dordrecht, Netherlands.

Clifford, W.K. (1876) 'Preliminary sketch of biquaternions' *Proc. London Math. Soc.* 4, 381-395.

Clifford, W.K. (1876) 'On the classification of Geometric algebras', published as Paper XLIII in *Mathematical Papers*, (ed. R. Tucker), Macmillan, London (1882).

Cnops, J. (1994) *Hurwitz pairs and applications of Möbius transformations*, Thesis, Universiteit Gent.

Elstrodt, J., Grunewald, F. and Mennicke, J. (1987) 'Vahlen's group of Clifford matrices and spin groups', *Math.Z.* 196, 369-390.

Fillmore, J. and Springer, A. (1990) 'Möbius groups over general fields using Clifford algebras associated with spheres', *Int. J. Theo. Phys.* 29, 225-246.

Freudenthal, H. (1951) *Oktaven, Ausnahmegruppen und Oktavengeometrie*, Math. Inst. der Rijk Univ. te Utrecht, Utrecht.

Fuchs (1963) *Partially ordered algebraic systems*, Pergamon, Oxford.

Gilmore, R. (1974) *Lie groups, Lie algebras and some of their applications*, Wiley-Interscience, New York.

Grassmann, H. (1844) *Die Wissenschaft der extensiven Grösse oder die Ausdehnungslehre, eine neue mathematishcen Disciplin*, Leipzig.

Graves, J. (1848) *Trans. Irish Academy* 21, p. 338.

Haantjes, J. (1937) 'Conformal representations of an *n*-dimensional euclidean space with a non-definite fundamental form on itself', *Proc. Ned. Akad. Wet. (Math)* 40, 700-705.

Hamilton, W.R. (1844) 'On quaternions, or on a new system of imaginaries in algebra', *Phil. Mag.* 25, 489-495, reprinted in *The mathematical papers of Sir William Rowan Hamilton*, Vol. III, *Algebra*, Cambridge University Press, London (1967), 106-110.

Hampson, A. (1969) *Clifford algebras* M.Sc. Thesis, University of Liverpool.

Helgason, S. (1962) *Differential Geometry and Symmetric Spaces*, Academic Press, New York.

Hermann, R. (1979) Appendix 'Kleinian mathematics from an advanced standpoint, A: Conformal and non-Euclidean geometry in R^3 from the Kleinian viewpoint', bound with F.Klein *Developments of Mathematics in the 19th Century*, translated by M. Ackerman, Math. Sci. Press, Brookline, Mass., 367-376.

Hopf H.(1931) 'Über die Abbildungen der dreidimensionalen Sphären auf die Kugelfläche', *Math Ann.* 104, 637-665.

Hopf H. (1935) 'Über die Abbildungen von Sphären auf Sphären niedrigerer Dimension', *Fundam. Math.* 25, 427-440.

Hurwitz, A. (1898) 'Über die Komposition der quadratischen Formen von beliebig vielen Variabeln', *Nachrichten von der Königlichen Gesellschaft der Wissenschaften zu Göttingen, Math. Phys. Kl.*, 308-316.

Hurwitz, A. (1923) 'Über die Komposition der quadratischen Formen', *Math. Ann.* 88, 294-298.

Jacobson, N. (1962) *Lie algebras*, Interscience, New York.

James, I.M. (1976) *The topology of Stiefel manifolds*, L.M.S. Lecture Note Series, 24, Cambridge University Press, Cambridge.

Kaplansky, I. (1963) 'Lie algebras', *Lectures on Modern Mathematics* (ed. T.L. Saaty), Vol. I, Wiley, New York.

Kervaire, M. (1958) 'Non-parallelizability of the n-sphere for $n > 7$', *Proc. natn. Acad. Sci. U.S.A.* 44, 280-283.

Kleinfeld, E. (1963) 'A characterisation of the Cayley numbers', *Studies in Modern Algebra* (ed. A.A. Albert), Vol. 2, Mathematical Association of America, Prentice-Hall, Englewood Cliffs, N.J.

Kuiper, N.H. (1949) 'On conformally-flat spaces in the large', *Ann. Math.* 50, 916-924.

Ławrynowicz, J. and Rembieliński, J. (1986) 'Pseudo-euclidean Hurwitz pairs and generalised Fueter equations', in J.S.R. Chisholm and A.K. Common (eds): *Clifford algebras and their applications in mathematical physics*, Reide, Dordrecht, Netherlands, 39-48.

Liouville, J. (1850) Appendix to Monge, G. *Application de l'analyse à la geométrie*, 5 éd. par Liouville.

Lipschitz, R. (1880) 'Principes d'un calcul algébriques qui contient comme espèces particulières le calcul des quantités imaginaires et des quaternions', *C.R. Acad. Sci. Paris* 91, 619-621, 660-664.

Lipschitz, R. (1886) *Untersuchungen über die Summen von Quadraten*, Max Cohen und Sohn, Bonn.

Loos, O. (1969) *Symmetric Spaces*, Vols. I and II, Benjamin, New York.

Lounesto, P. (1987) 'Cayley transform, outer exponential and spinor norm', Suppl. Rend. Circ. Mat. Palermo, Ser. II, 191-198.

Maks, J. (1989) *Modulo* $(1,1)$ *periodicity of Clifford algebras and the generalized (anti-)Möbius transformations*, Thesis, Technische Universiteit Delft.

Maks, J. (1992) 'Clifford algebras and Möbius transformations', in A. Micali et al. (eds), *Clifford algebras and their applications in mathematical physics*, Kluwer, Dordrecht, Netherlands.

Meusnier, J.-B.-M.-C. (1785) 'Mémoire sur la courbure des surfaces', *Mémoire Div. Sav.*, 10, 477-510.

Micali, A., Boudet, R. and Helmstetter, J. (1991) *Clifford algebras and their applications in mathematical physics*, Kluwer, Dordrecht, Netherlands.

Milnor, J. and Bott, R. (1958) 'On the parallelizability of the spheres' *Bull. Amer. Math. Soc.* 64, 87-89.

Montgomery, D. and Samelson, H. (1943) 'Transformation groups of spheres', *Ann. Math.* 44, 454-470.

Moufang, R. (1935) 'Zur Struktur von Alternativkörpern', *Math. Ann.* 110, 416-430.

Paige, L.J. (1961) 'Jordan algebras', *Studies in Modern Algebra*, ed. A.A. Albert, Vol. 2, Mathematical Association of America, Prentice-Hall, Engelwood Cliffs, N.J.

Pontrjagin, L.S. (1946) *Topological groups*, 2nd edition, Gordon and Breach, New York (1966).

Porteous, I.R. (1969) *Topological geometry*, 1st edition, Van Nostrand Reinhold, London.

Porteous, I.R. (1981) *Topological geometry*, 2nd edition, with additional material on triality, Cambridge University Press, Cambridge.

Porteous, I.R. (1993) 'Clifford algebra tables', in F. Brackx et al. (eds) *Clifford algebras and their applications in mathematical physics* Kluwer, Dordrecht, Netherlands, 13-22.
Radon, J. 'Lineare Scharen orthogonaler Matrizen', *Abh. math. Semin. Univ. Hamburg* 1, 1-14.
Randriamihamison L.-S. (1990) 'Paires de Hurwitz pseudo-euclidiennes en signature quelconque', *J. Phys. A: Math. Gen.* 23, 2729-2749.
Ryan, J. (1995) 'Some applications of conformal covariance in Clifford Analysis' in *Clifford Algebras in Analysis and Related Topics*, ed. J. Ryan, CRC Press, Studies in Advanced Mathematics, Boca Raton, FL, 128-155.
Segre, B. (1947) 'Gli automorphismi del corpo complesso ed un problema di Corrado Segre', *Atti Accad. naz. Lincei Rend.* (8) 3, 414-420.
Spurgeon, C. (1979) *Clifford algebras*, M.Sc. Thesis, University of Liverpool.
Steenrod, N.E. and Epstein, D.B.A. (1962) *Cohomology Operations*, Princeton University Press, Princeton, N.J.
Study, E. (1891) 'Von den Bewegungen und Umlegungen', *Math. Ann.* 39, 441-566.
Study, E. (1903) *Geometrie der Dynamen*, Teubner, Leipzig.
Study (1913) 'Grundlagen une Ziele der analytischen Kinematik', *Sitzungsberichte der Berliner Math. Gesellschaft* 12, 36-60.
Tits, J. (1959) 'Sur la trialité et certaines groupes qui s'en déduisent', *Inst. des Hautes Études Sci. Publ. Math.* 2, 13-60.
Trautman A. (1993) 'Geometric aspects of spinors', in F. Brackx et al. (eds) *Clifford algebras and their applications in mathematical physics*, Kluwer, Dordrecht, Netherlands 333-344.
Vahlen, K.Th. (1902) 'Über Bewegungen und complexe Zahlen', *Math. Ann.* 55, 585-593.
van der Waerden, B.L. (1985) *A history of algebra*, Springer-Verlag, Berlin.
Wall, C.T.C. (1968) 'Graded algebras, antiinvolutions, simple groups and symmetric spaces', *Bull. Amer. Math. Soc.* 74, 198-202.
Waterman, P.L. (1993) 'Möbius transformations in several dimensions', *Advances in Math.*, 101, 87-113.
Weyl, H. (1939) *The Classical Groups*, Princeton University Press, Princeton, N.J.

Index

LaVergne, TN USA
03 March 2011
218770LV00001B/2/P

9 780521 118026